Biophysical and Chemical Properties of Collagen: Biomedical Applications

Biophysical Society-IOP Series

About the Series

The Biophysical Society and IOP Publishing have forged a new publishing partnership in biophysics, bringing the world-leading expertise and domain knowledge of the Biophysical Society into the rapidly developing IOP ebooks program.

The program publishes textbooks, monographs, reviews, and handbooks covering all areas of biophysics research, applications, education, methods, computational tools, and techniques. Subjects of the collection will include: bioenergetics; bioengineering; biological fluorescence; biopolymers in vivo; cryo-electron microscopy; exocytosis and endocytosis; intrinsically disordered proteins; mechanobiology; membrane biophysics; membrane structure and assembly; molecular biophysics; motility and cytoskeleton; nanoscale biophysics; and permeation and transport.

Biophysical and Chemical Properties of Collagen: Biomedical Applications

John A M Ramshaw
Department of Surgery, St. Vincent's Hospital, University of Melbourne, Fitzroy, Australia and Collagen-61 (Consulting), Coburg, Australia

Veronica Glattauer
CSIRO Manufacturing, Clayton, Australia

IOP Publishing, Bristol, UK

ISBN 978-0-7503-2096-2 (ebook)
ISBN 978-0-7503-2094-8 (print)
ISBN 978-0-7503-2097-9 (myPrint)
ISBN 978-0-7503-2095-5 (mobi)

DOI 10.1088/978-0-7503-2096-2

Version: 20191101

IOP ebooks

British Library Cataloguing-in-Publication Data: A catalogue record for this book is available from the British Library.

Published by IOP Publishing, wholly owned by The Institute of Physics, London

IOP Publishing, Temple Circus, Temple Way, Bristol, BS1 6HG, UK

US Office: IOP Publishing, Inc., 190 North Independence Mall West, Suite 601, Philadelphia, PA 19106, USA

Cover image: Type IV collagen production shown by immunofluorescence (green) with cells (blue) stained by 4′,6-diamidino-2-phenylindole (DAPI), after culture of bovine testicular myoid cells on gelatin beads in spinner culture for 4 days. (Photograph by Veronica Glattauer, from studies with Dr Zerina Lokmic in the laboratory of Professor Lydia Sorokin).

Contents

Preface

This text is targeted as an introduction to the biophysics and chemistry of collagen and its use as a biomedical material in the rapidly changing field of biomedical device production, tissue engineering and regenerative medicine. This field is an amalgamation of a wide range of distinct disciplines.

The field has become huge, with the National Center for Biotechnology Information of the US National Library of Medicine database (PubMed) giving more than 14 000 entries for 'collagen and biomaterial' and more than 15 000 entries for 'collagen and tissue engineering'. And this does not count all the additional information found in the older chemical literature nor any of the many inventions described in the patent literature. And more work has been reported during the writing of this book!

Hence, the references cited are just a small selection, a mixture of older, original papers on topics, especially those by pioneers in the field, as well as more recent studies and reviews. The aim is to provide the reader with entry points to the broader literature, and to more recent papers, so that they can follow up and gain more detailed information on areas of particular interest.

The range of references may also be too limited for some of the present emerging areas of technology development. These include, for example, new fabrication approaches such as bioprinting, and production methods that include recombinant collagen, including designed collagens. Overall, it is clear that major developments will occur with innovative concepts for fabrication of composite materials that are based on collagen, where it provides critical biological functionality.

This book is focused on biomedical applications of collagens, so there are certain areas that have only been covered briefly. These include, for example, the large amount of data that covers collagen and its role in human health, including the issues that arise from mutations and other genetic changes that affect the collagen family. Also, the biosynthesis of collagen is only covered briefly, as only a limited amount of these data, such as proline hydroxylation, are relevant for biomaterials studies. This includes, for example, the information needed for recombinant collagen production.

The authors have worked together now for three decades. We have endeavored to be accurate, but undoubtedly some errors will have crept into the text. Communication and the exchange of ideas are key features to be encouraged in all areas of science. Some errors may arise from how we have defined terms. The definitions in this field change and are always under review. In 2018, a further Consensus Conference was held in Chengdu and these refined and revised definitions should be the ones to use in future.

Melbourne, 2019

Acknowledgments

The authors wish to thank many people for their roles in bringing this project to fruition.

They wish to thank their many present and former colleagues at CSIRO and collaborators elsewhere, including those in industry, for their inputs to the various studies described in this review. In particular, Dr Geoff Dumsday, Alan Kirkpatrick, Yong Peng, Violet Stoichevska, Tracy Tebb, Dr Jerome Werkmeister and Jacinta White at CSIRO, Professor Glenn Edwards, now at Charles Sturt University, Dr Anton Persikov and Dr Zhuoxin Yu when at University of Medicine and Dentistry of New Jersey, and especially Professor Barbara Brodsky when at University of Medicine and Dentistry of New Jersey and subsequently at Tufts University.

The authors are most grateful to their friends and colleagues in Chengdu, China, especially Professors Zingdong Zhang and Hai Lin, and Professor Mibel Aguilar of Monash University, for their support during the planning phase of this book.

We also wish to thank those who read sections of the manuscript during its preparation. These include Sue Barrett, Dr Joanne Caine, Dr Xiaojuan Hao, Dr Mark Hickey, Dr Louis Kyratzis, Dr Daniel Langley, Susan McMaster, Dr Mischa Mueller, Dr Ashley Murphy, Dr Tom Nebl, Paul Pasic, Jacinta Poole, Dr Lindsay Sparrow, Dr Helmut Thissen, Aditya Vashi, Dr Nick Welch and Jacinta White.

One of the authors (JAMR) wishes to thanks the University of Melbourne and Professors Peter Chung and Robert Kapsa and Ms Toni Ardino for their strong support.

The authors wish to thank the IOP staff, especially Jessica Fricchione, Daniel Heatley and Poppy Emerson and the production team for their support, advice and help during this project.

Finally, and most importantly, both authors would like to thank their families for their patience and support during the production of this book.

Author biographies

John A M Ramshaw

John A M Ramshaw is a chemist who studied at Van Mildert College, University of Durham, UK, where he was awarded BSc, PhD and DSc degrees. He recently retired as a chief research scientist at CSIRO and is now at the University of Melbourne. He has extensive experience on collagen and biomedical materials, through research leading to papers and patents, consulting and innovation. He has served on Government panels, and with the Australasian Society for Biomaterials and Tissue Engineering and the International Union of Societies for Biomaterial Science and Engineering. He was elected a Fellow of the Australian Academy of Technological Science and Engineering (2003) and won two CSIRO Research Medals (2005, 2012) as a key team member. He is a Fellow of the Royal Australian Chemical Institute (1997) and was awarded Fellow, Biomaterials Science and Engineering (2000).

Veronica Glattauer

Veronica Glattauer graduated from La Trobe University, in Melbourne, Australia, with a BBSci (Hons) degree in chemistry and biochemistry. She was later awarded a PGDip in Computer Science from Monash University. She has extensive scientific knowledge in the area of biomaterial science and engineering, extracellular matrix biology (principally collagen), protein chemistry and cell biology.

Glattauer is a Principal Research Scientist at CSIRO in the Biomedical Manufacturing Program. She leads a Cell Material Interactions team examining new biomedical material platforms and evaluating novel materials for medical devices and tissue engineering applications. Her research focuses on partnerships with Australian and international companies. She is a long-standing member and representative on the Executive Committee of the Australasian Society for Biomaterials and Tissue Engineering. She has received individual and team awards including two CSIRO Medals for Research Achievement (2005, 2012).

Abbreviations

ACI	Autologous chondrocyte transplantation
ACM	Acellular matrix
AGE	Advanced glycosylation end-products
Amino acids	Amino acids are in standard 3 and 1 letter code, with O=Hyp
ATP	Adenosine triphosphate
BAPN	β-aminopropionitrile
BCE	Before common era
BM-40	Basement membrane protein 40
BME	2-mercaptoethanol
BMP	Bone morphogenic protein
BP	Before present
BSE	Bovine spongiform encephalopathy
CD	Circular dichroism
CE	Common era
CL-domain	The collagen like domain in a bacterial collagen,
CMI	Cell-mediated immunity
CS	Chondroitin sulfate
DDR	Discoidin domain receptor
DSC	Differential scanning calorimetry
DTH	Delayed type hypersensitivity
DTT	Dithiothreitol
ECM	Extracellular matrix
EDC	1-ethyl-3(3-dimethylamino propyl)carbodiimide
EGF	Epidermal growth factor
ePTFE	Expanded poly(tetrafluoro ethylene)
FACIT	Fibril-associated collagen with an interrupted triple-helix
FLS	Fibrous long spacing
FMOC	Fluorenylmethyl oxycarbonyl
Fn	Fibronectin
FTIR	Fourier transform infrared spectroscopy
GA	Glutaraldehyde
GAG	Glycosaminoglycan
Gal	Galactosyl
Glc	Glucosyl
GPVI	Glycoprotein immune receptor
GT	Galactosyltransferase
GTT	Glucosyltransferase
HAp	Hydroxyapatite
HFIP	1,1,1,3,3,3-hexafluoro-2-propanol
HMDI	Hexamethylene diisocyanate
HyA	Hyaluronic acid
Hyl	Hydroxylysine
Hyp	Hydroxyproline
ILs	Ionic liquids
IMAC	Immobilised metal affinity chromatography
iPS	Induced pluripotent stem cell
LAIR	Leukocyte-associated immunoglobulin-like receptor

LbL	Layer-by-layer
LH	Lysyl hydroxylase
LOX	Lysyl oxidase
MAb	Monoclonal antibody
MIDAS	Metal ion-dependent adhesion site
MLBR	Major ligand binding region
MMP	Matrix metalloproteinase
mTG	Microbial transglutaminase
n.d.	Not determined
NC	Non-collagenous domain
NHS	*N*-hydroxysuccinimide
ORD	Optical rotary dispersion
OSCAR	Osteoclast-associated immunoglobulin-like receptor
OVP	Omniflow™ vascular prosthesis
P3H	Prolyl-3-hydroxylase
P4H	Prolyl-4-hydroxylase
P4HB	Poly-4-hydroxybutyrate
PBS	Phosphate-buffered saline
pC	C-terminal propeptide
PCL	Polycaprolactone
PEG	Poly(ethylene glycol)
PGA	Polyglycolic acid
PLA	Polylactic acid
PLGA	Poly(L-lactide-*co*-glycolide
PMMA	Polymethylmethacrylate
pN	*N*-terminal propeptide
PVA	Poly(vinyl alcohol)
PVP	Poly(vinylpyrrolidone)
RI	Refractive index
RTT	Rat tail tendon
SDS-PAGE	SDS-polyacylamide gel electrophoresis
SIS	Small intestine submucosa
SLS	Segment long spacing
SPARC	Secreted protein acidic and rich in cysteine
TCEP	Tris(2-carboxyethyl) phosphine
TCP	Tricalcium phosphate
T_d	Denaturation temperature
TEM	Transmission electron microscopy
TIMP	Tissue inhibitor of metalloproteinase
T_m	Melting temperature
TNBS	Trinitrobenzenesulfonic acid
TRL	Technology readiness level
tTG	Tissue transglutaminase
V-domain	A non-collagenous domain in a bacterial collagen
vWF	von Willebrand factor
Xaa	Amino acid residue after Gly in the repeating $(\text{Gly-Xaa-Yaa})_n$ triple helix
XPS	X-ray photoelectron spectroscopy
Yaa	Amino acid residue before Gly in the repeating $(\text{Gly-Xaa-Yaa})_n$ triple helix
Zaa	Amino acid residue interrupting a repeating $(\text{Gly-Xaa-Yaa})_n$ triple helix
σMet	L-selenomethionine

Disclaimer

Various commercial products and companies are mentioned in this book. These are examples only and other equivalent examples may exist. There is no endorsement or recommendation made or intended for those items that are mentioned and the authors cannot verify the efficacy of any material or process. Before using any products or processes, the reader is advised to independently ensure they are appropriate for the reader's needs, and that all aspects of safety have been considered.

Part I

Introduction

IOP Publishing

John A M Ramshaw and Veronica Glattauer

Chapter 1

Introduction

Collagens are recognised as important biomaterials. Devices made using collagen have been clinically successful and are well accepted by clinicians and by patients. Development of such devices requires a sound understanding of the structure, chemistry and biology of collagens. This introductory chapter includes a brief history of collagen and its application to materials, along with an outline of the topics covered in subsequent sections. The literature on collagen and its use as a biomedical material is now extensive, so this introduction also discusses the relative importance of different knowledge areas to the development of collagen-based products.

1.1 Overview

Collagen is an animal protein that is both critical and very abundant and is found throughout the body. It plays a large number of readily identifiable roles. It is the main component of skin, bone, tendon, ligament, cartilage and other specific tissues, where it provides strength and durability, coupled with flexibility, in addition to a broad range of specific biological interactions (see chapter 7). Collagen is found in the extracellular matrix of all multicellular animals, including sponges, invertebrates and vertebrates. It is possible that an early form of collagen was a key molecule whose role featured in the emergence of multicellular organisms [1]. Indeed, the presence of potential precursor forms has been observed recently in some, but not all, bacterial genomes [2].

The name 'collagen' is frequently used as a generic term for what in fact is a large family of proteins that are key structural components of the extracellular matrix. The 28 members of this diverse family are linked through a common structural element, the triple-helix (see chapter 2), as well as several shared biochemical themes [3]. The collagens are just a small part of the much larger family of members of the

doi:10.1088/978-0-7503-2096-2ch1

extracellular matrix, comprising well over 500 distinct genes, now referred to as the 'matrisome' [4]. This grouping was first introduced by George Martin and colleagues for a subgroup of components that form basement membranes [5]. The concept was extended by David Hulmes to include the full group of extracellular matrix components (as the matrixome, e.g., at FECTS, Sicily in 2004), before being further described by others [4]. Many of these other genes are associated with the biosynthesis and turnover of collagens (see chapter 3), as well as many other tissue components that are involved with collagen and its interactions (see chapter 7).

It did not take the complexity of the present day understanding of the matrisome and tissues for the value of collagen as a material, and particularly as a biomaterial, to be recognised. On the contrary, collagen in tissues has been recognised for millennia as a useful material, especially in the use of hides and skins for leather manufacture and also for gelatin production. Indeed, it is not clear when it was first recognised that hides and skins could be preserved against decay, by tanning the collagen, to give leather. Interestingly, it has been suggested that in prehistory, by the Chalcolithic period (or Copper Age) a variety of identifiable properties and applications for different sources of collagen products were recognised and considered, and that specific choices of material were made [6]. For example, the Tyrolean Iceman from the Ötzal Alps, a natural mummy from circa 5300 BP, wore and had implements made from five or more different animal sources, suggesting specific choices [6].

Equally, historically, it was well established that collagen tissues from all animal phyla were sources of gelatins (glues), through heating or other denaturation, although the resulting materials were used for making glues, paints and similar products rather than for food applications. In ancient times, gelatin was used as a medical material, with recipes from Philotas (1st century BCE, cited by Celsus in *De Medicina*, circa 40 CE) and Asclepiades (circa 80–110 CE), including fish-glue (gelatin) in materials that could 'glue wounds shut' and that were 'particularly useful in the treatment of skull fractures' [7]. It seems that a form of fish-glue may even have been used up until the 1950s to seal small wounds that did not need sutures [7]. At that time, it was recognised, for example by Pliny in *Naturalis Historia*, that the quality of gelatins varied between the sources of raw materials, for example between fish versus mammalian sources, and between different parts in an animal source [8]. Gelatin's role as a purified food product seems more recent, being used in royal courts from the middle ages [9], until it became more popular in the mid-19th century, possibly through improvements to isinglass [10, 11] and then to 'common gelatin' [9, 12], followed later by the development of quality standards for food manufacturing and applications (e.g., [13]; discussed by [9]).

Collagens have been useful as biomedical materials from early times. For example, Claudius Galen in circa 175 CE used intestine (gut) based, collagenous materials as sutures, with particular success, especially for the repair of severed tendons in gladiators [14], recommending sanitising the wound (and possibly the suture) with wine [15]. It is possible that fibres from sinew were used prior to this [16]. Early texts mention suturing, particularly the Edwin Smith Surgical Papyrus from around 3500 BP [17] which may be a copy of an earlier document from around 4500 BP [18], but the materials used are not described. It is possible, however, that if not collagen then either keratin (e.g., as horse hair), or wool, linen, or possibly silk was used; sericulture had been well established in China by the 4th millennium BP [19].

There seem to be few subsequent reports of collagen biomaterials until collagen tissues were examined, starting in the late 19th century. These materials included tubes made from demineralised chicken bone for use as drainage tubes [20]. These rubber-like tubes could have been quite useful in an era before synthetic polymer devices. Shortly after, other applications were also proposed for this material, including nerve repair and bone repair [21–23]. Later, foetal membranes were also examined for wound repair [24, 25], such as for a leg ulcer [24], and this material has continued to be developed up to the present day [26, 27]. However, it was several decades later, after these initial studies, before complex manufactured items were being produced using collagen. Initial products included dressings and haemostats, (e.g., *Brocatamp, Brocafil, Sangostop*) [28–30].

Subsequently, principally in the last 50–60 years, numerous successful collagen-based devices have emerged that have proved safe and effective in medical products and have achieved clinical and consumer acceptance. An advantage of collagen is that it can be readily adapted to meet a range of different clinical applications. Despite this excellent versatility, one consistent query that is often raised is whether collagen devices lead to any severe immune system anomalies. However, this has not been observed and the products continue to be regarded as safe. Hence, various exciting developments in collagen-based materials and their applications continue to emerge. Currently, this includes the development of collagen-based materials for applications in tissue engineering and cell therapy, and particularly novel composite materials. However, if further advances are to be made it is critical that the ever-increasing understanding of collagen science is fully integrated into scientific, technological and engineering solutions. This includes the understanding of structure and function, biochemistry and molecular and cellular biology of collagens and the host response to collagen materials.

In the following text, section II focuses on the structure and biochemistry of the collagen family. The authors feel that for researchers developing new collagen-based biomaterials, a broad understanding of these topics is essential. Although many critical developments occurred in the early days of collagen research, the total amount of literature for those that want to delve deeper into these topics is now very extensive. Collagen research has blossomed in many, diverse directions in recent years. However, for those who are interested, details on the history of collagen and gelatin can be found in a range of texts (see, for example, [31–35]). An area of significant progress in recent years has been the focus on collagen and matrix-based diseases, both inherited and acquired, that arise from changes in collagen structure and biology. The huge variety of disease conditions that have been elucidated reflect the wide number of collagen types, distributions and functions that have been established. Changes that alter the Gly residue found at every third residue in the collagen triple-helix are frequently associated with disease [36].

When considering the extent of information that is available in the literature, it is worth noting that for both collagen biology and structure, the amount of knowledge is greatest at the smallest scale, at an atomic level in many cases, while relatively little is known about higher structural features, such as packing of collagen into fibre bundles and tissues or about the development and control of function (see chapters 4 and 5). Yet it is these larger scale questions that are of particular relevance to collagen-based biomaterial and tissue engineering scaffold performance.

As section II aims to provide only a broad understanding of the structure and biochemistry of the collagen, there are a range of topics that are not covered in depth as the authors do not believe that they have immediate relevance to collagen-based biomaterials production and use. As such, further reading is encouraged for those who want to enquire further, and a selection of review papers are cited in the appropriate chapters. These areas include the complex biosynthetic and catabolic pathways for collagens, along with the large and complex nature of the structural genes, excepting where some information is required for the development of recombinant collagen materials. Also, the wide variety of collagen-based diseases, arising from collagen mutations or collagen processing or interactions, is not covered in detail and other texts are needed for this information (e.g., [37]). Embryonic tissue development is also not included, although some features that are emerging from these studies may be relevant to the host response to a biomedical implant and to the development of tissue engineering and regenerative medicine strategies.

Subsequently, section III discusses the chemistry and processes that are needed to take tissue collagen and convert it into functional biomedical materials. It was natural that collagens, with their vital roles in tissues, including structural and mechanical contributions, development, wound healing, platelet activation and angiogenesis, etc, would be seen as ideal biomedical materials. Despite the development of recombinant systems, animal derived materials still remain the main commercial source of the collagens and are likely to remain so for many applications.

Collagen-based biomaterials are often discussed as a single group whereas quite distinct formats are used. These include tissue-based biomaterials, where intact tissue is adapted so that the product meets the medical needs, and purified collagen, either as a comminuted product or as a soluble product, which may then be reformed into a particular shape or form. Whatever the format, key activities that are needed for biomedical material manufacture include preparation of the collagen into a pure, non-immunogenic form. This is followed by fabrication of the product into the various shapes and sizes of material that form the medical devices. These devices need stabilisation to either ensure their long-term durability or else to control the turnover in a device with an intended shorter lifetime.

Finally, in section IV, which is followed by some concluding remarks in section V, the application of collagen-based materials to specific clinical applications is considered. A selection of applications is discussed, including materials that have reached the market as accepted, regulated products, as well as prototype devices that provide new directions for clinical opportunities and products.

All medical implants, collagen ones included, are foreign to the host and as such lead to various tissue responses during the healing process. These vary depending on the nature of the material and its site of implant. Understanding these processes enables the performance of materials to be evaluated, as well as providing knowledge that can be used for the development of better materials. Of particular interest is the extent of surrounding fibrous tissue formation as this can often have a direct and detrimental bearing on device performance. However, most studies on implant performance have been cell-based and have focused on measuring negative or unwanted secondary cellular reactions and often limited to short-term assessments of biocompatibility rather than the longer-term performance of an implant.

References

[1] King N 2004 *Dev. Cell* **7** 313–25

[2] Rasmussen M, Jacobsson M and Björck L 2003 *J. Biol. Chem.* **278** 32313–16

[3] Ramshaw J A M and Brodsky B 2003 *Industrial Proteins in Perspective* ed W Y Aalbersberg *et al* (Amsterdam: Elsevier) pp 134–41

[4] Hynes R O and Naba A 2012 *Cold Spring Harb. Perspect. Biol.* **4** a004903

[5] Martin G R, Kleinman H K, Terranova V P, Ledbetter S and Hassell J R 1984 *Ciba Found. Symp.* **108** 197–212

[6] O'Sullivan N J, Teasdale M D, Mattiangeli V, Maixner F, Pinhasi R, Bradley D G and Zink A 2016 *Sci. Rep.* **6** 31279

[7] Scarborough J 2015 *Classical Philol.* **110** 54–65

[8] Ulrich R B 2008 *Roman Woodworking* (New Haven, CT: Yale University) p 71

[9] Schrieber R and Gareis H 2007 *Gelatine Handbook: Theory and Industrial Practice* (Weinheim: Wiley)

[10] Royle J F 1842 *Production of Isinglass along the Coasts of India* (London: Allen)

[11] Cooper P 1845 *US Patent* 4084

[12] Arney G A 1846 *GB Patent* 11045

[13] Bloom O T 1925 *US Patent* 1540979

[14] Mackenzie D 1973 *Med. Hist.* **17** 158–68

[15] Muffly T M, Tizzano A P and Walters M D 2011 *J. Roy. Soc. Med.* **104** 107–12

[16] Huber C, Eckstein F S, Halbeisen M and Carrel T P 2003 *Ann. Thorac. Surg.* **75** 1318–21

[17] Breastead J H 1930 *The Edwin Smith Surgical Papyrus, in Facsimilie and Hieroglyphic Transliteration with Translation and Commentary* (Chicago, IL: University of Chicago Oriental Inst.)

[18] Hughes J T 1988 *Paraplegia* **26** 71–82

[19] Barber E J W 1992 *Prehistoric Textiles: The Development of Cloth in the Neolithic and Bronze Ages with Special Reference to the Aegean* (Princeton, NJ: Princeton University Press) p 31

[20] Neuber G 1879 *Arch. Klin. Chir.* **24** 314–30

[21] Gluck T 1881 *Berliner Klinische Wochenschrift* **18** 554–57

[22] Vanlair C 1882 *Arch. Biol.* **3** 379–496

[23] Senn N 1889 *Am. J. Med. Sci.* **98** 219–47

[24] Davis J S 1910 *Johns Hopkins Hosp. Rep.* **15** 307–96

[25] Sabella N 1913 *Med. Rec. N Y* **83** 478–80

[26] Lim L S, Poh R W, Riau A K, Beuerman R W, Tan D and Mehta J S 2010 *Arch. Ophthalmol.* **128** 1303–10

[27] Lemke A, Ferguson J, Gross K, Penzenstadler C, Bradl M, Mayer R L, Gerner C, Redl H and Wolbank S 2018 *Acta Biomater.* **66** 335–49

[28] Feriz H 1940 *Surgery* **8** 654–61

[29] Feriz H 1941 *Surgery* **10** 326–35
[30] Schmitt F O 1985 *Annu. Rev. Biophys. Biophys. Chem.* **14** 1–22
[31] Alexander J 1923 *Glue and Gelatine* (New York: American Chemical Society)
[32] Gustavson K H 1956 *The Chemistry and Reactivity of Collagen* (New York: Academic)
[33] Ramachandran G N 1967 *Treatise on Collagen* vol 1 (New York: Academic)
[34] Fraser R D B and MacRae T P 1973 *Conformation in Fibrous Proteins* (New York: Academic)
[35] Ward A G and Courts A 1977 *The Science and Technology of Gelatin* (New York: Academic)
[36] Myllyharju J and Kivirikko K I 2001 *Ann. Med.* **33** 7–21
[37] Royce P M and Steinmann B 2002 *Connective Tissue and Its Heritable Disorders: Molecular, Genetic, and Medical Aspects* 2nd edn (New York: Wiley-Liss)

Part II

The structure and biochemistry of collagen

IOP Publishing

John A M Ramshaw and Veronica Glattauer

Chapter 2

The structure of collagen

The structure of collagen, like all proteins, defines its properties and its interactions with other molecules and cells in tissues. The chemical composition, and in particular its amino acid sequence (primary structure), provides the diversity of structure necessary for precise interactions while the higher orders of structure (secondary and tertiary structure) define and maintain topology that also provides specificity of interactions as well as providing bulk properties of assemblies, such as strength. All these, and other properties that arise from the collagen structure are important elements in the design and function of effective, collagen-based biomaterials.

2.1 Composition

The concept that all proteins are polymers of amino acids was proposed in 1902, by Fischer [1] and by Hofmeister [2]. They proposed that proteins were formed through bonds between the amino group of one amino acid and the carboxyl group of another, in a linear structure. The term amino acid itself had only emerged in 1898. Previously, throughout most of the 19th century, proteins, including gelatin that was derived from collagen, had been hydrolysed and individual amino acids isolated. Gelatin hydrolysis first gave rise to 'sugar of gelatin', named because of its sweet taste [3], which was later called glycine [4]. Hydrolysis was by boiling, initially using H_2SO_4, but later HCl was introduced [5], which remains the preference to the present. Thus by 1900, some 13 amino acids had been isolated from proteins [6]. The last of the 20 common amino acids to be discovered was threonine in 1935 [7]. The two amino acids derived by secondary modification of collagen, hydroxyproline (Hyp) and hydroxylysine (Hyl) were first identified in 1902 [8] and in 1921 [9] respectively. However, the realisation that secondary modification was involved (see chapter 3) and that they were not directly incorporated into protein came some decades later [10, 11]. Subsequently, collagen was also found to have a low (1%)

doi:10.1088/978-0-7503-2096-2ch2

total carbohydrate content that was shown to be galactose or glucosylgalactose bound to the hydroxyl groups of Hyl [12] (see chapter 3).

Throughout this period, determining the amino acid composition of proteins was a tedious and difficult process. Nevertheless, the distinct features of the amino acid composition of collagen and gelatin emerged, with the high contents of glycine and proline being the most noteworthy [13]. In the late 1950s the use of ion exchange resin columns allowed the development of an automated, rapid (overnight) and reliable system for amino acid analysis to be established [14, 15]. From this, reliable and reproducible data became available, for collagen [13], and subsequently for individual chains and for different collagen types (see below) (table 2.1).

These data confirmed the unusually high content of Gly, near 33%, and a high content of imino acids (Pro and Hyp), near 20% (table 2.1). Comparisons between the compositions of collagen and gelatin showed close similarity, but often gelatin

Table 2.1. The amino acid compositions of various collagen chains[a].

Amino acid	α1[I]	*α1[I] DNA*	α2[I]	*α2[I] DNA*	α1[II]	*α1[II] DNA*	α1[III]	*α1[III] DNA*
3-Hydroxyproline	1	*n.d.*	1	*n.d.*	2	*n.d.*	0	*n.d.*
4-Hydroxyproline	108	*n.d.*	93	*n.d.*	97	*n.d.*	125	*n.d.*
Aspartic acid	42	46	44	26	43	37	42	29
Asparagine	n.d.	12	n.d.	24	n.d.	14	n.d.	23
Threonine	16	18	19	19	23	22	13	15
Serine	34	39	30	34	25	29	39	44
Glutamic acid	73	49	68	45	89	59	71	49
Glutamine	n.d.	30	n.d.	23	n.d.	44	n.d.	25
Proline	124	240	113	207	120	249	107	240
Glycine	333	347	338	350	333	376	350	373
Alanine	115	121	102	109	103	112	96	97
Cysteine	0	0	0	0	0	0	2	2
Valine	21	21	35	40	18	20	14	15
Methionine	7	7	5	5	10	11	8	9
Isoleucine	6	7	14	18	9	13	13	16
Leucine	19	21	30	35	26	29	22	22
Tyrosine	1	5	4	5	2	2	3	6
Phenylalanine	12	15	12	12	13	16	8	9
Hydroxylysine	9	*n.d.*	12	*n.d.*	20	*n.d.*	5	*n.d.*
Lysine	26	38	18	32	15	44	30	40
Histidine	3	3	12	12	2	2	6	7
Arginine	50	53	50	57	50	58	46	48
Tryptophan	n.d.	0	n.d.	0	n.d.	0	n.d.	0
Gal-hydroxylysine	1	*n.d.*	1	*n.d.*	4	*n.d.*	—	*n.d.*
Glc-Gal-hydroxylysine	1	*n.d.*	2	*n.d.*	12	*n.d.*	—	*n.d.*

[a] Data determined by amino acid analysis (normal font: [13], expressed as residues per 1000) and from DNA sequences (italic font: taken from the NLM database, as total occurrences including telopeptides). Asn, Gln and Trp cannot be determined after acid hydrolysis. Alternative hydrolysis conditions are needed for sugar quantitation. DNA sequence data does not give values for Hyp, Hyl and sugar modified Hyl.

had a lower ionic point as a result of amide hydrolysis. Thus, amino acid hydrolysis data cannot be used to calculate an isoelectric point and the amino acid sequence data is needed, although if DNA derived data is used the residues that have been modified (Hyp, Hyl) are not identified.

2.2 X-ray diffraction studies

2.2.1 Fibre diffraction

It is now almost a century since the initial x-ray fibre diffraction studies on collagen (on nuchal ligament) were reported [16]. However, it took several decades before a sufficient understanding of helical diffraction and appropriate diffraction pattern photographs had emerged to allow a determination of the collagen structure (reviewed by [17]).

Initial progress on understanding the collagen structure was limited by the poor diffraction patterns that were available. Nevertheless, Astbury in 1933 [18] noted that the fibre diffraction pattern observed for rat tail tendon (RTT) was quite distinct to those seen for other fibrous proteins. The RTT pattern showed a strong axial reflection at 2.9 Å and a strong equatorial reflection between 11 Å and 16 Å [18]. The equatorial spacing depended on the extent of hydration of the sample [19]. Also, with great foresight, Astbury [18] suggested that the collagen structure included a 'Gly-Pro-Oxyproline' (oxyproline is now called Hyp) repeating unit, which would provide an axial height of around 2.8 Å per residue.

Various models for the collagen structure were proposed. These included several sheet structures of parallel chains hydrogen bonded into layers, but the density calculated from these models was not a good fit to the observed data [20]. Others included helical models, with the chains packed perpendicular in a helical array, but these models did not account well for the swelling data [20].

The major breakthrough, reported at a Faraday Society Meeting in London in 1953, was that stretching a collagen fibre led to a significantly improved diffraction pattern. This pattern appeared to be the transform of a rod-like structure with helical symmetry [20]. At that time, in the subsequent General Discussion, Mr Crick [*sic*] in the audience noted of the photograph that 'to the crystallographic eye it is quite overwhelming' as he recognised that from 'the theory of helical structure, ... the new picture is exactly what you expect'; also, he felt that 'the structure of collagen is likely to be worked out in a short period .... because this picture is rather an important one' [20]. This prediction proved to be correct.

In 1954, using a knowledge of composition and diffraction information, Ramachandran and Kartha proposed a triple-helical structure for collagen [21]. This suggested a three-stranded model in which each chain was in a left-handed helix. This model also placed Gly residues in the centre of the structure. The proposed structure left little space in the centre, and only Gly, the sole amino acid that lacks a C_β atom, was small enough to fit. Thus, an immediate consequence of this proposal was that these Gly residues would be part of a repetitive amino acid sequence pattern $(Gly\text{-}Xaa\text{-}Yaa)_n$ that has become recognised as characteristic of collagen molecules [22]. The proposed triple-helical structure was also supported by physico-chemical investigations that suggested that collagen molecules consisted of three polypeptide chains, each with length of about 1000 amino acids [23].

The three-chain 'triple-helical' concept has subsequently been generally accepted, although slight changes to the initial proposal by Ramachandran [24] and others [25] were needed to resolve all the stereochemical and hydrogen bonding issues.

The key issues that needed to be solved were that certain interatomic distances were very short, that a Pro residue could not be readily incorporated in the structure after a Gly residue and that the hydrogen bonding pattern was not clear. Significant progress was made when Rich and Crick [25] proposed two new models (termed RCI and RCII) that were stereochemically satisfactory (figure 2.1). Although developed from a poly-Gly model, in the new models all three chains have *trans* peptide bonds and a conformation that resembles the poly-L-proline II structure. This allowed Pro residues to be accommodated in the non-Gly positions without distortion. The differences between the new models focused on the arrangement of the hydrogen bonds. Various stereochemical considerations suggested preference for the RCII model [25].

Subsequently, a detailed description of the chain conformation of the kangaroo collagen molecule was derived by Fraser and colleagues using quantitative x-ray diffraction data collected from stretched tail tendon. The best solution, which had a crystallographic R factor of 0.272, very closely resembled the RCII structure (figure 2.1) and provided a possible interchain hydrogen bonding scheme. The RCI model proved unacceptable due to the large number of short contacts between non-bonded atoms [26], confirming the previous conclusions [25]. The new data also gave an estimated value for the unit twist of the molecular helix of 107.1° ± 0.6°,

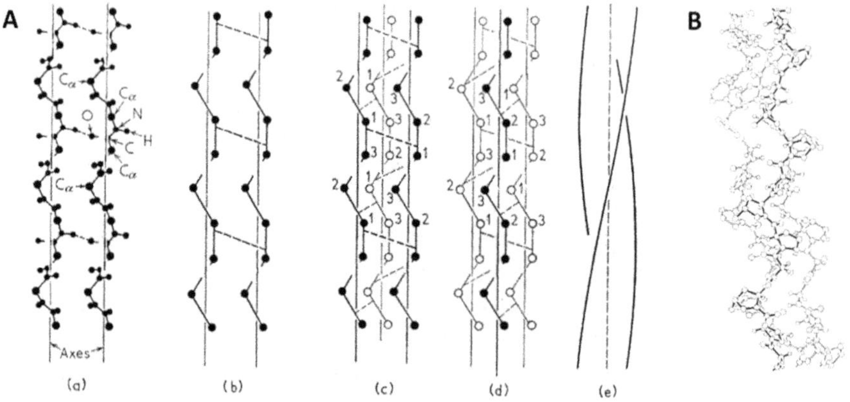

Figure 2.1. Diagrams illustrating the chain conformation of the collagen triple helix. **A** (a) Two strands of polyglycine with black dots representing various atoms and dashed lines as hydrogen bonds. (b) Two chains of the polyglycine II lattice in which only the *Ca* atoms and hydrogen bonds are shown. (c) The third chain shown with open circles makes a collagen RCI arrangement. (d) The third chain added giving rise to the collagen RCII model. (e) Solid lines representing the axes around which the chains are coiled. (Reprinted from [25] with permission from Elsevier.) **B** The 3-strand coiled coil structure determined by a linked-atom least-squares refinement procedure on x-ray data collected from dried kangaroo tail tendon. (Reprinted from [26] with permission from Elsevier.)

which is close to the value expected (108°) for a 10/3 helix, that is a helix with 10 units in 3 turns [26]. Of course, there there is no *a priori* reason why the collagen helix should be based on such a simple numerical format [22]. (In crystallographic notation, a 10/3 helix, with 3.3 units/turn is a 10_7 helix, while a 7/2 helix, with 3.5 units/turn is a 7_5 helix [27]). The availability of large collagen-like polypeptide structures [17] did not allow better data to be obtained.

Fibre diffraction studies had, therefore, characterised the defining feature of collagen, its triple-helical structure. This structure is present in every native collagen, whatever the source, and not just mammalian tissues. It comprises three polypeptide chains, each in a left-handed extended polyproline II-like helix (figure 2.1). The three chains, which are parallel and staggered by one residue with respect to each other, confirmed by NMR [28], are supercoiled about a common axis in a right-handed manner with interchain hydrogen bonds linking all three chains. The Yaa position on one chain is at the same axial level as the Xaa position on an adjacent chain, as a result of the stagger, so rotation of the side chains can bring them in close proximity and allow interactions. Every third residue in the amino acid sequence is a Gly residue, due to the close packing of the chains, while the extended conformation of the individual chains is stabilized by a high content of imino acids (Pro and Hyp) [22].

A consequence of this structure is that while all Gly residues are buried, all residues in the Xaa and Yaa positions have considerable exposure to solvent, with the Xaa position more exposed than the Yaa position. This allows for a high percentage of charged residues, up to 35%–40% of nonglycine residues, while potentially accommodating any of the 20 common amino acids without distortion.

2.2.2 Peptide diffraction

The collagen molecule is too big for detailed structural studies on the intact molecule. The availability of fragments and small peptide models for collagen has enabled very detailed structural information to be determined, down to an atomic scale. Initially, a model repeating tripeptide 30-mer $(\text{Pro-Pro-Gly})_{10}$ was crystallised [29]. However, the diffraction data indicated the collagen-like helices were aligned end-to-end in a fibre-like manner, rather than as crystalline packing of repeating units [30]. Nevertheless, structural data was obtained using a linked-atom least-square refinement. Interestingly, these data showed backbone conformations similar to those of dry stretched collagen [26], but with a different, 7/2 helical symmetry [30]. Although there is only a small structural difference between 10/3 and 7/2 helices, these changes can lead to major shifts in register over distances that include hundreds of amino acids.

Subsequently, the first crystal structure of a collagen peptide, a 30-mer peptide, was obtained by Brodsky and colleagues [31] and this has been followed by over 60 others, all at high resolution (table 2.2) [27]. This initial structure was a mimic of mild *Osteogenesis imperfecta* that included a mutation of a Gly to Ala [31]. The peptide chains were in extended left-handed polyproline II-like helices supercoiled in a right-handed manner. Of interest was that the helix in the unmutated zones was a

Table 2.2. Some examples of crystal structure data reported for collagen-like peptides (see [27]). Imino content (Pro and Hyp) is given for the non-Gly residues in the Xaa and Yaa positions of the collagen repeat. σMet is L-selenomethionine.

Structure	Resolution (Å)	Helical symmetry	Imino content (%)	
$(Pro\text{-}Hyp\text{-}Gly)_4$-Pro-Hyp-Ala-$(Pro\text{-}Hyp\text{-}Gly)_5$	1.9	7/2	100	[31]
$(Pro\text{-}Pro\text{-}Gly)_{10}$	1.8	7/2	100	[32]
$(Pro\text{-}Pro\text{-}Gly)_9$	1.33	7/2	100	[34]
$(Pro\text{-}Hyp\text{-}Gly)_{10}$	1.8	7/2	100	[33]
$(Gly\text{-}Pro\text{-}Hyp)_9$	1.45	7/2	100	[41]
$(Hyp\text{-}Hyp\text{-}Gly)_{10}$	1.5	7/2	100	[39]
$(Gly\text{-}Pro\text{-}Hyp)_3$-$(Gly\text{-}3Hyp\text{-}Hyp)_2$-$(Gly\text{-}Pro\text{-}Hyp)_4$	1.8	7/2	100	[38]
$(Pro\text{-}Hyp\text{-}Gly)_4$-Leu-Hyp-Gly-$(Pro\text{-}Hyp\text{-}Gly)_5$	1.6	7/2	95	[35]
$(Pro\text{-}Hyp\text{-}Gly)_3$-Pro-Arg-Gly-$(Pro\text{-}Hyp\text{-}Gly)_4$	1.45	7/2	94	[36]
$(Pro\text{-}Hyp\text{-}Gly)_4$-Glu-Lys-Gly-$(Pro\text{-}Hyp\text{-}Gly)_5$	1.75	7/2	90	[37]
$(Pro\text{-}Hyp\text{-}Gly)_4$-$(Leu\text{-}Hyp\text{-}Gly)_2$-$(Pro\text{-}Hyp\text{-}Gly)_4$	1.4	7/2	90	[35]
$(Gly\text{-}Pro\text{-}Hyp)_2$- Gly-Phe-Hyp-Gly-Glu-Arg-$(Gly\text{-}Pro\text{-}Hyp)_3$	1.8	7/2 and 10/3	78	[45]
$(Pro\text{-}Hyp\text{-}Gly)_3$-Ile-Thr-Gly-Ala-Arg-Gly-Leu-Ala-Gly-$(Pro\text{-}Hyp\text{-}Gly)_4$	2.0	7/2 and 10/3	70	[44]
Gly-Pro-Ile-Gly-Pro-Hyp-Gly-Pro-Arg-Gly-Asn-Arg-Gly-Glu-Arg-Gly-Ser-Glu-Gly-Ser-Hyp-Gly-His-Hyp-Gly-σMet-Hyp-Gly-Pro-Hyp-Gly-Pro-Hyp-Gly-Ala-Hyp-Gly-Pro-Cys-Cys-Gly-Gly	2.3	7/2 and 10/3	48	[46]

7/2 helix, not a 10/3 helix expected from the data of Fraser and colleagues [26]. Subsequently, a range of unsubstituted peptides including just Pro and Hyp were studied, for example $(Pro\text{-}Pro\text{-}Gly)_{10}$ and $(Pro\text{-}Hyp\text{-}Gly)_{10}$, where the Hyp was in the normal Yaa position [32–34] (table 2.2). Additional studies looked at similar constructs, but with substitutions of one [35, 36] or two [37, 38] amino acids, including substitution to provide a 3-Hyp in the Xaa position [38] (table 2.2). Studies also examined non-native variations in locating a 4-Hyp in the Xaa position [39, 40] (table 2.2). In many reports, interactions between adjacent molecules in the crystal are discussed. However, these data should be treated with caution as they may be artefacts of the crystal packing, where a rare interaction *in vivo* is the particular one that forms the crystal. All these studies also showed a 7/2 helix. There are some [41] who have seen this as the true, overall structure of the collagen triple-helix, whereas others suggest that it results from the very atypical composition of these peptides, with imino acid contents of 90% or more, that is not found in natural collagens [42, 43] (table 2.2). The conformational restraints arising from imino acids favour a tight helix [27]. Studies on other peptides, which have biologically relevant sequences that have a lower imino content [44–46], suggest that the 7/2 helix is indeed an artefact from the high imino acid content and that the helix is predominantly a 10/3 helix. However, this could include variations along its length, including changes arising from local helix dissociation and molecular packing [42, 43].

It is worth noting that the availability of peptide models for collagen has also proved of value for NMR studies. However, although determining structural information at a molecular level in solution became more feasible using NMR spectroscopy, especially through two-dimensional NMR spectroscopy, assignments proved difficult since the rod-like collagen molecule behaves like a globular protein with a radius similar to the axial length of the collagen peptide, giving rise to significantly broadened spectra, and difficulties arise from the repeating tripeptide sequences giving overlapping resonances. Nevertheless, the structural data obtained to date indicate that the collagen molecule in solution has a structure that is similar in general properties to those seen in crystalline form. For example, the atoms found to be close together in solution by NOE (i.e., less than 4 Å apart) are consistent with those expected for the triple-helical model, and other studies showed that the NH of Gly in $(Pro\text{-}Hyp\text{-}Gly)_{10}$ was very slow to exchange, again indicating the rigidity of the helix in solution [28].

2.2.3 Hydrogen bonding

The emergence of high-resolution crystal structures for collagen model peptides (table 2.2) allowed confirmation of the intramolecular hydrogen bonding patterns, which were seen to be fully consistent with the RCII assignments [25]. The interchain intermolecular amide Gly NH...O=C Pro bonds form the innermost and probably most stable H bonding set [31]. In addition, the presence of regular GlyCH...O–C bonds was confirmed [47] (figure 2.2).

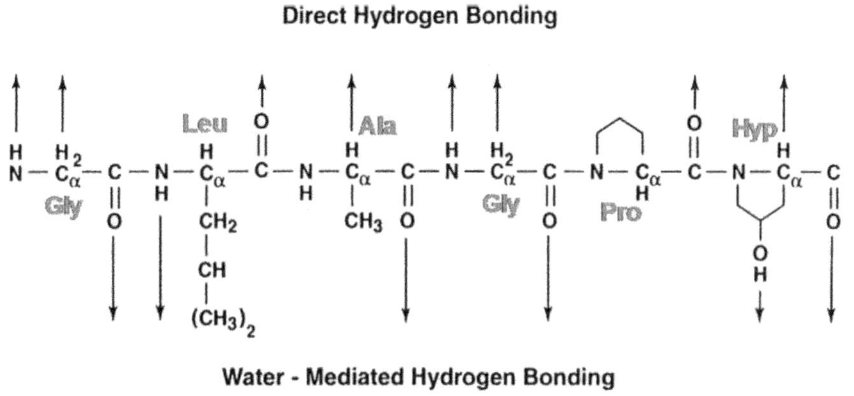

Figure 2.2. An illustration of the observed hydrogen bonding in a collagen triple-helix [44, 49] containing the sequence Gly-Leu-Ala-Gly-Pro-Hyp. Residues that have been shown to be involved in direct hydrogen bonding are shown with an upward pointing arrow (↑), while those involved in water-mediated hydrogen bonding are shown with a downward arrow (↓). From [50], reprinted with permission from *Springer Nature*.

Initially, there was no information concerning the H bonding pattern of an NH group in the Xaa position, since this position was always occupied by the imino acid Pro. Subsequently, in peptides where Pro did not occupy the Xaa position [37, 44], an additional interchain hydrogen bond, NH (Xaa)…CO (Gly), was observed, formed through one mediating water molecule.

To date, there are too few structures available for a definitive view of possible side chain interactions that are independent of any crystal packing considerations. Nevertheless, Arg side chains in the Yaa position were seen to make direct contacts to backbone carbonyl groups, which may account for the stabilizing effects of these residues [48]. In addition to demonstrating that sequence can influence local conformation affecting the helical pitch, the limited structural data from peptides containing non-Pro residues, also show the potential for hydrogen bonding and hydration variations as well.

2.2.4 Hydration

A topic that has advanced significantly through these peptide diffraction studies (table 2.2) is the knowledge of the hydration structure for collagens. Other studies had provided evidence that there was ordered water in fibrils that was distinct from bulk water [27, 51, 52]. The initial crystal structure that was solved [31] showed an extensive, complex and quite regular hydration network [53]. While the role of water molecules in collagen stabilisation has been questioned [54], they clearly do exist as an intrinsic feature of the triple-helix [53]. However, thermodynamic analyses by Privalov [55], indicated an unusually favourable enthalpy for collagen, which was interpreted as the dominance of hydrogen bonding, including hydration, in stabilization of the triple-helix; this has now been observed in the range of crystal structures. An extensive hydration network is part of the response to satisfy available backbone groups, and the role of Hyp in collagen stability may also be related to this hydration network.

Thus, 168 water molecules were localised in the structure for the 90 residues of the trimer and must confer a cumulative stability of many bonds. In the crystal structure, every OH of a Hyp and the C=O of the Gly and Yaa (Hyp) residue positions were all H-bonded to water, meaning that every possible H bonding donor and acceptor is satisfied (figure 2.3). Remarkably, the water molecules visualized in the crystal structure were even more intricate than had been proposed and 2, 3, 4 or 5 intervening water molecules connecting C=O groups or OH groups of hydroxyproline were observed (figure 2.3) [53]. Extended water bridges that appeared to relate to the lateral distance between adjacent molecules were less regular than the intrachain and interchain water network. It was expected that the intrachain hydrogen bonding would likely stabilize the conformation of individual chains, while interchain bonds between the three chains stabilise the molecule [27]. The structures show clearly that the basis for Hyp stabilization of the triple-helical structure lies in its pivotal position in this ordered water network. Further, in regions where there are no imino acids, the water-mediated hydrogen bonds contribute to local stability and are essential for maintaining the triple-helical conformation [27].

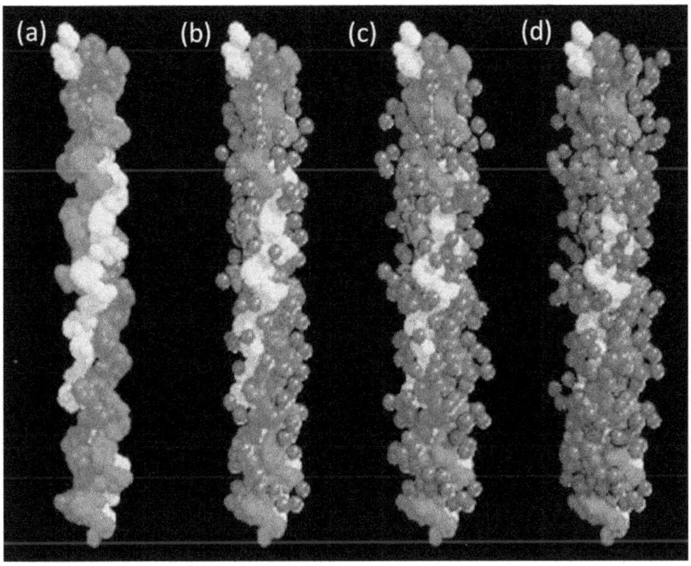

Figure 2.3. The defined hydration (water molecules in blue) that surrounds a collagen model peptide. (Reprinted from [53], with permission from Elsevier.)

2.2.5 The role of hydroxyproline

In animal collagens a key feature that adds significantly to stability is the modification of most Pro residues in the Yaa position to hydroxyproline (Hyp) [56] (see chapter 3). The uniqueness of Hyp to collagen structure and to its role in adding stability to the triple-helical structure was recognised early on [57]. Pro and Hyp groups are entropically favoured to adopt the triple-helix conformation [58]. The extent of the Pro hydroxylation is correlated to the stability required for the environmental temperature that the animal experiences [59, 60]. Hence, homeotherms usually have a high or full extent of hydroxylation, while poikilotherms, such as fish, generally have incomplete hydroxylation [59], although the extent may vary if higher environmental conditions are encountered [61]. As noted above, the role of Hyp in collagen stability may be related in part to the extent of the hydration network. The role of Hyp in providing triple-helical stability was also shown through cell culture experiments with hydroxylation inhibitors. This showed that chicken type I collagen had a thermal stability reduced by about 15 °C when there was no hydroxylation [62]. Similar data was also obtained from un-hydroxylated recombinant plant collagen [63], which also showed a critical role for Hyp in fibril formation [63].

On the other hand, the determination of the higher stability of a fluoroproline containing peptide $(\text{Gly-Pro-fluoro-Pro-Gly})_{10}$, led to a suggestion that Hyp residues stabilize the helix through inductive and stereoelectronic effects and not through a hydration network [54]. It was suggested that inductive effects were influencing the puckering of the imide ring structure [64], and crystal data has shown that Pro residues in the Xaa and Yaa positions systematically present different backbone and side chain conformations, with those in the Xaa position oriented downwards and those of the Yaa position upwards [65]. Later, however, it became evident that stereochemical considerations based on conformational preferences were probably sufficient to provide a direct explanation for stabilizing effect of Hyp, fluoro-Pro and a range of other analogs in the Y-position, as well as the destabilising effect of Hyp in the Xaa position [27, 65].

2.3 Different collagen types

Until the late 1960s, some 15 years after a triple-helical structure of collagen had first been proposed [21], only one collagen (type I) was thought to exist. In 1969, biochemical evidence was presented showing that a second, distinct collagen type was present in cartilage [66, 67]. Subsequently, and especially as the human genome analysis progressed, a wide range of other collagen types has steadily been discovered and characterised, such that 28 genetically distinct types have now been described, arising from over 45 different genes [68].

It is probable that other types of collagen had been observed prior to the description of type II collagen, but the significance was not then clear. For example, the tissue network of reticulin had been described from 1888 [69], and again in 1903 [70], but while it was of particular interest to histologists, biochemical data was not available. It did seem to be different from collagen by various criteria [71] although, more recently, has been interpreted as including type III collagen [72]. Also, Bazin and Delauney [73] first noticed that granulation tissue appeared to contain a type of collagen that was distinct from type I collagen and similar to the type often associated with embryonic collagens in skin. This collagen was later characterised as type III collagen [74, 75].

All collagen molecules comprise three chains. In some collagens the molecules are homotrimers with three identical chains, such as type II or type III collagens, $\alpha1[II]_3$, $\alpha1[III]_3$. Some of the other collagens are heterotrimers [68]. In some cases, there are two chains of one structure and one of another. For example, type I collagen normally has the form of $\alpha1[I]_2\alpha2[I]$, although type I trimer $\alpha1[I]_3$ has been reported [76]. Similarly, the common structure of type V collagen is $\alpha1[V]_2\alpha2[V]$ [77]. However, in this case a heterotrimer with three different chains is also found, $\alpha1[V]\alpha2[V]\alpha3[V]$ [78].

The different collagen types fall into a selection of structural and functional subgroups [68, 79] (figure 2.4). The most abundant are the fibril forming, interstitial collagens), with type I collagen comprising about 90% of the body's collagen. Clearly, this had made it the collagen of choice for biomedical material developments. Type II collagen is found in cartilage, while type III collagen is associated with extensible structures such as blood vessels and the skin. Minor fibrillar collagens include types V and XI. Most remaining collagen types are non-fibrillar. In addition to the fibrillar collagens, another notable group comprises the Fibril Associated Collagens with Interrupted Triple Helices (FACIT collagens) [80] that

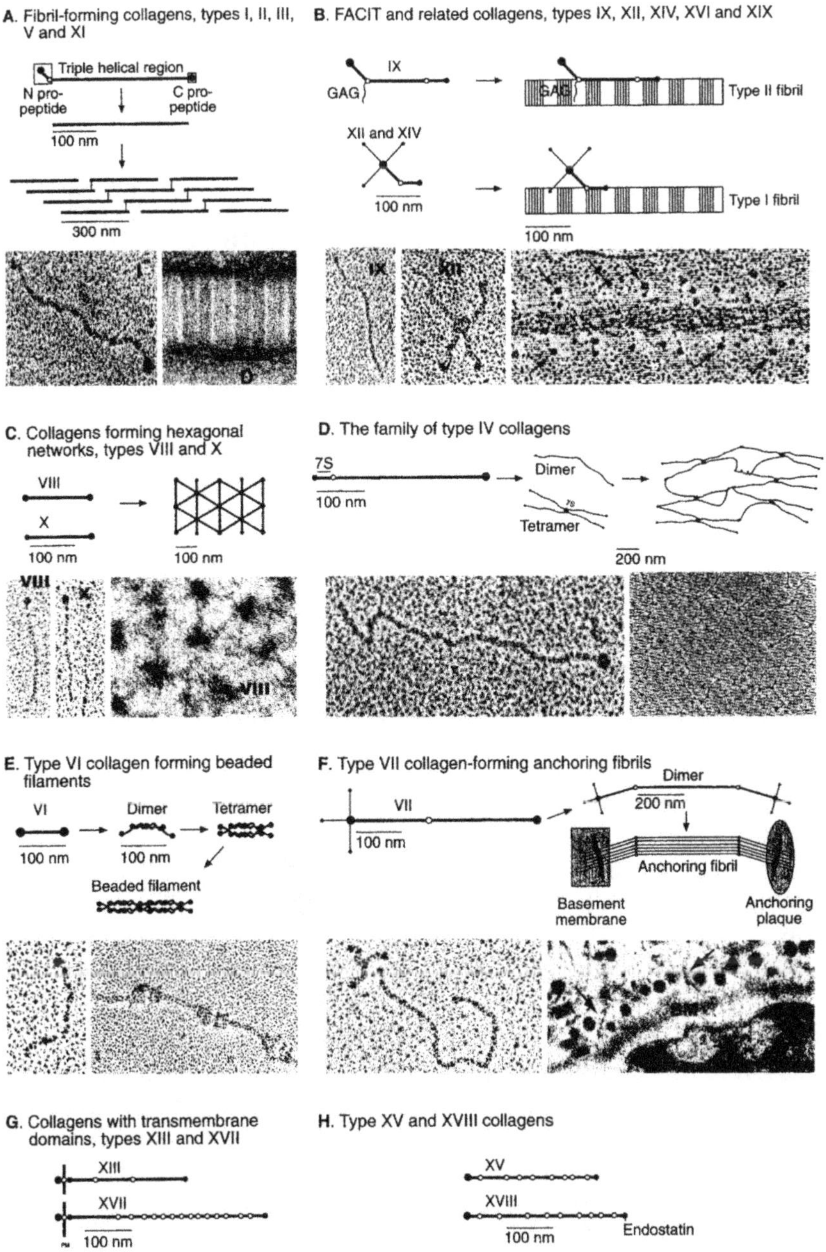

Figure 2.4. Schematic illustrations of the variations in the shapes, forms, and packing of various collagen types, representing different groups of collagens. Circles indicate non-collagenous structural domains. (Republished with permission of John Wiley & Sons, Inc., from [83], through the Copyright Clearance Center Inc.)

Table 2.3. Examples of triple-helical interruptions in various collagen sequences. (Reprint from [84], with permission from Elsevier.)

Interruption	Motif	Type	Sequence example	Collagen
One deletion (Xaa/Yaa)	Gly-Zaa-Gly	G1G	GSP-GVP-**GI-G**PP-GAR	α1[IV]
Double deletion	Gly-Gly	G0G	GEK-GPP-**G -G**PA-GPP	
One insertion	Gly-$(Zaa)_3$-Gly	G3G	GLP-GRP-**GHIK-G**VK-GDI	α2[IV]
Gly deletion	Gly-$(Zaa)_4$-Gly	G4G	GFP-GFP-**GLDMP-G**PK-GDK	α1[IV]
Gly substitution	Gly-$(Zaa)_5$-Gly	G5G	GPP-GPP-**GSAALP-G**SK-GDT	α2[IV]
Double Gly deletion	Gly-$(Zaa)_6$-Gly	G6G	GPK-GEP-**GKIVPLP G**PP-GAE	α1[IV]
Long insertion	Gly-$(Zaa)_n$-Gly	G*N*G	GAK-GEP-**GEFYFDLRLK-G**DK-GDP	α1[IV]
Tandem interruption	Gly-$(Zaa)_n$-Gly-$(Zaa)_m$-Gly	G*N*G*M*G	GER-GLP-**GEVL**G**AQP-G**PR-GDA	α2[IV]

includes types IX, XII, XIV, XVI and XIX collagens. Others similarities are the Multiple Triple Helix domains with Interruptions (Multiplexins) including types XV and XVIII, and the Membrane Associated Collagens with Interrupted Triple Helices (MACIT) collagens that includes types XIII and XVII [81] (figure 2.4). In some cases, the collagen has not been isolated from native tissue, but has only been produced by recombinant methods.

A number of these collagens, notably type IV and the FACIT collagens have interruptions along the triple helix (table 2.3). These interruptions take a variety of forms and lengths. However, as the same structure forms have persisted, for example in type IV collagen from hydra to humans [82], they may well have an important role, such as forming binding domains or allowing essential flexibility, but broadly speaking their roles are not yet fully understood.

The best studied collagens by far are the various human collagens [68] that show a full range of distinct and different structures and functions (figure 2.4). However, researchers recognised early on that collagens were distributed throughout the animal kingdom, particularly from studies on gelatin production, although it was recognised that collagens from species that were more primitive than fish were much harder to identify.

Subsequent studies have shown that in addition to collagens that are equivalent to the interstitial collagens in species from sponges to humans [85] and to basement membrane type IV collagen [82, 86], collagens from more primitive species include numerous other novel structures that have been characterised [60]. Some of the more unusual include, for example, the extracorporeal collagen that forms the 'silk' of the cocoons of a small group of hymenopteran insects, sawflies [87]. Another, is the collagen containing composite protein that comprises the byssal threads of marine mussels, which contain collagen, resilin-like and silk fibroin-like components [88]. Among the most diverse and largest set of structures identified are the numerous collagens from the nematode, *Caenorhabditis elegans*, which has been used extensively as a model organism. It contains, perhaps, more than 170 collagen genes, with many being small structures potentially involved in the cuticle structure while some are larger and show similarities to mammalian collagens [89–91].

2.4 Other proteins with a triple-helix

2.4.1 Animal proteins

In addition to the collagen family, a range of other proteins that are not part of the extracellular matrix have also been characterised and shown to contain a collagen-like triple helix [22], typically with the same structural characteristics of those in the collagen family (table 2.4); for further details see [68]. It is of interest that many of these proteins, for example C1q and the lung surfactant proteins, play various roles in the host defence systems. In several cases, various functional domains are present that can be useful for incorporation into designed recombinant proteins, for example the heparin binding domains in the acetylcholinesterase collagen-like tail [92].

2.4.2 Bacterial collagens

Moderately recently, around 2000, a new group of collagens from bacteria started to be described [93–96]. This was unexpected, as until then, collagens were generally thought to be found only in multicellular animals and to require the presence of Hyp. So far, no direct studies have been carried out on any of the collagen-like proteins extracted from their natural bacteria, but several have been expressed in *E. coli* and studied as recombinant proteins. They frequently seem to be associated with the outer membrane of the organisms, but are quite distinct from the mammalian transmembrane collagens (XIII, XVII, XXIII and XXV) [68]. The roles of these bacterial collagens seem diverse, although several can interact with mammalian proteins and may play important roles in bacterial interactions [97].

These bacterial proteins were identified as collagen-like proteins as they contained the collagen consensus sequence, $(\text{Gly-Xaa-Yaa})_n$, and an early example was shown to form into a triple helix, and to lack Hyp [96], as has proved to be the case for others in this group. The sizes of the collagens that were predicted from initial analyses of the 25 genomes that contained collagen-like sequences in 136 eubacterial genomes [98] were from seven triplet repeats to over 700 triplets. However,

Table 2.4. Examples of proteins that contain a collagen-like triple helix, but are not included in the collagen family.

Acetylcholinesterase collagen-like tail	Ficolins 1, 2, and 3
Adiponectin	Gliomedin
Complement C1q (incl. subunits A, B, C)	Macrophage receptor (MARCO)
Ectodysplasin A	Mannose-binding protein C
EMI domain-containing protein 1 Emilins 1 and 2	Pulmonary surfactant-associated proteins A1, A2 and D

those with very few triplets, where $n < 20$, were considered unlikely to form stable triple-helices [99]. Nevertheless, the data indicated that the sequences comprised an average of 76 triplets, suggesting that a good number of triple-helices may exist. However, in practice, only a limited number, ~8, of these collagens have been confirmed to date as forming stable triple-helical structures [100], and all consist of a single uninterrupted collagen triple-helical domain. The number of genomic sequences in databases has increased substantially since the study in 2003 [98].

The most interesting feature of these collagens is that they are stable around mammalian body temperature, even without the presence of Hyp [100, 101]. Initially, this high stability was thought to be necessary for the effectiveness of pathogens, but non-pathogenic species also showed collagens of similar stability [99]. This stability arises from the sequence features in the Xaa and Yaa positions. For example, there is still a high proline content, often in the Xaa position. Other stabilising features, that had been established from peptide studies [102] (see chapter 6) were also present, including arginine residues in the Yaa position [48], some hydrophobic segments and a high number of ion pairs that contribute significantly to the triple helix stability [103] (figure 2.5). In these examples, the collagen-like domains vary from 117 to 246 amino acids, with calculated pI values from about 4.7 to 9.3. All have additional N- and C-terminal domains of various sizes. Of great interest is that all have high thermal stabilities, ranging from 35.0 to 38.8 °C (table 2.5).

Structurally, these sequences generally comprise a non-collagenous registration domain (V) adjacent to the collagen-like domain (CL), at either the N- or C-terminal [99]. Also, there is an attachment domain, but for recombinant production this can be removed from the construct. One notable feature of many of these bacterial collagens that is not seen in animal collagens is the frequent presence of repeating amino acid sequence patterns, such as $(GPKGEP)_n$ in *M.* sp 4–46. Another key difference is that when compared with animal interstitial collagens, the non-animal, bacterial collagens do not readily form ordered aggregates [104]. This may reflect the absence of Hyp, which is implicated in the self-association process [63]. However, the high calorimetric enthalpy observed for *S. pyogenes* Scl2 collagen [104] suggested that a high degree of hydrogen bonding mediated by hydration is still present.

These bacterial collagens, are no longer curiosities, but represent an opportunity for approaching basic science problems in collagen and for new, safe materials for use as medical materials. As potential biomaterials they have been shown to be non-cytotoxic and not immunogenic, and acute toxicity and sub-chronic toxicity were also absent [107].

```
GQD GRN GER GEQ GPT GPT GPA GPR GLQ GLQ GLQ GER GEQ GPT GPA GPR
GLQ GER GEQ GPT GLA GKA GEA GAK GET GPA GPQ GPR GEQ GPQ GLP GKD
GEA GAQ GPA GPM GPA GER GEK GEP GTQ GAK GDR GET GPV GPR GER GEA
GPA GKD GER GPV GPA GKD GQN GQD GLP GKD GKD GQN GKD GLP GKD GKD
GQN GKD GLP GKD GKD GQD GKD GLP GKD GKD GLP GKD GKD GQP GKP
```

Figure 2.5. The sequence of the triple-helical domain of *Streptococcus pyogenes* Scl2 collagen [96], with stabilising peptide sequences indicated for Pro in the Xaa and Yaa positions, Arg in the Yaa position, charged pairs involving Asp and Lys and an Xaa position Leu cluster [103].

Table 2.5. Examples of bacterial collagen domains known to form a stable triple-helix.

Bacterium	[a]Collagen length (aa)	Calculated pI	T_m (°C)	
Streptococcus pyogenes Scl1	150	5.1	36.4	[96]
Streptococcus pyogenes Scl2	237	5.4	37.6	[96]
Bacillus anthracis	228	3.1	37.0	[105]
Legionella pneumophila	105	5.3	n.d.	[106]
Clostridium perfringens	189	4.7	38.8	[99]
Solibacter usitatus	246	5.6	38.5	[99]
Rhodopseudomonas palustris	117	9.3	37.0	[99]
Methylobacterium sp. 4–46	147	8.6	35.0	[99]

[a] The numbers of residues given are for a specific reported isolate, but extensive polymorphisms are present for many, if not all species. n.d. Not determined. (Reprinted from [100] with permission from Elsevier.)

More recently, a group of triple-helical collagens has been reported in *E. coli* [108]. These collagens are particularly stable, with an example showing a T_m of 42 °C. These collagens are not found in the common laboratory strains, but are found in various pathogenic strains associated with phage particles [108].

2.5 Protein sequences

The ability to determine protein sequences comes from the pioneering studies of Sanger working on insulin [109] and Edman working on sequential degradation chemistry [110]. For collagen, the initial studies using partial acid hydrolysates suggested a regular structure in which Gly was potentially present as every third residue. However, the long chain length of over 1000 amino acids in fibrillar collagens made for particular difficulties. Further progress emerged from the separation of the individual chains of collagen by ion exchange chromatography [111]. These purified chains could then be chemically cleaved by CNBr and the smaller fragments separated. CNBr cleaves specifically after the limited number of Met residues present in collagen chains (see table 2.1). The characteristic CNBr fragment patterns of different collagen chains separated by SDS-polyacylamide gel electrophoresis (SDS-PAGE) were useful to identify different collagen types (figure 2.6). They were suitable for further enzymatic cleavage to give fragments for manual sequencing [112], or could be used in the automated protein sequencing instruments that had been developed [113].

The first complete chain sequence, for the bovine α1[I] chain, was achieved in 1972, and other sequences for chains from types II, III and V collagens soon emerged (see [114, 115]). These emerging data showed the locations and degree of occupancy of various secondary modifications, including Hyp and Hyl [116], 3-Hyp [117], carbohydrates [118] and crosslink sites [112, 119]. These are all key structural features that were not available from subsequent DNA sequence studies. These protein sequence data also allowed the distribution of different amino acids to be examined [119, 120] most noteably for acidic, basic and hydrophobic groups, which allowed an excellent correlation between sequence and the banding patterns

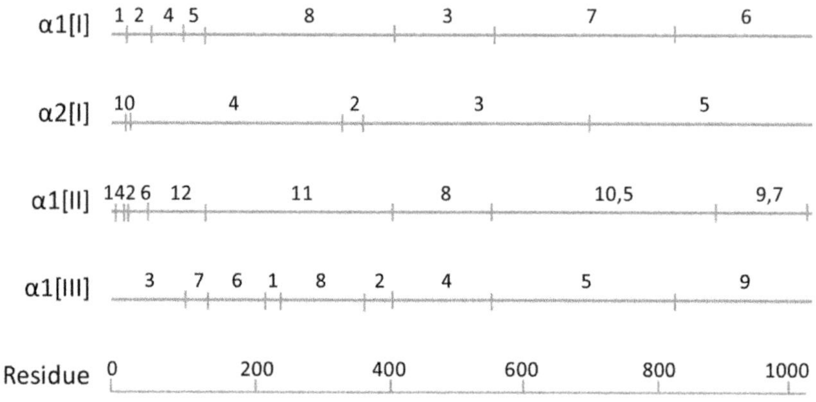

Figure 2.6. CNBr fragment arrangement of various human collagen types.

observed for fibril and aggregate structures (see chapter 4) in transmission electron microscopy (TEM) studies [119, 121]. The sequence data also allowed evaluation of the distributions of amino acid residues within individual triplets, which form the basis for specific interactions, showing that only a small proportion of potential triplets occur frequently in collagen triple-helices, and that there are strong, non-random preferences for certain residues for either Xaa or Yaa positions (table 2.6). Structure and sequence data indicated that there were, for example, steric constraints in the Yaa position, such that certain amino acids, such as Phe, Tyr, or Leu are rarely found in the Yaa position (table 2.6) (see chapter 6) [122].

The difficulties of determining protein sequences have now disappeared as a result of the development of DNA sequencing technology, which has massively increased the available sequence database, for different collagen types, including those of low abundance and from different species. These data are now readily accessible through, for example, the US National Library of Medicine Database, as well as from other databases in Europe and Asia. In addition to providing sequence data for both the triple-helical and other structural parts of collagens, these DNA data provided details on the complexity of the collagen gene, including defining the various coding and non-coding sequences as well as specific DNA sequence elements that are involved in the control of biosynthesis. Clearly, the availability of the DNA sequences and clones derived from these studies have provided the fundamental basis for expression studies of recombinant collagens (see chapter 9).

Table 2.6. Examples for positional preferences of certain amino acids in several collagens.

Type	III								Type	Ala		Leu		Phe	
	Xaa	Yaa		Xaa	Yaa		Xaa	Yaa		Xaa	Yaa	Xaa	Yaa	Xaa	Yaa
Ala	46	45	Phe	8	0	Arg	8	40	I	60	59	18	1	12	0
Thr	7	8	Leu	19	2	Lys	12	26	III	47	42	19	2	8	0
			Glu	46	2	Asp	10	17	IV	17	17	60	2	31	1

References

[1] Fischer E 1902 *Ber. Dt. Chem. Ges.* **35** 1095–106
[2] Hofmeister F 1902 *Ergebn. Physiol.* **1** 759–802
[3] Braconnot H 1820 *Ann. Chim. Phys.* **13** 113–25
[4] Berzelius J 1848 *Jahres-Bericht über die Fortschritte der Chemie und Mineralogie* vol 47 (Tübigen: Laupp) p 654
[5] Bopp F 1849 *Justus Liebigs Annln. Chem.* **69** 16–37
[6] Vickery H B and Schmidt C L A 1931 *Chem. Rev.* **9** 169–318
[7] McCoy R H, Meyer C E and Rose W C 1935 *J. Biol. Chem.* **112** 283–302
[8] Fischer E 1902 *Ber. Dt. Chem. Ges.* **35** 2660–65
[9] Van Slyke D D and Hiller A 1921 *Proc. Natl Acad. Sci. USA* **7** 185–6
[10] Stetten M R 1949 *J. Biol. Chem.* **181** 31–7
[11] Sinex F M and Van Slyke D D 1957 *Fed. Proc.* **16** 250
[12] Butler W T and Cunningham L W 1966 *J. Biol. Chem.* **241** 3882–88
[13] Eastoe J E 1967 *Treatise on Collagen* vol 1 ed G N Ramachandran (New York: Academic) pp 1–72
[14] Moore S and Stein W H 1956 *Adv. Protein Chem.* **11** 191–236
[15] Moore S, Spackman D H and Stein W H 1958 *Fed. Proc.* **17** 1107–15
[16] Herzog R O and Jancke W 1921 *Festschrift der Kaiser Wilhelm Gesellschaft zur Forderung der Wissenschaften* (Berlin: Springer) pp 118–21
[17] Fraser R D B and MacRae T P 1973 *Conformation in Fibrous Proteins* (New York: Academic)
[18] Astbury W T 1933 *Trans. Faraday Soc.* **29** 193–211
[19] Küntzel A and Prakke F 1933 *Biochem. Ztschr.* **267** 243–96
[20] Cowan P M, North A C T and Randall J T 1953 *Nature and Structure of Collagen* ed J T Randall and S Fitton-Jackson (London: Butterworths Scientific) pp 241–49
[21] Ramachandran G N and Kartha G 1954 *Nature* **174** 269–70
[22] Brodsky B and Ramshaw J A M 1997 *Matrix Biol.* **15** 545–54
[23] Boedtker H and Doty P 1956 *J. Am. Chem. Soc.* **78** 4267–80
[24] Ramachandran G N and Kartha G 1955 *Nature* **176** 593–95
[25] Rich A and Crick F H C 1961 *J. Mol. Biol.* **3** 483–506
[26] Fraser R D B, MacRae T P and Suzuki E 1979 *J. Mol. Biol.* **129** 463–81
[27] Bella J 2016 *Biochem. J.* **473** 1001–25
[28] Li M H, Fan P, Brodsky B and Baum J 1993 *Biochemistry* **32** 7377–87
[29] Sakakibara S, Kishida Y, Okuyama K, Tanaka N, Ashida T and Kakudo M 1972 *J. Mol. Biol.* **65** 371–73
[30] Okuyama K, Okuyama K, Arnott S, Takayanagi M and Kakudo M 1981 *J. Mol. Biol.* **152** 427–43
[31] Bella J, Eaton M, Brodsky B and Berman H M 1994 *Science* **266** 75–81
[32] Nagarajan V, Kamitori S and Okuyama K 1998 *J. Biochem.* **124** 1117–23
[33] Nagarajan V, Kamitori S and Okuyama K 1999 *J. Biochem.* **125** 310–18
[34] Hongo C, Noguchi K, Okuyama K, Tanaka Y and Nishino N 2005 *J. Biochem.* **138** 135–44
[35] Okuyama K, Narita H, Kawaguchi T, Noguchi K, Tanaka Y and Nishino N 2007 *Biopolymers* **86** 212–21
[36] Okuyama K, Haga M, Noguchi K and Tanaka T 2014 *Biopolymers* **101** 1000–9

[37] Kramer R Z, Venugopal M G, Bella J, Mayville P, Brodsky B and Berman H M 2000 *J. Mol. Biol.* **301** 1191–205

[38] Schumacher M A, Mizuno K and Bächinger H-P 2006 *J. Biol. Chem.* **281** 27566–74

[39] Kawahara K, Nishi Y, Nakamura S, Uchiyama S, Nishiuchi Y, Nakazawa T, Ohkubo T and Kobayashi Y 2005 *Biochemistry* **44** 15812–22

[40] Okuyama K, Hongo C, Wu G, Mizuno K, Noguchi K, Ebisuzaki S, Tanaka Y, Nishino N and Bächinger H P 2009 *Biopolymers* **91** 361–72

[41] Okuyama K, Miyama K, Mizuno K and Bächinger H P 2012 *Biopolymers* **97** 607–16

[42] Bella J 2010 *J. Struct. Biol.* **170** 377–91

[43] Orgel J P R O, Persikov A V and Antipova O 2014 *PLoS One* **9** e89519

[44] Kramer R Z, Bella J, Mayville P, Brodsky B and Berman H M 1999 *Nat. Struct. Biol.* **6** 454–57

[45] Emsley J, Knight C G, Farndale R W and Barnes M J 2004 *J. Mol. Biol.* **335** 1019–028

[46] Boudko S P, Engel J, Okuyama K, Mizuno K, Bächinger H P and Schumacher M A 2008 *J. Biol. Chem.* **283** 32580–89

[47] Bella J and Berman H M 1996 *J. Mol. Biol.* **264** 734–42

[48] Yang W, Chan V C, Kirkpatrick A, Ramshaw J A M and Brodsky B 1997 *J. Biol. Chem.* **272** 28837–40

[49] Kramer R Z, Bella J, Brodsky B and Berman H M 2001 *J. Mol. Biol.* **311** 131–47

[50] Brodsky B 1999 *Proc. Indian Acad. Sci. (Chem. Sci.)* **111** 13–8

[51] Nomura S, Hiltner A, Lando J B and Baer E 1977 *Biopolymers* **16** 231–46

[52] Peto S, Gillis P and Henri V P 1990 *Biophys. J.* **57** 71–84

[53] Bella J, Brodsky B and Berman H M 1995 *Structure* **3** 893–906

[54] Holmgren S K, Taylor K M, Bretscher L E and Raines R T 1998 *Nature* **392** 66–667

[55] Privalov P L 1982 *Adv. Protein Chem.* **35** 1–104

[56] Myllyharju J 2003 *Matrix Biol.* **22** 15–24

[57] Gustavson K H 1955 *Nature* **175** 70–4

[58] Josse J and Harrington W F 1964 *J. Mol. Biol.* **9** 269–87

[59] Rigby B J and Robinson M S 1975 *Nature* **253** 277–9

[60] Mathews M B 1975 *Connective Tissue* (Berlin: Springer)

[61] Cohen-Solal L, Castanet J, Meunier F J and Glimcher M J 1986 *Comp. Biochem. Physiol. B-Biochem. Mol. Biol.* **83** 483–6

[62] Rosenbloom J, Harsch M and Jimenez S 1973 *Arch. Biochem. Biophys.* **158** 478–84

[63] Perret S, Merle C, Bernocco S, Berland P, Garrone R, Hulmes D J S, Theisen M and Ruggiero F 2001 *J. Biol. Chem.* **276** 43693–98

[64] Bretscher L E, Jenkins C L, Taylor K M, DeRider M L and Raines R T 2001 *J. Am. Chem. Soc.* **123** 777–8

[65] Vitagliano L, Berisio R, Mastrangelo A, Mazzarella L and Zagari A 2001 *Protein Sci.* **10** 2627–32

[66] Miller E J and Matukas V J 1969 *Proc. Natl Acad. Sci. USA* **64** 1264–68

[67] Miller E J 1971 *Biochemistry* **10** 1652–59

[68] Ricard-Blum S 2011 *Cold Spring Harb. Persp. Biol.* **3** a004978

[69] Mall F P 1888 *Anat. Anz.* **3** 397–401

[70] Mallory F B 1903 *J. Med. Res.* **10** 334–41

[71] Kramer H and Little K 1953 *Nature and Structure of Collagen* ed J T Randall and S Fitton-Jackson (London: Butterworths Scientific) pp 33–43

[72] Strum J M, Gartner L P and Hiatt J L 2007 *Cell Biology and Histology* (Hagerstwon, MD: Lippincott Williams & Wilkins) p 83
[73] Bazin S and Delaunay A 1964 *Am. Inst. Pasteur* **107** 163–72
[74] Epstein E H 1974 *J. Biol. Chem.* **249** 3225–31
[75] Chung E and Miller E J 1974 *Science* **183** 1200–201
[76] Moro L and Smith B D 1977 *Arch. Biochem. Biophys.* **182** 33–41
[77] Sage H, Pritzl P and Bornstein P 1981 *Biochemistry* **20** 3778–84
[78] Rhodes R K and Miller E J 1981 *Coll. Rel. Res.* **1** 337–43
[79] Sherman V R, Yang W and Meyers M A 2015 *J. Mech. Behav. Biomed. Mat.* **52** 22–50
[80] Shaw L M and Olsen B R 1991 *Trends Biochem. Sci.* **16** 191–4
[81] Tu H, Huhtala P, Lee H-M, Adams J C and Pihlajaniemi T 2015 *BMC Evol. Biol.* **15** 281
[82] Fowler S J, Jose S, Zhang X, Deutzmann R, Sarras M P and Boot-Handford R P 2000 *J. Biol. Chem.* **275** 39589–99
[83] Ramshaw J A M, Glattauer V and Werkmeister J A 2003 *Encyclopaedia of Polymer Science and Technology* 3rd edn vol 5 ed J Kroschwitz (New York: Wiley) pp 603–32
[84] Bella J 2014 *J. Struct. Biol.* **186** 438–50
[85] Exposito J Y, Valcourt U, Cluzel C and Lethias C 2010 *Int. J. Mol. Sci.* **11** 407–26
[86] Fidler A L, Darris C E, Chetyrkin S V, Pedchenko V K, Boudko S P, Brown K L, Gray-Jerome W, Hudson J K, Rokas A and Hudson B G 2017 *Elife* **6** e24176
[87] Sutherland T D, Peng Y Y, Trueman H E, Weisman S, Okada S, Walker A A, Sriskantha A, White J F and Huson M G *et al* 2013 *Sci. Rep.* **3** 2864
[88] Qin X X and Waite J H 1995 *J. Exper. Biol.* **198** 633–44
[89] Cox G N, Kramer J M and Hirsh D 1984 *Mol. Cell. Biol.* **4** 2389–95
[90] Kramer J M 1994 *FASEB J.* **8** 329–36
[91] Page A P and Johnstone I L 2007 *WormBook* **2007** 1–15
[92] Doss-Pepe E, Deprez P, Silva T, Inestrosa N C, Kirkpatrick A, Ramshaw J A M and Brodsky B 2004 *Biochim. Biophys. Acta* **1698** 187–95
[93] Engel J and Bachinger H-P 1999 *Proc. Indian Acad. Sci. (Chem. Sci.)* **111** 81–6
[94] Lukomski S, Nakashima K, Abdi I, Cipriano V J, Ireland R M, Reid S D, Adams G G and Musser J M 2000 *Infect. Immun.* **68** 6542–53
[95] Rasmussen M, Edén A and Björck L 2000 *Infect. Immun.* **68** 6370–77
[96] Xu Y, Keene D R, Bujnicki J M, Höök M and Lukomski S 2002 *J. Biol. Chem.* **277** 27312–18
[97] Caswell C C, Han R, Hovis K M, Ciborowski P, Keene D R, Marconi R T and Lukomski S 2008 *Mol. Microbiol.* **67** 584–96
[98] Rasmussen M, Jacobsson M and Bjorck L 2003 *J. Biol. Chem.* **278** 32313–16
[99] Xu C, Yu Z, Inouye M, Brodsky B and Mirochnitchenko O 2010 *Biomacromolecules* **11** 348–56
[100] Yu Z, An B, Ramshaw J A M and Brodsky B 2014 *J. Struct. Biol.* **186** 451–61
[101] Mohs A, Silva T, Yoshida T, Amin R, Lukomski S, Inouye M and Brodsky B 2007 *J. Biol. Chem.* **282** 29757–65
[102] Persikov A V, Ramshaw J A M and Brodsky B 2005 *J. Biol. Chem.* **280** 19343–49
[103] Persikov A V, Ramshaw J A M, Kirkpatrick A and Brodsky B 2005 *Biochemistry* **44** 1414–22
[104] Yoshizumi A, Yu Z, Silva T, Thiagarajan G, Ramshaw J A M, Inouye M and Brodsky B 2009 *Protein Sci.* **18** 1241–51

[105] Boydston J A, Chen P, Steichen C T and Turnbough C L 2005 *J. Bacteriol.* **187** 5310–17

[106] Vandersmissen L, De Buck E, Saels V, Coil D A and Anné J 2010 *FEMS Microbiol. Lett.* **306** 168–76

[107] Peng Y Y, Yoshizumi A, Danon S J, Glattauer V, Prokopenko O, Mirochnitchenko O, Yu Z, Inouye M and Werkmeister J A *et al* 2010 *Biomaterials* **31** 2755–761

[108] Ghosh N, McKillop T J, Jowitt T A, Howard M, Davies H, Holmes D F, Roberts I S and Bella J 2012 *PLoS One* **7** e37872

[109] Sanger F 1988 *Annu. Rev. Biochem.* **57** 1–28

[110] Edman P 1970 *Mol. Biol. Biochem. Biophys.* **8** 211–55

[111] Piez K A, Eigner E A and Lewis M S 1963 *Biochemistry* **2** 58–66

[112] Bornstein P and Piez K A 1966 *Biochemistry* **5** 3460–73

[113] Edman P and Begg G 1967 *Eur. J. Biochem.* **1** 80–91

[114] Fietzek P P and Kühn K 1976 *Int. Rev. Connect. Tissue Res.* **7** 1–60

[115] Kadler K 1995 *Protein Profile* **2** 491–619

[116] Wendt P, Fietzek P P and Kühn K 1972 *FEBS Lett.* **26** 69–73

[117] Gryder R M, Lamon M and Adams E 1975 *J. Biol. Chem.* **250** 2470–74

[118] Butler W T 1970 *Biochemistry* **9** 44–50

[119] Fietzek P P and Kühn K 1975 *Mol. Cell. Biochem.* **8** 141–57

[120] Hofmann H, Fietzek P P and Kühn K 1980 *J. Mol. Biol.* **141** 293–314

[121] Meek K M, Chapman J A and Hardcastle R A 1979 *J. Biol. Chem.* **254** 10710–714

[122] Ramshaw J A M, Shah N K and Brodsky B 1998 *J. Struct. Biol.* **122** 86–91

IOP Publishing

John A M Ramshaw and Veronica Glattauer

Chapter 3

Biosynthesis and biodegradation of collagen

It is important to have a broad understanding of collagen biosynthesis when considering collagen as a biomedical material. In particular, the processes of conversion of procollagens to tissue forms and the key secondary modification reactions that affect molecular and tissue stability are important. Many of these steps may need to be replicated if recombinant systems are used. Understanding the biodegradation and remodelling of tissue is important for the biocompatibility of implants, even though variations to the native processes may apply for heavily stabilised materials.

3.1 Biosynthetic pathway

The biosynthesis of collagen (figure 3.1) has been determined principally through studies on interstitial collagens, but the broad principles are assumed to apply to other collagens as well. The majority of the intracellular processes are no doubt common to all, but the secretion and matrix formation and associated processing may vary between collagens. Also, there will potentially be additional processing steps for certain collagens, for example for collagen types IX, XII, XIV, XV, XVIII that also have bound glycosaminoglycan chains [1, 2].

When considering collagen as a biomedical material, it is the processes that happen during secretion, aggregation and crosslinking (figure 3.1) that are the most relevant. However, an understanding of the preceding steps, especially certain of the secondary modification reactions such as proline hydroxylation, is important for those trying to replicate collagens using recombinant methods (see chapter 9). Many of the other processes, particularly the transcription and translation stages, are common to all protein biosynthesis and are therefore covered in detail in most, if not all, recent biochemistry and molecular biology text books [3–5].

doi:10.1088/978-0-7503-2096-2ch3

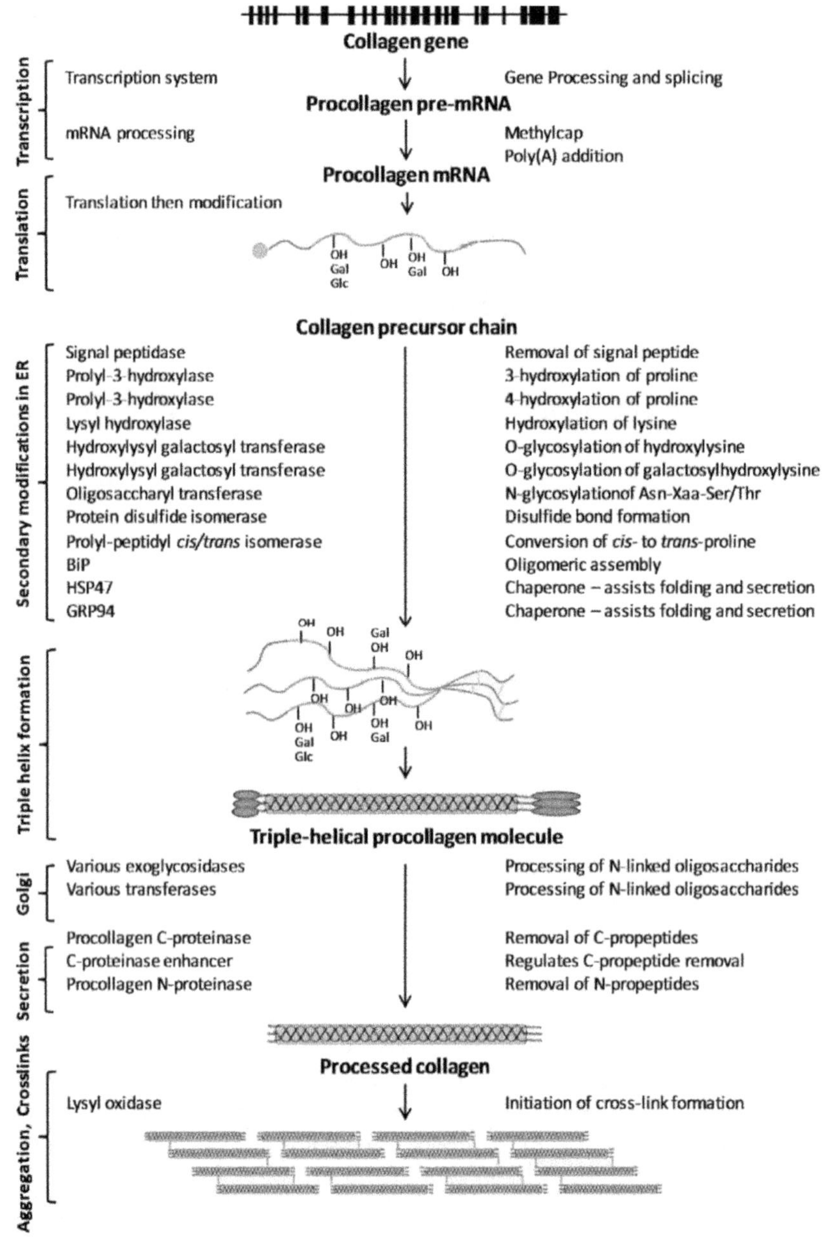

Figure 3.1. An overview of some of the key steps and enzyme and chaperone involvement in the biosynthesis pathway for collagen. (Republished with permission of John Wiley & Sons, Inc., from [19], through the Copyright Clearance Center.)

The majority of information on collagen biosynthesis has been derived from studies on collagen types I and III [6, 7]. These and the other fibril-forming collagens, and most other collagens [2] (figure 2.4), all have a globular *N*-propeptide and C-propeptide flanking the triple-helix domain. All collagens are also synthesised with an N-terminal signal sequence that directs subsequent processing. Since collagens are targeted for the extracellular matrix, much of the biosynthetic process takes place in the endoplasmic reticulum, followed by transport through the Golgi and secretion from the cell (figure 3.1). The folding of fibril-forming collagens is initiated when three appropriate globular C-propeptides recognise each other and associate, and then become disulfide bonded to form stable trimers. The recognition domain structure [8] has been shown to drive chain selection and the correct associations that are present in homotrimeric and heterotrimeric collagen structures. This domain also includes coiled-coil sequence interactions [9]. The next structural section is the C-telopeptide region followed by the C-terminal triple-helix nucleating zone. This region is typically rich in a Gly-Pro-Hyp and similar sequences [10] (Hyp = hydroxyproline).

During the biosynthetic process, collagen chains interact with various enzymes and molecules that assist in folding and are involved in secondary modifications. Some key reactions, notable proline hydroxylation and lysine hydroxylation only occur during this intermediate unfolded stage [11]. As the triple-helix formation propagates from the C-terminus to the N-terminus, in a zipper-like fashion, *cis–trans* isomerization of the imino acids is the rate limiting step in folding [12] and a *cis–trans* prolyl isomerase accelerates this isomerization *in vivo* [13], allowing proper triple-helix folding. After folding, the secondary modification reactions to give Hyp and hydroxylysine (Hyl) no longer occur [11].

Once the collagen has folded into its triple-helical structure, associated with the endoplasmic reticulum, a particular molecular chaperone, heat shock protein 47 (Hsp47), binds to the helix [14]. This binding may increase the collagen thermal stability and prevent any premature intracellaular aggregation of the collagen. After progressing to the Golgi, Hsp47 is released and some initial lateral aggregation of procollagen molecules is observed [15]. Collagen molecules leave the cell in secretory vesicles, accompanied by specific protease cleavage of the C-propeptide and frequently the N-propeptide. Some retention of N-propeptide may be associated with subsequent control of fibril formation and size [16]. The entire biosynthetic process takes about 15 min for type I collagen [17]. However, during the biosynthetic process, a significant quantity of collagen, perhaps 15%, is never fully synthesised and is broken down intracellularly [18].

The collagen molecules, which typically now contain only small non-helical telopeptides at each end of the large central triple-helix, spontaneously self-associate, possibly in an intermediate aggregate form [15] and progress to form fibrils with the collagen molecules associated in a staggered manner (see chapters 2 and 4). This process needs further resolution, but early forms of fibrils with a D axial period are found in cell crypts [20] prior to formation of mature, crosslinked fibrils. The specific crosslinking, another key biosynthetic step that imparts stability and strength to tissues, occurs in the fibril structure via the action of lysyl oxidase (LOX) (see below).

In part, during secretion and generally prior to aggregation and crosslinking, various collagen intermediates are present that can be purified as individual intermediate products in small quantities if proteolysis is inhibited. In certain animals, such as sheep and calves, deficiencies in the processing to remove the pro-domains leads to an accumulation of precursors in the skin that allow larger yields of precursors to be purified [21]. These include procollagen, where both the N- and C-terminal telopeptides are still attached [22] and pN-collagen, where only the N-propeptide is still attached and the C-propeptide has been removed [23]. The converse, where just the C-propeptide is retained is not seen. The yields of the precursor molecules are typically low, except for type III pN-collagen which accumulates in certain tissues such as skin, possibly as part of the regulation process for fibril growth. This precursor can be readily isolated and purified if proteolysis is inhibited [24].

The formation of crosslinks eventually makes the collagen insoluble (see chapter 5). In the initial stages some forms can be reversed in acid conditions. The key parts of the collagen structure that are involved in crosslinking are the telopeptides (figure 3.1). Depending on type, the N-terminal telopeptide is about 14 residues long while the C-terminal is about 26 residues long [25]. A lysine group in a telopeptide can be modified by LOX (figure 3.2), possibly once the collagen is associated with a fibril, to give a modified amino acid with a terminal aldehyde side chain, allysine. This residue can then participate in crosslinking [26] (see chapter 5). In addition, certain lysine residues in the telopeptides and in the triple-helical domain are converted by lysyl hydroxylase to Hyl. The initial crosslinks that form can be acid labile, but during maturation this reversibility is lost. For example, aldol condensation products involving adjacent telopeptides are reversible and so are some of the initial crosslinks involving Lys, but not those involving Hyl [26] (see chapter 5).

3.2 Selected secondary modification enzymes in collagen biosynthesis

3.2.1 Prolyl hydroxylase

Proline hydroxylation was first identified in the biosynthesis of collagen [27, 28] and until moderately recently was seen as a fairly specialised post-translational modification that was restricted to this family of proteins and other collagen-like entities [29]. Thus, the hydroxylation of certain Pro residues in unfolded collagen chains is important for both the stability and the interactions of the final triple helix. Hyp residues make a critical contribution to the stability of the triple helix through stereoelectronic effects [30] and/or hydration [31], and also appear essential for collagen self-association [32] and for some receptor interactions (see chapter 7).

The main reaction is that of prolyl-4-hydroxylase (P4H), leading to 4-Hyp (figure 3.2). This enzyme is one of the expanding family of 2-oxoglutarate-dependent dioxygenases, of which there are now around 70 members [29, 33], with a wide range of functions [34], many of which are associated with hypoxia dependent pathways [35]. Mammalian P4H is located in the endoplasmic reticulum and consists of an $\alpha 2\beta 2$ tetramer containing 2 distinct subunits. In some more primitive organisms, however, an active $\alpha\beta$ dimer is found [36]. The α-subunit, with a MW of about 59 kDa, is responsible for substrate binding and the catalytic activity, while the slightly smaller, MW 55 kDa, β-subunit is responsible for localising the enzyme within the endoplasmic reticulum. The β-subunit is identical to protein disulfide

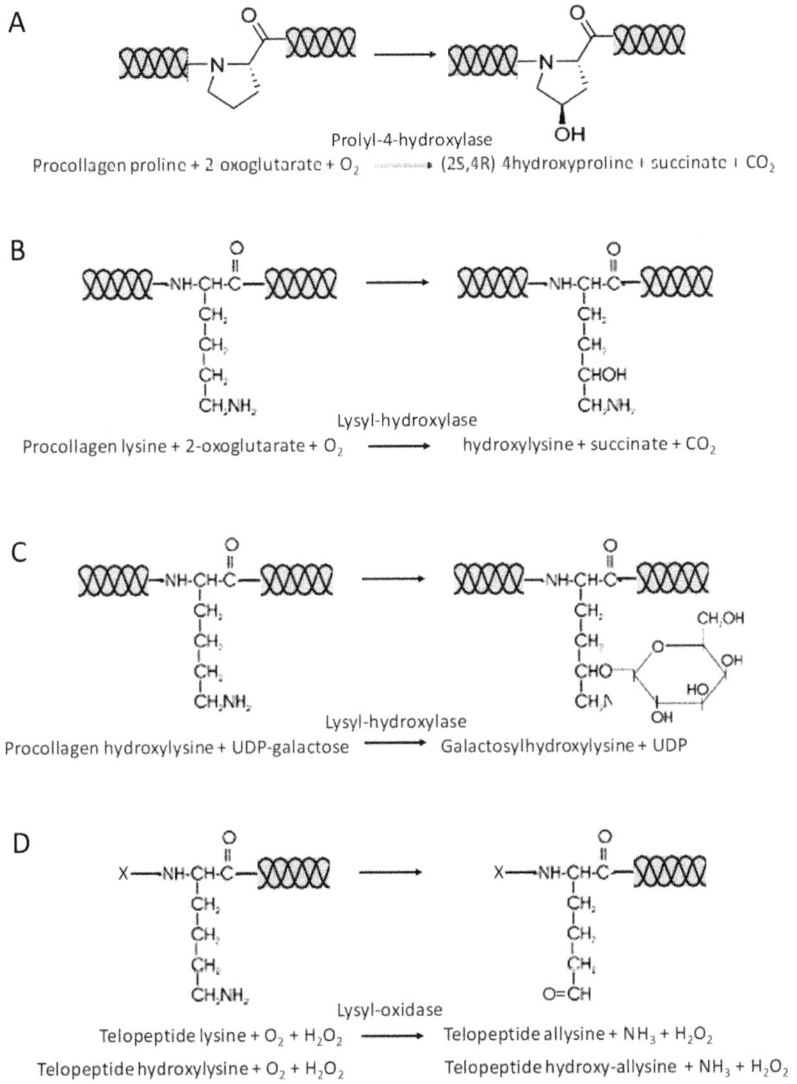

Figure 3.2. Selected secondary modification reactions involved in collagen biosynthesis. (A) Proline hydroxylation, (B) Lysine hydroxylation, (C) Glycosylation following lysine hydroxylation, and (D) Lysine oxidation.

isomerase which, in addition to this function, also acts as a chaperone during collagen folding [37].

P4H is an iron-containing enzyme. A proposed mechanism involves sequential binding of 2-oxoglutarate and then substrate to the active site, followed by oxygen. The decarboxylation of the 2-oxoglutarate gives a ferryl intermediate to give a 2-electron oxidation, in this case giving hydroxylation while ascorbic acid is utilised as a critical cofactor to reduce the iron site [33]. Ascorbic acid (vitamin C) is essential for proper production of collagen and a deficiency reduces the levels and stability of Hyp (and Hyl) leading to scurvy [38]. P4H has a substrate specificity to the Pro in an

Xaa-Pro-Gly motif which limits its activity to Pro residues in the Yaa position in the (Gly-Xaa-Yaa) repeating sequence of the collagen chain. P4H activities are found throughout the animal kingdom [39], and also in plants where a much broader specificity is observed [40].

Another prolyl hydroxylase that is also associated with collagen is prolyl-3-hydroxylase (P3H) [41]. This enzyme has a still more stringent specificity to the individual, unfolded collagen chain, recognising the Pro in a Gly-Pro-Hyp-Gly sequence, such that typically only one residue and a few partially occupied sites are found in a wide variety of interstitial collagen chains [42], although many more, 10–15, are present in type IV collagen [43]. The role of the 3-hydroxyproline residue is not fully understood, but it has been suggested that it is involved in local destabilisation to assist fibril assembly and crosslinking [42], with its biological relevance confirmed through the association of reduced activity with various abnormal tissue states [44, 45].

3.2.2 Lysyl hydroxylase

The hydroxylation of the penultimate carbon atom of certain Lys residues during collagen biosynthesis is catalysed by lysyl hydroxylase (LH) activity. Various LH activities (see below) have now been identified in vertebrates that, like P4H, are also part of the 2-oxoglutarate-dependent dioxygenases family and have similar functional requirements, requiring ferrous iron, oxygen, 2-oxoglutarate, and ascorbate [33]. The main LH1 activity reacts only with the unfolded helical domain, whereas LH2 also acts upon the specific telopeptide Lys residues [46]. The LH3 activity interestingly comprises three different activities [47], as in addition to LH activity, it also shows galactosyltransferase (GT), and glucosyltransferase (GGT) activities that function on modified Hyl resides in the helical domain [48] (figure 3.2).

The LH activities occur in the endoplasmic reticulum before triple helix folding. All three forms catalyse to conversion of Lys to 5-Hyl at Xaa-Lys-Gly- triplets in the collagenous domain. LH2, in addition acts upon Xaa-Lys-Ala- or -Xaa-Lys-Ser- sequences in the telopeptide regions of the fibril-forming procollagens [46]. The extent of hydroxylation varies between collagen types and tissues, with types I and III collagens containing very few Hyl residues while higher levels are found in types IV and VI collagens [47]. Certain Hyl residues may then be glycosylated, to form galactosyl-Hyl and galactosyl-glucosyl-Hyl moieties that appear to be modifications that are unique to collagenous sequences. Compared to Hyp formation, which has clear cut importance for stability and interactions, the understanding of a role of Lys hydroxylation (other than in crosslinking) and subsequent glycosylation is less well advanced.

3.2.3 Lysyl oxidase

LOX activity is critical in the formation of functional extracellular matrix as it is responsible for the first step in crosslinking of collagen, thus providing tissue strength and structural integrity to connective tissues [49]. It was originally discovered due to the adverse effects on connective tissues that were observed when people and animals consumed seeds from the Lathyrus family that contain naturally occurring LOX inhibitors. More recently, it has been recognised that in addition to its key role in collagen and elastin crosslinking and normal development, LOX also plays a role in fibrotic disorders [50], in cancer through enhancing cell proliferation, invasion and other processes and has been implicated in many other diseases and disorders [51, 52].

LOX is a Cu^{++} and lysyl tyrosylquinone (LTQ) dependent amine oxidase. It is a member of a family of five oxidases, LOX-1, 2, 3, and 4 and LOX-like (LOXL), that all show a similar catalytic site at a conserved C-terminal region that includes the key copper binding domain as well as LTQ and cytokine receptor-like (CRL) domains [53]. Mature, active LOX has a MW of 32 kDa [53]. All members can catalyse the oxidation of the Lys and Hyl side chains of collagens in the telopeptide regions of fibrillar collagens, leading to the formation of peptidyl α-aminoadipic-δ-semialdehyde residues or peptidyl-δ-hydroxy-α-aminoadipic-δ-semialdehyde residues, also known as allysine and hydroxyallysine, respectively (see figure 3.2). These resulting aldehydes are highly reactive and lead to the formation of the native collagen crosslinks, apparently without any subsequent enzyme involvement.

Substrate preferences may exist for the various forms and enzyme specificities are not clear cut. However, Hyl residues are more nucleophilic and chemically reactive than non-modified Lys residues at physiologic pH values, and so could, in principle, be better substrates for LOX and more likely to participate in the formation of crosslinks [53].

3.3 Degradation of collagen in tissues

The degradation of collagen in native tissues is a key part of the continual remodelling and rebuilding processes that occur. The enzymes involved include matrix metalloproteinases (MMPs), which are an extended family of more than 20 members of Zn^{2+} dependent proteases [54]. However, only a limited number of these MMPs may be involved in turnover of interstitial collagens. They are involved in complex pathways, and are tightly regulated by various systems [55], including activation of proenzymes and interactions with specific tissue inhibitors of metalloproteinases (TIMPs) [56].

It has been known for a long time that the collagen triple-helix is resistant to degradation by proteases, such as trypsin, while gelatin is readily digested [57]. An early report of a collagenase [58] discussed an enzyme from ox pancreas. However, some thought that its activity was due to bacterial contamination, as bacterial collagenases were known to exist [59]. Also, there was doubt as to whether is was a collagenase, that being an enzyme that can efficiently hydrolyse an intact triple-helix. Later, a specific collagenalytic activity was detected in tadpole tails [60], and later this activity, now called MMP-1, was purified and characterised [61].

Each of the MMPs appears to have a different specificity profile, with some degrading gelatin, rather than the triple helix [54]. The 'classical' collagenases, MMP-1, 8 and 13, break down triple-helical collagen fibrils through cleavage at a unique site in a collagen molecule located about ¾ of the molecular length from the

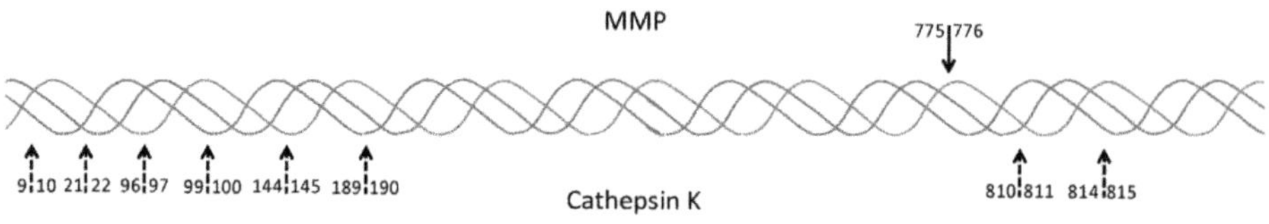

Figure 3.3. A schematic representation of the sites of cleavage of type I collagen by MMP and Cathepsin K. (Data from [54]).

N-terminus [54] (figure 3.3). The cleavage site in type I collagen α1 chain occurs at the -Gly(775)-Ile(776) sequence (figure 3.3) and it is thought that an imino acid deficient sequence in this zone contributes to the specificity [62]. The relative importance of residues near the cleavage site have been assessed using recombinant and peptide constructs (see chapter 6). The effectiveness of these MMPs varies between collagen types. So, MMP-1 has best activity against type III collagen, MMP-8 has best activity to type I collagen and MMP-13 to type II collagen. A gelatinase activity is found for MMP-2 and MMP-9, while MMP-12 cleaves at the -Gly-Ile- site (figure 3.3) as well as many other sequences [54]. The cleavage rate for individual molecules is dependent on species and is also much greater than for fibrils [63], so an associated cleavage of telopeptides may also be required [64].

The three-dimensional structures of certain MMP's have been determined by x-ray crystallography and by NMR [65]. The structures comprise a hemopexin domain connected to a linker sequence, both of which are involved in specificity, and also a separate catalytic (cleavage) domain [65].

The other proteases that are probably important are the cathepsins, notably cathepsin K where several sites on collagen have been observed (figure 3.3) [54].

Degradation of collagen has been implicated in various disease processes, such as collagen degradation in cartilage in osteoarthritis, as well as in collagen degradation as part of tumour growth and metastasis. Inhibitors to MMP enzymes have been explored for therapeutic applications, especially for cancer therapies, but this approach has been complicated by the redundant, alternative pathways that exist and by the observation that some enzymes have essential beneficial functions that should not be blocked [54, 66, 67].

Collagen-based biomaterials, even after chemical modifications and crosslinking, would potentially be subject to most, if not all, of the same degradation enzymes. However, the specific collagenalytic pathways of native collagen may no longer represent the path(s) used if the cleavage sites are modified or sterically hindered. In such cases it is possible that less specific cathepsins or other cell or tissue proteases play increased and more significant roles. A further examination of models that have been developed in other disciplines for collagen degradation (for example, [68]) may be valuable and could possibly be adapted to biomaterial degradation.

References

[1] Li D Q, Clark C C and Myers J C 2000 *J. Biol. Chem.* **275** 22339–47

[2] Ricard-Blum S 2011 *Cold Spring Harb. Perspect. Biol.* **3** a004978

[3] Grisham C M 2016 *Biochemistry* (Boston, MA: Cengage Learning)
[4] Cox M and Nelson D 2017 *Lehninger Principles of Biochemistry, International Edition* (Stuttgart: Macmillan)
[5] Snape A 2018 *Biochemistry and Molecular Biology* (Oxford: Oxford University Press)
[6] Engel J and Prockop D J 1991 *Annu. Rev. Biophys. Biophys. Chem.* **20** 137–52
[7] Bulleid N J, Wilson R and Lees J F 1996 *Biochem. J.* **317** 195–202
[8] Sharma U, Carrique L, Vadon-Le Goff S, Mariano N, Georges R N, Delolme F, Koivunen P, Myllyharju J and Moali C *et al* 2017 *Nat. Commun.* **8** 14671
[9] McAlinden A, Smith T A, Sandell L J, Ficheux D, Parry D A D and Hulmes D J S 2003 *J. Biol. Chem.* **278** 42200–07
[10] Xu Y, Bhate M and Brodsky B 2002 *Biochemistry* **41** 8143–51
[11] Pihlajaniemi T, Myllylä R and Kivirikko K I 1991 *J. Hepatol.* **13** S2–7
[12] Bruckner P and Eikenberry E F 1984 *Eur. J. Biochem.* **140** 391–95
[13] Steinmann B, Bruckner P and Superti-Furga A 1991 *J. Biol. Chem.* **266** 1299–303
[14] Thomson C A and Ananthanarayanan V S 2000 *Biochem. J.* **349** 877–83
[15] Hulmes D J S, Bruns R R and Gross J 1983 *Proc. Natl Acad. Sci. USA* **80** 388–92
[16] Hulmes D J S, Kadler K E, Mould A P, Hojima Y, Holmes D F, Cummings C, Chapman J A and Prockop D J 1989 *J. Mol. Biol.* **210** 337–45
[17] Vuust J and Piez K A 1972 *J. Biol. Chem.* **247** 856–62
[18] Bienkowski R S 1984 *Coll. Relat. Res.* **4** 399–411
[19] Ramshaw J A M, Glattauer V and Werkmeister J A 2003 *Encyclopaedia of Polymer Science and Technology* 3rd edn vol 5 ed J Kroschwitz (New York: Wiley) pp 603–32
[20] Birk D E, Zycband E I, Winkelmann D A and Trelstad R L 1989 *Proc. Natl. Acad. Sci. USA* **86** 4549–53
[21] Lenaers A, Ansay M, Nusgens B V and Lapiere C M 1971 *Eur. J. Biochem.* **23** 533–43
[22] Hoffmann H P, Olsen B R, Chen H T and Prockop D J 1976 *Proc. Natl Acad. Sci. USA* **73** 4304–08
[23] Fleischmajer R, Timpl R, Tuderman L, Raisher L, Wiestner M, Perlish J S and Graves P N 1981 *Proc. Natl. Acad. Sci. USA* **78** 7360–64
[24] Lenaers A and Lapiere C M 1975 *Biochim. Biophys. Acta* **400** 121–31
[25] Kadler K 1995 *Protein Profile* **2** 491–619
[26] Yamauchi M and Mechanic G L 1988 *Collagen* vol 1 ed M E Nimni (Boca Raton, FL: CRC Press) pp 157–72
[27] Hutton J J, Kaplan A and Udenfriend S 1967 *Arch. Biochem. Biophys.* **121** 384–91
[28] Hutton J J, Tappel A L and Udenfriend S 1967 *Arch. Biochem. Biophys.* **118** 231–40
[29] Zurlo G, Guo J, Takada M, Wei W and Zhang Q 2016 *Biochim. Biophys. Acta* **1866** 208–20
[30] Bretscher L E, Jenkins C L, Taylor K M, DeRider M L and Raines R T 2001 *J. Am. Chem. Soc.* **31** 777–78
[31] Bella J, Brodsky B and Berman H M 1995 *Structure* **3** 893–906
[32] Perret S, Eble J A, Siljander P R, Merle C, Farndale R W, Theisen M and Ruggiero F 2003 *J. Biol. Chem.* **278** 29873–79
[33] Markolovic S, Wilkins S E and Schofield C J 2015 *J. Biol. Chem.* **290** 20712–22
[34] Loenarz C and Schofield C J 2011 *Trends Biochem. Sci.* **36** 7–18
[35] Taylor C T, Doherty G, Fallon P G and Cummins E P 2016 *J. Clin. Invest.* **126** 3716–24
[36] Veijola J, Koivunen P, Annunen P, Pihlajaniemi T and Kivirikko K I 1994 *J. Biol. Chem.* **269** 26746–53

[37] Wilson R, Lees J F and Bulleid N J 1998 *J. Biol. Chem.* **273** 9637–43

[38] Kipp D E, McElvain M, Kimmel D B, Akhter M P, Robinson R G and Lukert B P 1996 *Bone* **18** 281–8

[39] Pozzolini M, Scarfì S, Mussino F, Ferrando S, Gallus L and Giovine M 2015 *Mar. Biotechnol.* **17** 393–407

[40] Robinson D G, Andreae M and Blankenstein P 1990 *Organisation and Assembly of Plant and Animal Extracellular Matrix* ed W S Adair and R P Mecham (San Diego, CA: Academic Press-Harcourt Brace Jovanovich) pp 283–99

[41] Risteli J, Tryggvason K and Kivirikko K I 1977 *Eur. J. Biochem.* **73** 485–92

[42] Weis M A, Hudson D M, Kim L, Scott M, Wu J J and Eyre D R 2010 *J. Biol. Chem.* **285** 2580–90

[43] Kefalides N A 1973 *Int. Rev. Connect. Tissue Res.* **6** 63–104

[44] Vranka J A, Pokidysheva E, Hayashi L, Zientek K, Mizuno K, Ishikawa Y, Maddox K, Tufa S and Keene D R *et al* 2010 *J. Biol. Chem.* **285** 17253–62

[45] Hudson D M, Werther R, Weis M, Wu J J and Eyre D R 2014 *PLoS One* **9** e93467

[46] Takaluoma K, Lantto J and Myllyharju J 2007 *Matrix Biol.* **26** 396–403

[47] Myllylä R, Wang C, Heikkinen J, Juffer A, Lampela O, Risteli M, Ruotsalainen H, Salo A and Sipilä L 2007 *J. Cell. Physiol.* **212** 323–9

[48] Ruotsalainen H, Sipilä L, Vapola M, Sormunen R, Salo A M, Uitto L, Mercer D K, Robins S P and Risteli M *et al* 2006 *J. Cell Sci.* **119** 625–35

[49] Kagan H M and Trackman P C 1991 *Am. J. Resp. Cell Mol. Biol.* **5** 206–10

[50] Erler J T, Bennewith K L, Nicolau M, Dornhöfer N, Kong C, Le Q T, Chi J T, Jeffrey S S and Giaccia A J 2006 *Nature* **440** 1222–26

[51] Kumari S, Panda T K and Pradhan T 2017 *Indian J. Clin. Biochem.* **32** 134–41.

[52] Wang T H, Hsia S M and Shieh T M 2017 *Int. J. Mol. Sci.* **18** E62

[53] Trackman P C 2016 *Matrix Biol.* **52–54** 7–18

[54] Amar S, Smith L and Fields G B 2017 *Biochim. Biophys. Acta* **1864** 1940–51

[55] Yamamoto K, Murphy G and Troeberg L 2015 *Matrix Biol.* **44** 255–63

[56] Murphy G 2011 *Genome Biol.* **12** 233

[57] Bogue R H 1922 *The Chemistry and Technology of Gelatin and Glue* (New York: MaGraw-Hill) p 51

[58] Ssadikow W S 1927 *Biochem. Ztschr.* **181** 267–83

[59] Mall F P 1896 *Johns Hopkins Hosp. Rep.* **1** 171–208

[60] Gross J and Lapiere C M 1962 *Proc. Natl Acad. Sci. USA* **48** 1014–22

[61] Nagai Y, Lapiere C M and Gross J 1966 *Biochemistry* **5** 3123–30

[62] Fields G B 1991 *J. Theor. Biol.* **153** 585–602

[63] Welgus H G, Burgeson R E, Wootton J A, Minor R R, Fliszar C and Jeffrey J J 1985 *J. Biol. Chem.* **260** 1052–59

[64] Perumal S, Antipova O and Orgel J P R O 2008 *Proc. Natl Acad. Sci. USA* **105** 2824–29

[65] Fields G B 2014 *Adv. Protein Chem. Struct. Biol.* **97** 37–48

[66] Jabłońska-Trypuć A, Matejczyk M and Rosochacki S 2016 *J. Enzyme Inhib. Med. Chem.* **31** 177–83

[67] Turunen S P, Tatti-Bugaeva O and Lehti K 2017 *Biochim. Biophys. Acta* **1864** 1974–88

[68] Collins M J, Riley M S, Child A M and Turner-Walker G 1995 *J. Archaeol. Sci.* **22** 175–83

IOP Publishing

Biophysical and Chemical Properties of Collagen: Biomedical Applications

John A M Ramshaw and Veronica Glattauer

Chapter 4

Collagen assemblies

While there is a considerable body of knowledge on collagens at a molecular level, as structures increase in size, such as in fibrils or collagen networks, the amount of knowledge on higher order systems becomes steadily less. Nevertheless, a range of insights into the formation of these structures has emerged and opportunities to control these processes could enable better responses to biomaterial implants to be obtained.

4.1 Ordered collagen structures

Various collagen types form ordered aggregates. Most notable of these are the interstitial collagens that form the major structural fibrils of tissue (figure 4.1). Others, particularly the network forming collagens, also form aggregated structures (see figure 2.4). These include type IV collagen (figure 4.1) where a net-like structure is present that was originally suggested as a chicken wire-like structure [1], and is a key element of the basement membrane. Type VI collagen forms into fibres [2], which in some instances assemble into much larger sheet-like structures (figure 4.1). Types VIII and X collagen both form hexagonal array structures [3, 4] (figure 4.1). Type VII collagen appears to be arranged in bundles of dimers, as anchoring fibrils proximal to a basement membrane but it is not fully clear whether the molecules are in an ordered array or just clustered into the location [5]. With the exception of the fibrillar, interstitial collagens, the paths toward the formation of these various structures is often poorly understood, at best. Other collagen types and their structures are discussed in the literature [6].

4.1.1 Interstitial collagen fibrils

For type I collagen, the development of transmission electron microscopy (TEM) during the 1930s eventually led to images for native collagen and reconstituted

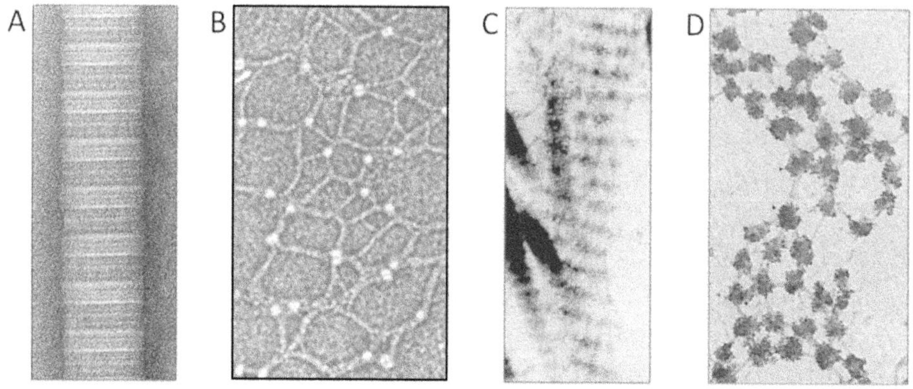

Figure 4.1. Ordered collagen structures. (A) TEM image of an interstitial collagen fibril (photo courtesy of Jacinta White), (B) Scanning electron microscopy (SEM) image of type IV collagen from mouse EHS tumour (photo from [11], republished with permission of Rockefeller University Press, with permission conveyed through Copyright Clearance Center, Inc.), (C) A TEM image of a sheet structure of type VI structure, adjacent to collagen fibrils (photo courtesy of Jacinta White) and (D) Immuno-TEM image of the hexagonal array of type VIII collagen from bovine corneal endothelial cell culture (photo from [4], republished with permission of Rockefeller University Press, with permission conveyed through Copyright Clearance Center, Inc.).

collagen fibrils [7]. These images clearly showed a simple banded pattern with repeating striations at 600 Å to 700 Å. This spacing, called the D-period, was equivalent to the 640 Å period seen in contemporary low angle x-ray studies on various dry collagen samples [8]. A variety of later studies, using native wet tendon collagen and reconstituted type I and III fibrils gave the generally accepted 670 Å D-period, while skin, and other extensible tissues that contain both type I and III collagens gave shorter periods of around 652–655 Å [9, 10], suggesting that a different packing mode may be present.

Examination of the cross-striation period and molecular length from other packing variations seen by electron microscopy, the segment long spacing (SLS) and fibrous long spacing (FLS) formats, led to a proposal that the collagen fibril comprised a parallel array of molecules staggered by one-quarter [12]. Subsequently, after temporary difficulties arising from the erroneous belief that the α_1 and α_2 chains of type I collagen were not the same size [13], the staggered organisation model was refined to give 4.4 D-periods per molecular length with a gap following, prior to the start of the next axially aligned collagen molecule [14] (figure 4.2). It is the differential stain uptake between the overlap (less) and gap (more) zones that gives the major banded pattern seen by TEM, with greater detail within the bands arising from variations in charged/hydrophobic residue positions [15].

A wide range of proposals have been made regarding the arrangement of type I collagen molecules within a fibril. For hydrated rat tail tendon, x-ray diffraction reflections indicated the presence of a three-dimensional lattice [17, 18]. A possible tetragonal cell was proposed [19], which built on the concept that collagen comprised microfibrillar, 5-stranded ropes [20] (figure 4.3). Various other microfibril models had also been proposed, including those with 2, 4, 6, 7 and 8 strands (see [21]). Later

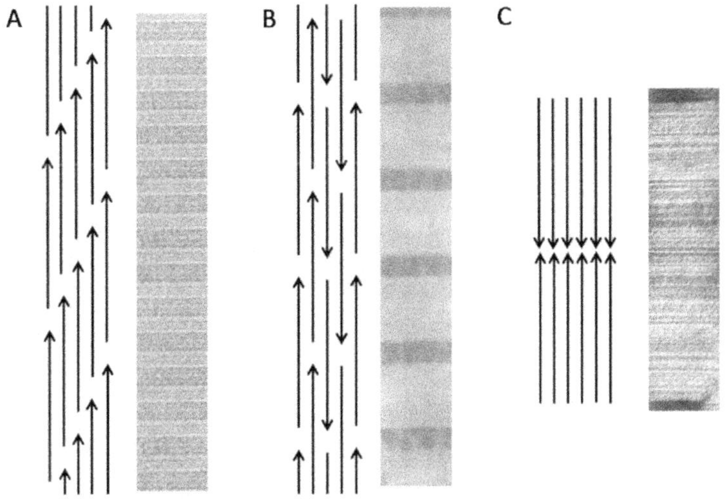

Figure 4.2. TEM of alternative forms of interstitial collagen packing. (A) Native fibril structure of sheep type I collagen (photo courtesy of Jacinta White), (B) FLS IV packing (photo reprinted from [16], with permission from Elsevier), (C) SLS packing of bovine type V collagen (photo courtesy of the late Dr Peter Tulloch). Individual collagen molecules (arrows) are ~290 nm.

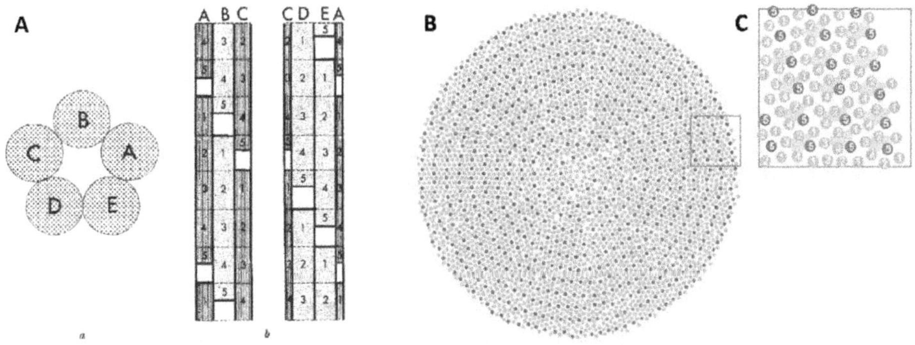

Figure 4.3. Molecular packing of collagen in fibrils. (A) A 5-stranded microfibril model (from [20], reprinted by permission from Springer Nature). (B) Transverse section of a radial packing model (based on [22] showing molecules in a section of thickness equal to the D repeat), with the gap segments in black [22]. (C) Enlarged view of selected area in (B) showing molecules grouped together as micro-(penta-)-fibrils (in orange) [22]. (B and C: Reprinted from [22] with permission from Elsevier.)

studies, using phosphotungstic acid enhanced patterns, suggested that other lattice options should be explored [18]. After a proposal of a hexagonal lattice that did not fully explain all reflections, a quasi-hexagonal lattice was proposed [23] (figure 4.3) that allowed a detailed indexing of all the observed x-ray reflections [18].

This work led to a model for rat tail tendon collagen (figure 4.4) [23] that indicated a path inclined slightly relative to the fibre axis of collagen molecules. Later, it was shown that the molecules were not entirely straight [24]. It was

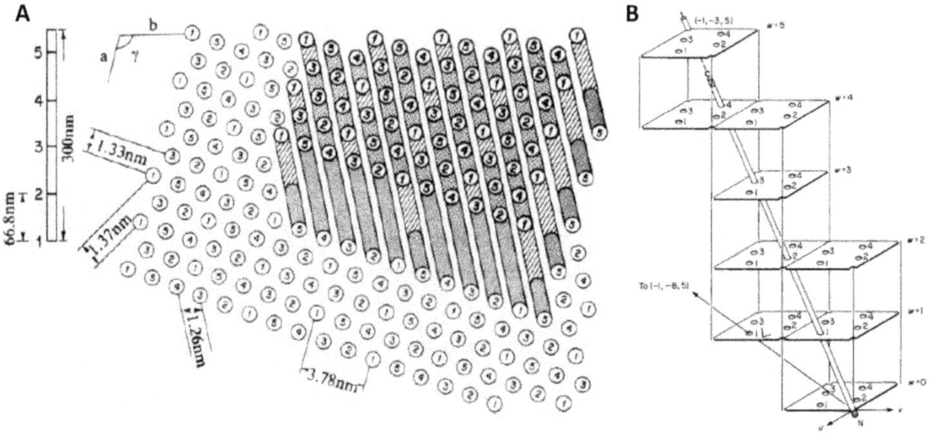

Figure 4.4. Molecular packing of collagen in fibrils. (A) The quasi-hexagonal model for packing ([21], reprinted by permission from Springer Nature). (B) An example of a permitted molecular path of a collagen molecule. The figure is not drawn to scale, but is contracted substantially in the *z*-dimension (reprinted from [23], with permission from Elsevier).

suggested that the x-ray data define the positions of the collagen molecules in the overlap region of the fibril, but give less information on the gap region where crystallinity seems low or absent [24]. The required connectivity between collagen segments in the overlap regions defines a path for the connecting segments, showing that the collagen is not a rigid, straight rod-like molecule, but has a degree of flexibility [24]. Additional models have emerged that extend these concepts to the larger scale packing of a collagen fibril [22] (figure 4.3).

The vast majority of data collected on collagen fibril organisation has been from rat tail tendon, as the degree of crystallinity in this tissue gives by far the best diffraction patterns. However, it is possible that not all fibrils are created equal. For example, native tissues containing type III collagen, which is possibly an integral component of heterotypic fibrils, give diffraction patterns with a shorter D-period (see chapter 2) [9]. Studies on cartilage type II collagen (and associated collagens), based on scanning-TEM mass mapping, suggested a different, 10+4 microfibril structure for thin, ≈4 nm cartilage fibrils, in which a core of 4 is surrounded by 10 molecules [25]. Other studies, using lamprey notochord, suggest that a molecular packing difference may exist compared to type I collagen, although this may reflect different telopeptide arrangements in the different collagen types, which may also have implications for crosslinking [26].

4.1.2 Other ordered structures for interstitial collagens

Early studies on reconstituted aggregates of soluble collagen revealed a selection of non-native forms, which gave further insights into collagen structure and evolution.

FLS aggregates

A group of at least four centrosymmetric structures have been observed [16, 27] (figure 4.2) that were identified as being laboratory derived rather than of natural

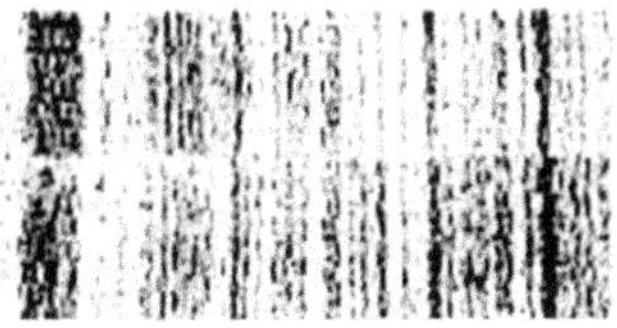

Figure 4.5. Comparison between the SLS aggregates of bovine and fish type I collagens, showing the similarity of the staining patterns. ([32] Reprinted by permission from Springer Nature). Each molecular aggregate is about 280 nm long.

origin [28]. These distinct structures arose from precipitation of collagen from solutions in citrate [29], and later with 'bovine mucoprotein' [28] and negatively charged compounds such as chondroitin sulfate [27].

SLS aggregates

The packing found in SLS aggregates is quite distinct to all other formats. It was found that mixing collagen solutions in 0.5% acetic acid with adenosine triphosphate (ATP) led to a rapid and reliable opalescent precipitate that consisted of sheet-like arrays of collagen molecules arranged parallel to each other and in register [30]. A number of alternatives to ATP, such as triphosphate and perdisulfate ions have also been found to produce SLS aggregates [31], but ATP is still preferred. The TEM studies showed that these structures were the length of a collagen molecule, and that the staining showed a fine structure [30] (figures 4.2 and 4.5). Subsequent studies showed an excellent correlation between this staining pattern and the location of charged residues the amino acid sequence (see [15]). Each of the interstitial collagens can form SLS aggregates. Most interestingly, comparisons between SLS aggregates for type I collagen from various species, including fish and mammals [32], showed a remarkable similarity in staining patterns (figure 4.5), indicating a high degree of sequence conservation over evolutionary history.

4.1.3 Ordered structures for other collagen types

Most data on collagen biosynthesis and the subsequent assembly of molecules to give functional tissues has been obtained for the interstitial collagen and particularly type I collagen (see chapters 2 and 3). This is also true for the higher order structures, such as collagen fibrils.

Type IV collagen also forms into a higher order structure in the formation of the basement membrane. Initially, a net-like structure was proposed [1]. In this proposal, the 7S domain of the collagen, which had been identified as an N-terminal cross-linked region joining different molecules, two in one direction and two in the opposite direction, gave one junctional domain while crosslinking between non-collagenous C-terminal domains (NC1) provided a further junctional element. Subsequently, an unusual crosslink was identified in the NC1 dimer, S-hydroxylysyl-methionine, derived from adjacent Met and Hyl residues [33]. Images of basement membranes (figure 4.1) do not show a regular arrangement. A net-like structure is still seen, but the arms of these structures are less regular, and much less than the 7S to NC1 chain length. This suggests that some lateral aggregation of the type IV collagen chains occurs and extensive image analyses support this new model [11, 34]. The various interruptions to the Gly-Xaa-Yaa repeat in collagen type IV [35], which have been maintained throughout the animal kingdom [36], may be important for allowing changes in organisation.

X-ray diffraction studies have provided some structural information but have not fully resolved the molecular arrangement [37]. The features of the pattern of highly stretched lens capsule samples suggest a staggered liquid crystal-like arrangement [37], with removal of GAG components giving a clarified pattern allowing meridional reflections which index as orders of 21 nm to be resolved [38]. The mechanical, stress–strain properties of mature lens capsules after GAG removal showed a less stiff material, while no significant change in mechanical properties was observed upon reduction of disulfide bonds [39]. However, at high concentrations of reducing agent that may be expected to reduce the NC1 sulfilimine bond, a less stiff material was obtained, with diffraction patterns suggesting nanometre scale changes [39].

A range of interactions are important for type IV collagen (see chapter 7) and seem an integral part of the network structure. The most important of these are with various laminins, nidogen [40], which is bound to laminin [41], and a heparin sulfate proteoglycan, e.g., perlecan. In addition, a range of other molecules may be intimately associated with this structure [42]. During embryonic development, type IV collagen is not immediately essential as the first basement membrane component, but becomes essential later in development to ensure the structural integrity of the basement membrane; laminin first forms a network and collagen type IV subsequently augments this structure [43]. The details are not clear, and there are considerably more biological data than structural evidence. Notwithstanding, structural models have been proposed that have combined the data that is available (figure 4.6) [44].

Type VI collagen comprises six distinct chains [45] that assemble into $\alpha_1\alpha_2\alpha_X$ heterotrimer triple helices [46]. Assembly of larger structures is initiated intracellularly, where two individual molecules assemble into disulfide-bonded antiparallel dimers. These dimers in turn align to form tetramers that are also stabilised by disulfide bonding. After secretion into the extracellular environment other organised

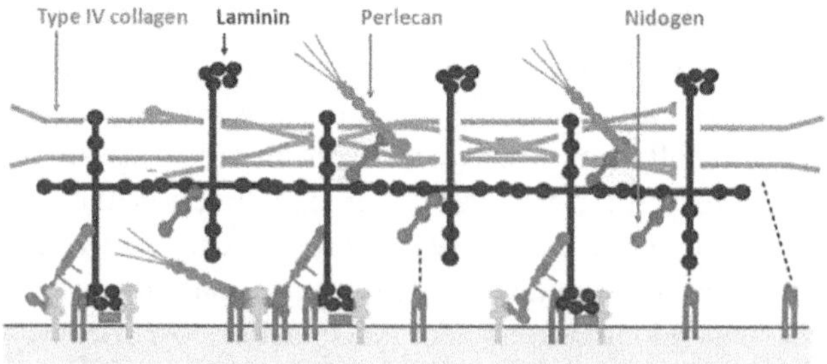

Figure 4.6. A model for the involvement of collagen type IV in the structure of a basement membrane. (Reprinted from [44] with permission from Elsevier.)

structures are formed by end-to-end non-covalent association to give beaded, moniliform filaments [2] (figure 4.7), that may themselves then aggregate into laterally associated sheets (figure 4.7).

A further distinct packing system, a hexagonal-like array, has been observed for type VIII and X collagens. These are minor collagens that are found in specialised tissues, such as Descemet's membrane and hypertrophic cartilage respectively. For type VIII collagen this array is visible by TEM (Figure 4.1) [4]. For type X collagen, TEM images of hexagonal arrays have also been recorded [48]. This study also indicated that individual molecules aggregate through the C-terminal NC1 domain and that lateral association of triple-helical domains leads to the network structure [48].

4.2 Fibrillogenesis

Although several collagens assemble into ordered structures only one of these has been extensively studied. This is the formation of fibrils by the interstitial collagens. The abundance, together with the structural and functional roles of these fibrils, clearly explains this focus.

4.2.1 Interstitial collagen fibril formation

There is an abundance of information at the molecular level on the structure and the mechanisms of biosynthesis of collagen (see chapter 3), but as the structures become larger, such as in fibrils, fibre bundles and tissues (figure 4.8) the amount of information on the mechanisms by which they are formed and how these processes are controlled becomes steadily less and less. Yet, these steps are the key to forming functional tissues.

TEM data suggested that the initial steps of collagen fibril formation, nucleation, that initiates fibril formation may happen within extracellular compartments

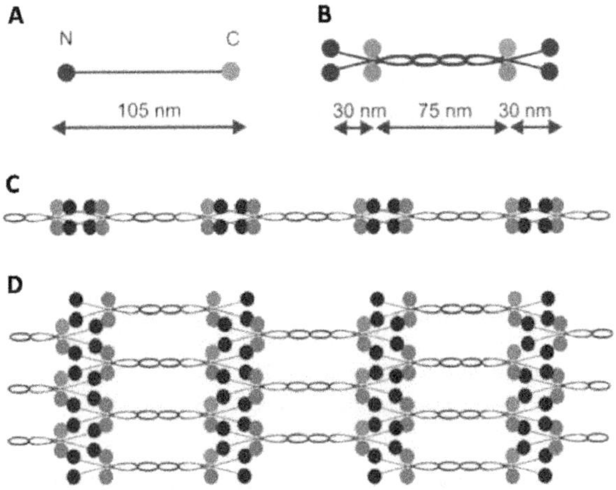

Figure 4.7. The chain arrangement in type VI collagen: (A) individual molecule, (B) tetrameric form, (C) beaded filament, and (D) lateral aggregate. (Reprinted from [47] with permission from Elsevier.)

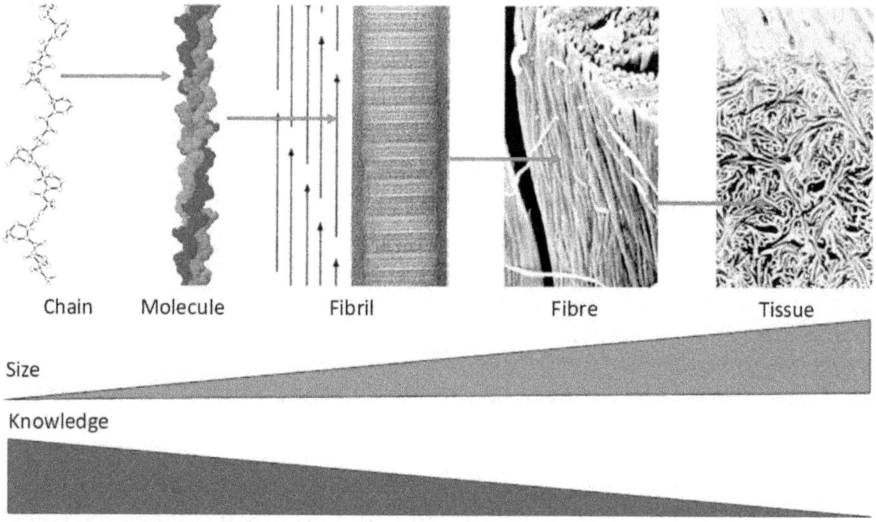

Figure 4.8. Structural steps in the establishment of tissues from collagen chains, indicating the relative levels of the scale against the knowledge base. (The image of a single chain is from [49], reprinted with permission from Elsevier, and the SEM image is from [50], reprinted with permission from Elsevier.)

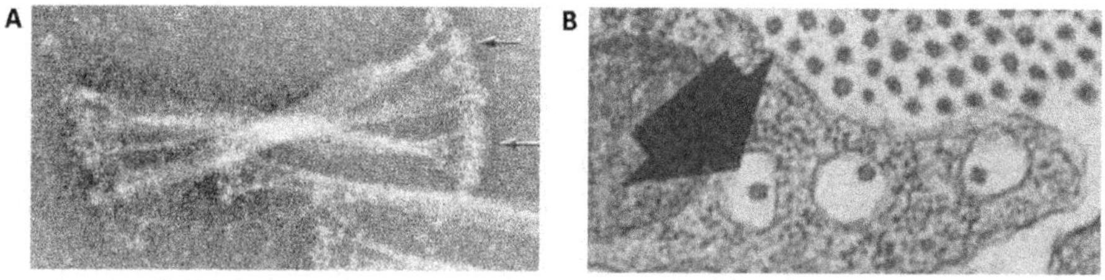

Figure 4.9. Potential steps in fibril formation. (A) Formation of SLS-like structures prior to fibril assembly, (from [52], with permission from authors (by DJSH)) and (B) individual fibrils forming within crypts within a cell (arrow), (from [53], republished with permission of John Wiley & Sons through the Copyright Clearance Center).

(crypts) at the cell surface [51], and that the early structures may develop from SLS type precursor aggregates [52] (figure 4.9).

Thermal initiation

Studies had shown that acetic acid soluble collagen, which had intact telopeptides, could readily reconstitute into a fibrillar mass comprised of D-periodic fibrils when solutions at neutral pH were warmed to 37 °C [54]. This can be followed by observing the increase in opacity with time. This shows a lag period with little change, followed by rapid growth in opacity, until a maximum opacity with no further increase is seen [54]. Decreasing protein concentration did not affect the duration of the lag period but did reduce the rate at which opacity developed. Also, increasing ionic strength by NaCl, from 0.14 to 0.40, shortened the lag period and increased the rate of opacity gain. Additions of urea retarded both steps and dropping the pH to less than pH 5 gave a precipitate rather than a fibrous gel [54].

Subsequent studies showed that treating acid soluble collagen with pepsin, which would remove the telopeptides, led to a major increase in the lag phase and the rate of increase in the following opacity. These data demonstrated the important role of the telopeptides in the nucleation step of fibril formation and in subsequent fibril growth [55]. Further studies confirmed the importance of the telopeptides in effective fibrillogenesis and showed a role in the initiation of the process and in subsequent regulation steps [56]. Light scattering data suggested that the initial nucleation aggregates were small, ⩾1500 nm long and <8 nm in diameter, and then grow linearly to a critical length at which point lateral association leads to mature fibrils [57]. On the other hand, other evidence suggested that a molecular conformation change in the collagen monomer from flexible to rigid was a critical step [58], although the molecular mechanism is not fully clear.

Enzyme initiation

The emerging availability of pN- pC- and procollagen and of the N- and C-terminal procollagen proteases enabled a different approach to study fibrillogenesis to be undertaken [59–61]. Thus, mixing procollagen with both proteases leads to the formation of structures with a distinct, needle-like shape that is distinct from fibrils isolated from tissues [59]. TEM examination of these structures showed the characteristic D-periodic banding pattern in TEM, as well as all having a single polarity reversal where the N to C oriented fibrils from each end merge (figure 4.10) [59], which contrasted with the fibrils formed *in vitro* from acid soluble collagen, which would contain small oligomers, that are unipolar and do not have this polarity reversal [60, 61]. The needle structures showed tapered ends, with one relatively blunt end and the other more pointed [59, 61], and fibril growth was seen to occur from only the sharper end, lengthening the segment away from the reversal point (figure 4.10) [59–61].

Embryonic tissues provide a useful source of newly synthesised fibrils. Initial studies on chick tendons, however, showed limitations in using this source. Fibril breakage during extraction was very high, such that out of around 6 × 10 fragments only 13 intact fibrils with both ends were identified, the longest being around 14 μm long [62]. Nevertheless, these 13 fibrils were very useful; five showed what had been the expected, unidirectional fibrils with one N-terminal and one C-terminal end. The

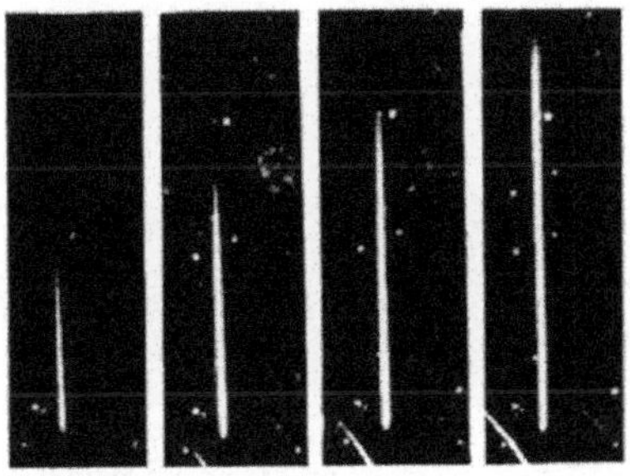

Figure 4.10. *In vitro* fibrils formed by C-proteinase action on pC-collagen at increasing time intervals ([59] reprinted with permission from Portland Press ©1990).

other eight, however were unusual with both ends as N-terminal sequences [62]. Following the standard D-period along these eight fibrils revealed a reversal zone, covering a small number of D-periods. In light of this discovery, examining fragments showed a preponderance of N-terminal tips [62]. A similar phenomenon was also seen for fibrils isolated from embryonic chick cornea and dermis, as well as tendon. All showed an asymmetry along their length, with a longer, sharper end and a shorter blunt end either side of the reversal zone. Although the fibrils from different tissues had different diameters, the question of how growth occurs remained a major question [63]. Some collagen fibrils in a primitive species, an echinoderm, also showed a similar pattern [64].

New fibrils formed in a wound healing model

A biomaterial implant necessarily creates a wound. So, a model based on a subcutaneous implant of a piece of expanded polytetrafluoroethylene (ePTFE) tubing in sheep has been used to examine new fibril formation and collagen deposition. The void inside the tube readily allows isolation of fibrils at different time points that do not contain any older pre-existing, mature host fibrils [65]. The porous wall of the tube allows new deposition of various collagen types to be examined [66]. Histology data indicated that there was no chronic or unwanted response in the wound healing process in this model and *in situ* hybridisation and immunohistology showed that normal collagen synthesis and deposition was occurring, consistent with a normal wound repair process [67].

Examination of the tube core showed no fibrils in 6-day samples after implant, but from 14 days onwards extensive, new collagen tissue along with fibroblasts was seen, which continued to increase up to a 28-day sample (figure 4.11). At both 14 and 28 days, *in situ* hybridisation showed extensive type I collagen synthesis, while immunostaining showed the presence of both type I and III collagens throughout the samples [65]. TEM of 14-day samples showed many small collagen fibrils with an average diameter of 48 nm (n = 47) and an average length 9.2 μm. The more extensive tissue in 28-day samples showed that isolated fibrils had an average diameter of 48 nm (n = 100) with an average length of 28.1 μm. Within each group

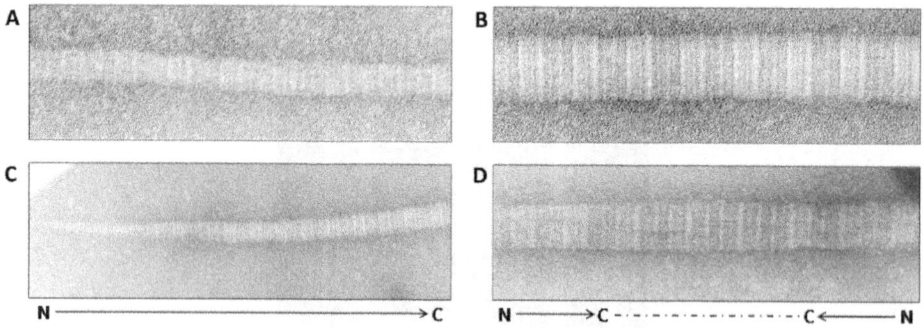

Figure 4.11. TEM images of newly formed collagen fibrils from a wound healing model. (A) 30 nm diameter at 14 days, (B) 68 nm diameter at 28 days, (C) the tip of a new fibril, and (D) a region of polarity reversal. (Photographs courtesy of Jacinta White of CSIRO.)

the diameters of the shorter fibres were similar to the longer fibres, suggesting a preliminary population that may later grow into longer and wider fibrils [63]. In both cases, possibly very recent fibrils about just 1 μm long were seen [65]. Fibril lengths are probably under estimates due to the preparation method used. For 14- and 28-day samples at least 80% of fibrils showed a single polarity reversal with two N-terminal ends [65], equivalent to those seen by other approaches.

It is probable that a specific mechanism is used to add additional single collagen molecules onto a developing fibril. Experimental data have shown that peptides containing sequences for the telopeptides can inhibit fibril formation in an *in vitro* model, with C-telopeptide mimics being particularly effective [68]. These studies also showed that the telopeptide regions show binding to a specific region of the triple helix, residues 776 to 822. This was also supported by molecular modelling studies that suggested that residues 781–794 was the most probable binding zone [68]. Further molecular modelling studies have suggested that the N-telopeptide is involved in the final stages of registration and docking of a newly added molecule to a growing fibril [69].

4.2.2 Regulation of collagen fibril formation

Regulation of collagen fibril formation and accumulation is clearly a complex and multi-layered process. A wide range of systems and molecules have been proposed as regulators of collagen fibril formation and growth. These include a potential function for cleaved propeptides, such as the N-propeptide, which has been proposed as being part of a feed-back system [70]. Other metabolic regulation may occur through the action of a glucose-dependent insulinotropic polypeptide [71]. The retention of the N-propeptide was suggested as a steric mechanism for control of fibril growth [72]. This has been demonstrated for type V collagen [73]. Other collagen regulation systems that may act via steric control include the interactions with FACIT collagens [74, 75]. Molecules that regulate fibril growth and interactions through binding to fibrils include proteoglycans, particularly the small interstitial leucine-rich repeat proteoglycans including decorin, biglycan, lumican and fibromodulin [76]. For biomaterials, especially collagen-based biomaterials, the opportunity to include these regulators as part of the formulation for release on implant may allow control of unwanted or excess tissue formation. The complexity of regulation has further depth if the role of molecules such as hevin, which regulates decorin production, are considered [77].

4.2.3 Heterotypic collagen fibrils

Most studies on fibril initiation have studied tissues (except the enzyme-initiated studies) while structural studies and fibrillogenesis models have focused on tendons and type I collagens. Yet many tissues are a complex mixture of several collagen types.

Immuno-TEM studies on embryonic tissues showed that type III pN-collagen was present along with type I collagen in newly formed fibrils [78], possibly indicating a role for the pN-collagen in regulating fibril development.

Subsequently, various other extracellular matrix (ECM) components, such as decorin, lumican, fibromodulin and hyaluronan, have also been suggested as being regulators of fibril growth [79–81].

Dual labelling immuno-TEM of newly formed fibrils from a wound healing model [65], showed that all the fibrils contained both type I and type III collagens [65] (figure 4.12), although whether either was as a precursor form was not determined. Strangely, staining for either collagen was absent in small diameter fibrils, typically less than 40 nm in diameter, and was also absent in the smaller diameter tapered ends of fibrils where the diameter had fallen to less than 40 nm [65].

Type V collagen has also been shown to be an important and integral component of newly formed native fibrils. It is widespread, being found in most (all) tissues, where it is typically a minor component, around 2%–5%, although in a highly specialised tissue, the cornea, it can be as much as 20% [82]. Most initial studies looked at type V collagen in the cornea, but more recently other tissues, including ligament and tendon have been studied [83].

In vitro experiments illustrate the interaction of type V with type I collagen [84]. Purified type V collagen forms thin fibrils, generally lacking in apparent periodicity, but when type I collagen is present forms into heterotypic fibrils [84]. Those formed with more type V collagen had smaller diameters, with this effect requiring the presence of the attached type V N-propeptide domain [84].

In the cornea, immuno-TEM showed that both types I and V collagens are codistributed within the same fibril [85]. However, fibril disruption was needed to visualise the type V collagen, suggesting that it was present as an inner core structure [85]. Type V collagen is key to the initiation of fibril formation. Type V deficient mice show deficient collagen fibril formation, with a resultant lethal phenotype [86, 87]. Retention of the N-terminal domain of the type V collagen, which can be seen by immuno-TEM independent of the helical domain on the surface of newly formed fibrils, acts to regulate fibril formation and aggregation [82, 87]. In tendons, type V collagen is associated with cells, consistent with a role in new tissue deposition [83].

Immunohistology shows that type V collagen is deposited early in new tissue formation, but then seems to disappear as additional type I and III collagen

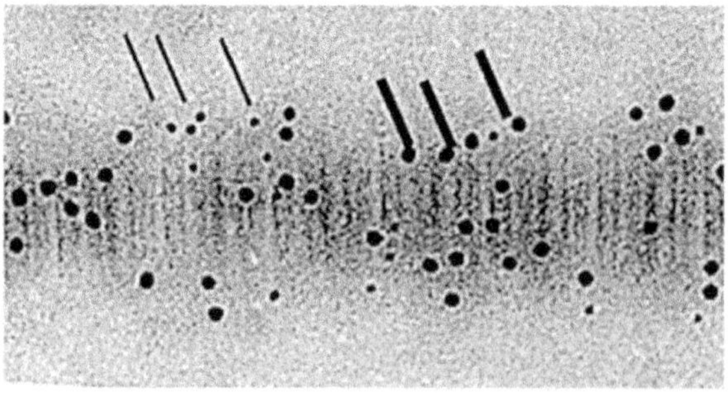

Figure 4.12. Dual immuno-labelling using gold labels of a 28-day fibril isolated from a wound healing model, showing the co-localisation in the same fibril of collagen types I and III. Type I collagen I (large black dots, examples indicated by wide lines) and type III collagen (small black dots, indicated by thin lines). (Reprinted from [65] with permission from Elsevier.)

accumulate [66]. However, enzyme treatment of the new tissue shows that the collagen has become masked rather than lost by resorbtion, confirming the suggestion that type V collagen is present as part of an initial core filament [66].

4.2.4 Interstitial collagen fibril growth

Polarity reversals suggests that axial aggregation occurs; this appears as a single event in early fibril formation and fibrils with two or more sites have not been observed. Nor have stable lateral associations of fibrils been observed in the early developmental stage. So, the steps and control mechanisms that lead to macroscopic, visible fibre bundles in tissues remain uncertain.

Various options have been discussed. For example, formation of longer fibrils may occur by the fusion of small (1–10 μm) segments [88]. Also, lateral association of fibrils to give thicker fibrils [89] has not been discounted. The actual mechanism may also depend on the particular tissue environment, with tendon showing differences to cornea and dermis [63]. At this time, even the final structure of fibrils, such as their length, is not clear [90].

4.2.5 Crimp in collagen fibrils

Crimp is widespread and is present in a wide variety of native tissues. It is a property of fibre-forming collagens [91]. The wavy structure has immediate implications for the mechanical properties of tissues. For example, the collagen crimp characteristics of different tendons are strongly tissue type specific, indicating that crimp specialisation is crucial in the respective mechanical function [92].

The most convenient, and probably the best, means of observing crimp is using polarising light microscopy [93, 94] (figure 4.13). Crimp can also be studied with SEM [95], and by TEM using appropriately stabilised tissues examined at a lower resolution than is normally used in identifying fine structure detail [91, 96].

Various studies show that crimp is due to an underlying periodically varying orientation of a continuous birefringent material, in a zig-zag rather than wave-like

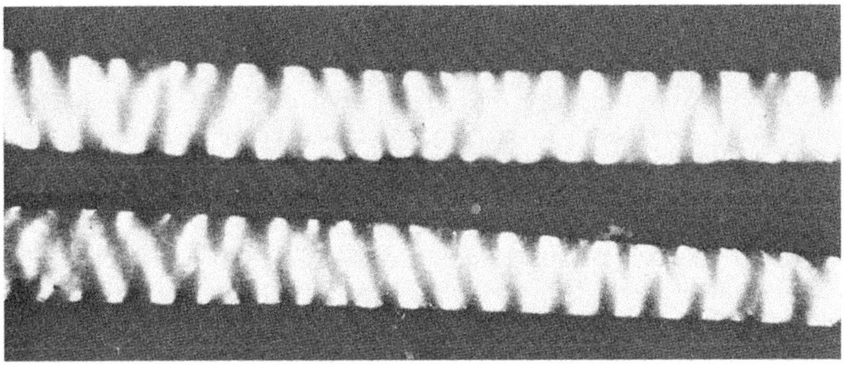

Figure 4.13. A polarised light micrograph of mature rat tail tendon collagen fibrils, showing the crimp period of around 100 μm. (From [94], republished with permission of Taylor & Francis Group LLC Books, through the Copyright Clearance Centre Inc.)

organisation, where the periodicity ranges from a few microns up to fractions of a millimetre [91]. However, this is not necessarily constant and can change in tendons, for example, as a result of exercise [97].

Crimp is probably present in all structures that are subject to tensile forces, not just in tendons. This is consistent with it being absent from corneal fibres [98] although it is present in other ocular tissues including the corneoscleral shell [99]. In more complex hierarchical structures, such as intestines, crimped sheets can be resolved that are in successive overlapping layers, oriented at an alternating angle of about 60° relative to adjacent layers [95]. In intervertebral disks, the fibre directions of adjacent layers are also alternately tilted, from a vertical axis, again with an interlamellar angle of around 60° [91]. A further higher order of structure may exist, a left-handed helical twist arising from the crimp orientation [100].

The development of crimp occurs during embryonic tendon development [101] and is clear in young animals [102]. Cells seem to play a central role in crimp formation, for example tenocytes in tendons, through a contraction process [103]. For rat tail tendons in animals of increasing age, up to 3 years, the crimp angle decreases as the tail length and the crimp periodicity increase, suggesting that the crimp number did not increase during this timeframe [93].

References

[1] Timpl R, Wiedemann H, van Delden V, Furthmayr H and Kühn K 1981 *Eur. J. Biochem.* **120** 203–11
[2] Bruns R R, Press W, Engvall E, Timpl R and Gross J 1986 *J. Cell Biol.* **103** 393–404
[3] Jakus M A 1956 *J. Biophys. Biochem. Cytol.* **2** 243–52
[4] Sawada H, Konomi H and Hirosawa K 1990 *J. Cell Biol.* **110** 219–27
[5] Keene D R, Sakai L Y, Lunstrum G P, Morris N P and Burgeson R E 1987 *J. Cell Biol.* **104** 611–21
[6] Ricard-Blum S 2011 Cold Spring Harb *Perspect. Biol.* **3** a004978
[7] Hall C E, Jakus M A and Schmitt F O 1942 *J. Am. Chem. Soc.* **64** 1234
[8] Bear R S 1942 *J. Am. Chem. Soc.* **64** 727
[9] Brodsky B, Eikenberry E F and Cassidy K 1980 *Biochim. Biophys. Acta* **621** 162–66
[10] Brodsky B and Ramshaw J A M 1994 *Int. J. Biol. Macromol.* **16** 27–30
[11] Yurchenco P D and Ruben G C 1987 *J. Cell Biol.* **105** 2559–68
[12] Schmitt F O, Gross J and Highberger J H 1955 *Exp. Cell Res.* **3** 326–34
[13] Petruska J A and Hodge A J 1964 *Proc. Natl Acad. Sci. USA* **51** 871–76
[14] Hodge A J 1967 *Treatise on Collagen* vol 1 ed G N Ramachandran (New York: Academic) pp 185–205
[15] Chapman J A 1974 *Connect. Tissue Res.* **2** 137–50
[16] Doyle B B, Hukins D W L, Hulmes D J S, Miller A and Woodhead-Galloway J 1975 *J. Mol. Biol.* **91** 79–99
[17] North A C T, Cowan P M and Randall J T 1954 *Nature* **174** 1142–43
[18] Fraser R D B and MacRae T P 1981 *Int. J. Biol. Macromol.* **3** 193–200
[19] Miller A and Parry D A D 1973 *J. Mol. Biol.* **75** 441–47
[20] Smith J W 1968 *Nature* **219** 157–8
[21] Hulmes D J S and Miller A 1979 *Nature* **282** 878–80

[22] Hulmes D J S 2002 *J. Struct. Biol.* **137** 2–10
[23] Fraser R D B, MacRae T P, Miller A and Suzuki E 1983 *J. Mol. Biol.* **167** 497–521
[24] Fraser R D B, MacRae T P, Miller A and Suzuki E 1987 *J. Mol. Biol.* **193** 115–25
[25] Holmes D F and Kadler K E 2006 *Proc. Natl Acad. Sci. USA* **103** 17249–54
[26] Antipova O and Orgel J P R O 2010 *J. Biol. Chem.* **285** 7087–96
[27] Armitage P M and Chapman J A 1971 *Nat. New Biol.* **229** 151–2
[28] Highberger J H, Gross J and Schmitt F O 1951 *Proc. Natl Acad. Sci. USA* **37** 286–91
[29] Highberger J H, Gross J and Schmitt F O 1950 *J. Am. Chem. Soc.* **72** 3321–2
[30] Schmitt F O, Gross J and Highberger J H 1953 *Proc. Natl. Acad. Sci. USA* **39** 459–70
[31] Bowden J K and Chapman J A 1972 *Connect. Tissue Res.* **1** 109–12
[32] Mathews M B 1975 *Connective Tissue. Macromolecular Structure and Evolution* (Heidelberg: Springer)
[33] Vanacore R M, Friedman D B, Ham A J, Sundaramoorthy M and Hudson B G 2005 *J. Biol. Chem.* **280** 29300–10
[34] Yurchenco P D and Ruben G C 1988 *Am. J. Pathol.* **132** 278–91
[35] Vogeli G, Horn E, Carter J and Kaytes P S 1986 *FEBS Lett.* **206** 29–32
[36] Fowler S J, Jose S, Zhang X, Deutzmann R, Sarras M P and Boot-Handford R P 2000 *J. Biol. Chem.* **275** 39589–99
[37] Gathercole L J, Barnard K and Atkins E D T 1989 *Int. J. Biol. Macromol.* **11** 335–8
[38] Barnard K and Gathercole L J 1991 *Int. J. Biol. Macromol.* **13** 359–65
[39] Dyksterhuis L B, White J F, Hickey M, Kirby N, Mudie S, Hawley A, Vashi A, Nigro J, Werkmeister J A and Ramshaw J A M 2011 *Biophys. J.* **100** 2077–83
[40] Yurchenco P D and Patton B L 2009 *Curr. Pharm. Des.* **15** 1277–94
[41] Stetefeld J, Mayer U, Timpl R and Huber R 1996 *J. Mol. Biol.* **257** 644–57
[42] Engvall E 1995 *Int. J. Dev. Biol.* **39** 781–7
[43] Pöschl E, Schlötzer-Schrehardt U, Brachvogel B, Saito K, Ninomiya Y and Mayer U 2004 *Development* **131** 1619–28
[44] Yurchenco P D, Amenta P S and Patton B L 2004 *Matrix Biol.* **22** 521–38
[45] Cescon M, Gattazzo F, Chen P and Bonaldo P 2015 *J. Cell Sci.* **128** 3525–31
[46] Gara S K, Grumati P, Urciuolo A, Bonaldo P, Kobbe B, Koch M, Paulsson M and Wagener R 2008 *J. Biol. Chem.* **283** 10658–70
[47] Knupp C, Pinali C, Munro P M, Gruber H E, Sherratt M J, Baldock C and Squire J M 2006 *J. Struct. Biol.* **154** 312–26
[48] Kwan A P, Cummings C E, Chapman J A and Grant M E 1991 *J. Cell Biol.* **114** 597–604
[49] Fraser R D B, MacRae T P and Suzuki E 1979 *J. Mol. Biol.* **129** 463–81
[50] Ramshaw J A M, Werkmeister J A and Peters D E 1990 *Current Perspectives on Implantable Devices* vol 2 ed D F Williams (London: JAI Press) pp 151–220
[51] Birk D E and Trelstad R L 1984 *J. Cell Biol.* **99** 2024–33
[52] Bruns R R, Hulmes D J S, Therrien S F and Gross J 1979 *Proc. Natl Acad. Sci. USA* **76** 313–17
[53] Birk D E and Zycband E 1994 *J. Anat.* **184** 457–63
[54] Gross J and Kirk D 1958 *J. Biol. Chem.* **233** 355–60
[55] Bannister D W and Burns A B 1972 *Biochem. J.* **129** 677–81
[56] Gelman R A, Poppke D C and Piez K A 1979 *J. Biol. Chem.* **254** 11741–5
[57] Gelman R A and Piez K A 1980 *J. Biol. Chem.* **255** 8098–102

[58] Bernengo J C, Ronziere M C, Bezot P, Bezot C, Herbage D and Veis A 1983 *J. Biol. Chem.* **258** 1001–6

[59] Kadler K E, Hojima Y and Prockop D J 1990 *Biochem. J.* **268** 339–43

[60] Kadler K E, Holmes D F, Trotter J A and Chapman J A 1996 *Biochem. J.* **316** 1–11

[61] Kadler K E 2017 *Int. J. Exp. Pathol.* **98** 4–16

[62] Holmes D F, Lowe M P and Chapman J A 1994 *J. Mol. Biol.* **235** 80–3

[63] Birk D E, Hahn R A, Linsenmayer C Y and Zycband E I 1996 *Matrix Biol.* **15** 111–8

[64] Thurmond F A and Trotter J A 1994 *J. Mol. Biol.* **235** 73–9

[65] White J F, Werkmeister J A, Darby I A, Bisucci T, Birk D E and Ramshaw J A M 2002 *J. Struct. Biol.* **137** 23–30

[66] White J F, Werkmeister J A, Bisucci T, Darby I A and Ramshaw J A M 2014 *J. Biomed. Mater. Res. A.* **102** 3550–5

[67] Clark R A F 1996 *The Molecular and Cellular Biology of Wound Repair* 2nd edn (New York: Plenum)

[68] Prockop D J and Fertala A 1998 *J. Biol. Chem.* **273** 15598–604

[69] Malone J P, George A and Veis A 2004 *Proteins* **54** 206–15

[70] Hörlein D, McPherson J, Goh S H and Bornstein P 1981 *Proc. Natl Acad. Sci. USA* **78** 6163–7

[71] Mieczkowska A, Bouvard B, Chappard D and Mabilleau G 2015 *Bone* **74** 29–36

[72] Hulmes D J S 1983 *Coll. Related Res.* **3** 317–21

[73] Linsenmayer T F, Gibney E, Igoe F, Gordon M K, Fitch J M, Fessler L I and Birk D E 1993 *J. Cell Biol.* **121** 1181–9

[74] Chen S, Mienaltowski M J and Birk D E 2015 *Exp. Eye Res.* **133** 69–80

[75] Zvackova I, Matalova E and Lesot H 2017 *Front. Physiol.* **8** 554

[76] Chen S and Birk D E 2013 *FEBS J.* **280** 2120–37

[77] Sullivan M M, Barker T H, Funk S E, Karchin A, Seo N S, Höök M, Sanders J, Starcher B and Wight T N *et al* 2006 *J. Biol. Chem.* **281** 27621–32

[78] Fleischmajer R, Olsen B R, Timpl R, Perlish J S and Lovelace O 1983 *Proc. Natl Acad. Sci. USA* **80** 3354–8

[79] Nurminskaya M V and Birk D E 1996 *Biochem. J.* **317** 785–9

[80] Scott J E and Parry D A D 1992 *Int. J. Biol. Macromol.* **14** 292–3

[81] Neame P J, Kay C J, McQuillan D J, Beales M P and Hassell J R 2000 *Cell. Mol. Life Sci.* **57** 859–63

[82] Birk D E 2001 *Micron* **32** 223–37

[83] Smith S M, Zhang G and Birk D E 2014 *Matrix Biol.* **33** 47–53

[84] Birk D E, Fitch J M, Babiarz J P, Doane K J and Linsenmayer T F 1990 *J. Cell Sci.* **95** 649–57

[85] Birk D E, Fitch J M, Babiarz J P and Linsenmayer T F 1988 *J. Cell Biol.* **106** 999–1008

[86] Wenstrup R J, Florer J B, Brunskill E W, Bell S M, Chervoneva I and Birk D E 2004 *J. Biol. Chem.* **279** 53331–7

[87] Sun M, Chen S, Adams S M, Florer J B, Liu H, Kao W W, Wenstrup R J and Birk D E 2011 *J. Cell Sci.* **124** 4096–105

[88] Birk D E, Nurminskaya M V and Zycband E I 1995 *Dev. Dyn.* **202** 229–43

[89] Parkinson J, Kadler K E and Brass A 1994 *Phys. Rev. E Stat. Phys. Plasmas. Fluids Relat. Interdiscip. Topics* **50** 2963–6

[90] Craig A S, Birtles M J, Conway J F and Parry D A D 1989 *Connect. Tissue Res.* **19** 51–62

[91] Gathercole L J and Keller A 1991 *Matrix* **11** 214–34

[92] Spiesz E M, Thorpe C T, Thurner P J and Screen H R C 2018 *Acta Biomater.* **70** 281–92

[93] Diamant J, Keller A, Baer E, Litt M and Arridge R G 1972 *Proc. Roy. Soc. Lond. B Biol. Sci.* **180** 293–315

[94] Parry D A D and Craig A S 1988 *Collagen* vol 2 ed M E Nimni (Boca Raton, FL: CRC) pp 1–24

[95] Orberg J W, Klein L and Hiltner A 1982 *Connect. Tissue Res.* **9** 187–93

[96] Dlugosz J, Gathercole L J and Keller A 1978 *Micron* **9** 71–82

[97] Franchi M, Torricelli P, Giavaresi G and Fini M 2013 *Connect. Tissue Res.* **54** 267–74

[98] Gathercole L J and Keller A 1975 *Structure of Fibrous Biopolymers, Colston Papers no. 26* ed E D T Atkins and A Keller (London: Butterworths) pp 153–87

[99] Jan N J, Brazile B L, Hu D, Grube G, Wallace J, Gogola A and Sigal I A 2018 *Exp. Eye Res.* **172** 159–70

[100] Franchi M, Ottani V, Stagni R and Ruggeri A 2010 *J. Anat.* **216** 301–9

[101] Shah J S, Palacios E and Palacios L 1982 *Dev. Biol.* **94** 499–504

[102] Gathercole L J and Keller A 1978 *Micron* **9** 83–9

[103] Herchenhan A, Kalson N S, Holmes D F, Hill P, Kadler K E and Margetts L 2012 *Biomech. Model. Mechanobiol.* **11** 449–59

Chapter 5

Tissue arrangement

Tissues are the largest resulting structural arrangements of collagens and are the areas where the mechanisms of assembly and control are least understood. For biomaterials, and particularly collagen-based biomaterials, where the functional material is macroscopic in nature, observing and controlling the response to implanted materials is important. The form and nature of the new, associated tissue will significantly influence the performance of a device.

5.1 Formation of new tissue

There is a significant literature on tissue formation during development and during wound repair [1]. Both processes are complex, involving the migration of specific cell types in an ordered fashion. This includes the central role of the macrophage and the various cytokines and growth factors that regulate other cell types. However, as a consequence, much of this literature is concerned with these cellular changes and only a broad description of new extracellular matrix (ECM), often lacking details on the specific composition, including the particular type of collagen being synthesised [2].

Data on the host response to biomaterials has also been limited. The fibrous capsule that can form around devices has received attention, but the response to porous materials is less common, with, for example, reports of predominant type I production in corneal models [3, 4].

Other studies have followed the deposition of new collagens within a porous collagen-based biomaterial, the biosynthetic Omniflow™ Vascular Prosthesis (chapter 15). These studies showed the infiltration of type III into the wall of the device, for up to four years, while during the same time, the type III collagen of the device itself was maintained [5]. Also, these studies showed that type V collagen was a precursor to larger fibril formation and infiltration [6], while type VI collagen was also among the first collagens to be deposited accompanying cell infiltration [7].

doi:10.1088/978-0-7503-2096-2ch5

Similar changes were seen in clinical samples as well as in these animal model studies [8].

More recently, a study using monoclonal antibodies (MAb's) to different collagen types has tracked the order and extent of invasion of different collagen types into the porous walls of an expanded polytetrafluoroethylene (ePTFE) tube [9]. These data showed, by *in situ* hybridisation, the rapid accumulation of type I collagen around the device. Immunohistology showed the deposition of different collagen types within the porous walls (figure 5.1). Types V and VI collagens form the first stable matrix, and are then followed by the main interstitial collagens, types I and III. Type V collagen seems to disappear but is being masked by the deposition of the principal collagens that can be removed by brief proteolysis. Type XII appears limited to the fibrous capsule, while type XIV is an integral constituent of the new invading ECM [9].

5.2 Native crosslinking

A key aspect of new tissue formation is the development of crosslinking to stabilise the structure, and give it structural and functional integrity. It was recognised, for example, from studies on the effects of lathyrogens (see chapter 8), that the high tensile strength and mechanical stability are due to the presence of covalent crosslinks between collagen molecules [10]. Thus, the formation of specific covalent crosslinks between collagen molecules stabilises the collagen fibrils in tissues [11].

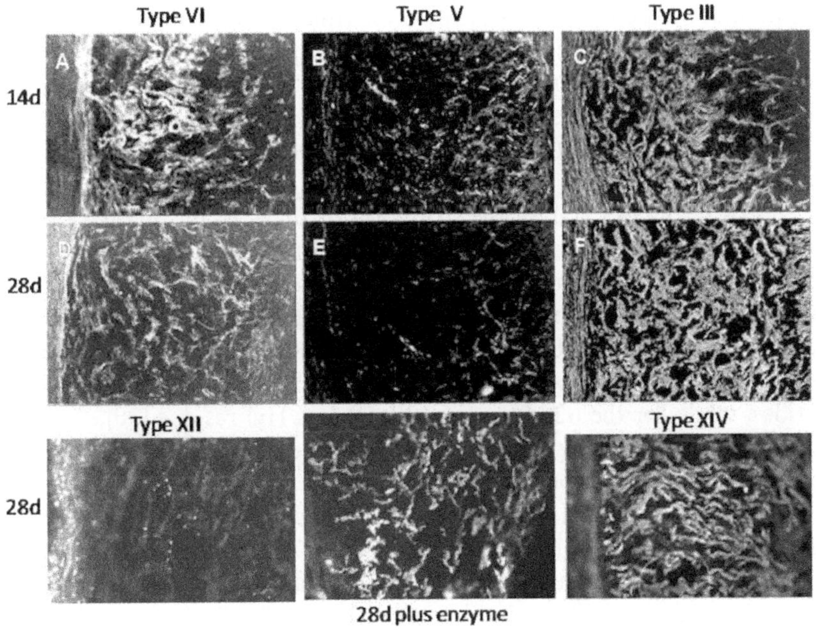

Figure 5.1. Immunofluorescent tracking of new collagen deposition into the wall pores of an ePTFE biomaterial implant into sheep dermis, following different collagen types at different time points. (Adapted from [9], republished with permission of John Wiley & Son Inc., through the Copyright Clearance Center Inc.)

Initially, these crosslinks are enzyme-mediated and are found at specific locations in the structure. With increasing age, a range of non-enzymatic covalent crosslinks, the advanced glycosylation end-products (AGE) accumulate and lead to stiffening of collagen tissues at low strains [12]. These AGE crosslinks result from sugars (such as glucose and ribose) in the body. Of interest, the extent of this AGE crosslinking is greater in diabetic patients where the glucose levels are typically higher, even with medication [13].

The nature and extent of collagen crosslinking also impacts on the *in vivo* turnover of collagens. Collagens are generally seen as having a very long half-life, and the AGE formation will increase this further in older people [14], for example, by restricting matrix metallo-proteinase (MMP) activities [15]. This is not just the case for interstitial collagens but also happens for type IV collagen, with implications for glomerular function [16]. It is possible that other biologically important binding sites could also be affected, even by initial monofunctional reaction with Lys prior to any crosslink formation. This increase in crosslinking can lead to other adverse indications, including mechanical issues such as arterial wall stiffening [17]. Long half-lifes for skin and cartilage have been reported, around 20 years and many decades respectively [14], while in contrast the collagen turnover in periodontal ligament has been reported as being only days into a couple of weeks [18] dependent on location.

5.2.1 Enzyme-initiated crosslinking

The enzyme-initiated crosslinking of the interstitial collagens involves the action of the lysyl oxidase (LOX) family of enzymes (see chapter 3). It has been studied most extensively for type I collagen, but it is probable that this is a good model for other interstitial collagens. This is supported by analyses that show crosslinking using the same mechanism between type I and type III collagen molecules, also showing their joint presence in fibrils [19].

These LOX enzymes catalyse the oxidation of the Lys and hydroxylysine (Hyl) side chains of collagens in the telopeptide regions of fibrilar collagens. The resulting products are peptidyl α-aminoadipic-δ-semialdehyde (allysine) and peptidyl-δ-hydroxy-α-aminoadipic-δ-semialdehyde (hydroxyallysine) residues (see figure 3.2) at specific locations in the telopeptides. For the α1-chain of type I collagen the residues that are modified are Lys/Hyl residues 9^N and 16^C in the N- and C-telopeptides, while for the α2-chain it is the 5^N Lys/Hyl residue [20].

These aldehydes are highly reactive and do not require any enzyme(s) for further reactions. Adjacent telopeptide aldehydes can undergo an aldol condensation reaction leading to the formation of an intramolecular crosslink. More important are the intermolecular crosslinks. These arise from the condensation reaction of an aldehyde with an amino group on an adjacent collagen molecule. For the α1-chain of type I collagen, the residues that react are Lys/Hyl residues at positions 87 and 930 in the triple-helix, while for the α2-chain it is the Lys/Hyl residues at positions 87 and 933 in the triple-helix (figure 5.2) [20].

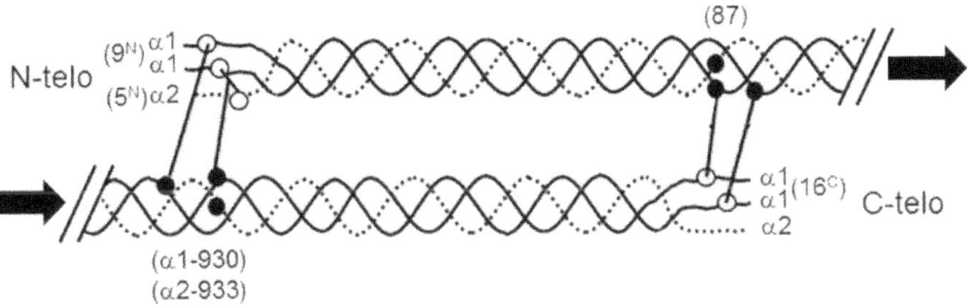

Figure 5.2. Locations of intermolecular crosslinks for type I collagen. (From [20], reprinted with permission from Portland Press ©2012.)

These initial crosslinks are termed the 'reducible' crosslinks, and the types vary between different tissues. They have unsaturated structures that can be reduced, normally using sodium borohydride [21] or a similar borohydride or borotritide compound. These reduced derivatives are stable to acid hydrolysis and can be used for analysis of the number and proportion of different crosslink types in a given tissue [22, 23]. A variety of column chromatography approaches have been developed for analysis of these crosslinks [24] and also for multi-functional crosslinks [25]. Various reducible crosslinks have been observed (figure 5.3) and their structures determined (figure 5.3) [11].

The initial crosslinks that form are aldimines and these are reversible under acid conditions, and are typically found in skin and tendon. However, those involving a telopeptide Hyl undergo an Amadori rearrangement giving keto-imines that are not reversible, and which are typically found in bone and cartilage [26].

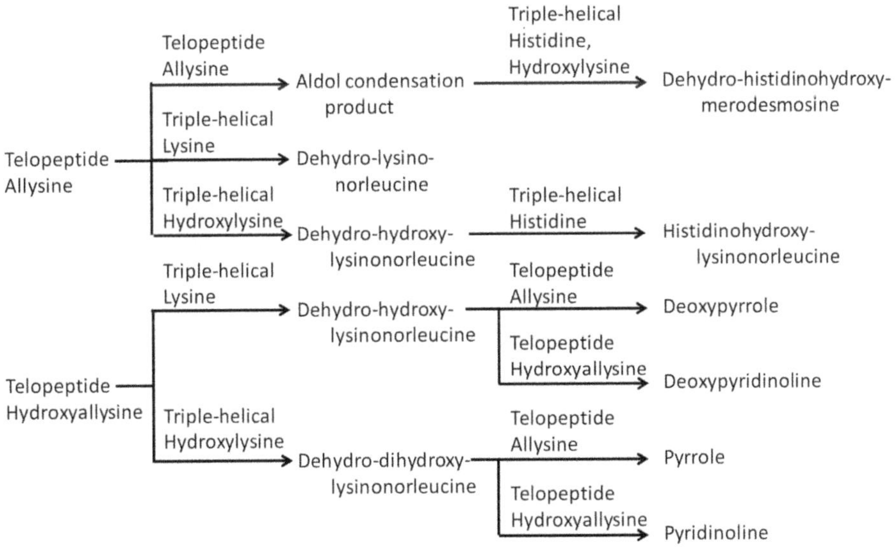

Figure 5.3. A schematic of the pathways that lead from aldehydes (left) to the 'reducible' crosslinks (centre) and thence to mature crosslinks (right).

Subsequently, the initial reducible crosslinks undergo a series of non-enzymatic, spontaneous reactions, to give the mature crosslinks (figure 5.3). Several have been identified and characterised (figure 5.3), but it is not clear if there are others present. The processes of mature crosslink formation may, in some cases, include the involvement of oxygen [27]. Notwithstanding, this is an area where there has been disagreement and debate.

Beyond these crosslinks in interstitial collagens, an unusual, probably unique crosslink has been reported for type IV collagen [28]. This is a sulfilimine crosslink, and its formation is catalysed by peroxidasin. It forms between a Met residue and a Lys residue in adjacent NC1 domains of this collagen [28]. Type IV collagen also contains the reducible crosslink dehydro-dihydroxy-lysinonorleucine [29].

Little else has been reported for crosslinking systems in other collagens. Type VIII contains an acid labile crosslink that is not β-aminopropionitrile (BAPN) sensitive, suggesting that it is not formed via lysyl oxidase activity [30]. Conversely, the extraction of type X collagen was enhanced in the presence of BAPN, so may contain aldehyde-based crosslinking [31].

5.2.2 Non-enzymatic crosslinking

The 'browning reaction', and specifically the Maillard reaction, is well known to the food industry, where heat, in combination with a reducing sugar and protein, can accelerate the browning process [32]. It had been suggested, in 1981, that this type of reaction could occur at body temperature, i.e. with no additional heating, and lead to the modification and crosslinking of collagen with age [33]. Almost a decade later one such crosslink, an imidazo[4,5-*b*]pyridinium molecule, that was called pentosidine, was isolated and characterised and model systems used to confirm its identity [34]. This crosslink, arising from ribose, leads to a characteristic fluorescence [33]. Subsequently, a range of other products, including some that involve reaction with glucose have been identified, although it is probable that other products will emerge. These crosslinks can provide bifunctional and trifunctional crosslinking.

The initial reaction in the formation of the AGE crosslinks is between a reducing sugar (glucose, ribose) and the ε-amino groups of a Lys residue to form a reversible and relatively unstable Schiff base, which undergoes an Amadori rearrangement to give a more stable intermediate. This intermediate, over time, participates in further reactions, to form AGE (advanced Maillard) products (figure 5.4).

The sites where such reactions could occur are limited. Although each type I collagen chain has over 30 Lys residues and over 70 Arg residues (see chapter 2), the number that are adjacent so as to allow a Lys and an Arg crosslink is limited. The number of possible crosslink sites with a Lys and Arg will depend on collagen structure and organisation (see chapters 2 and 4). There are possibly around 24 sites for intramolecular reactions [36], and fewer, around 14 sites for the potentially more important intermolecular crosslinks [37].

Beyond glucosepane (figure 5.4) at least 15 other AGE-related moieties that are associated with collagen have been also been identified in tissues over the last 20 years, as well as some in further non-collagenous entities [38]. In addition to

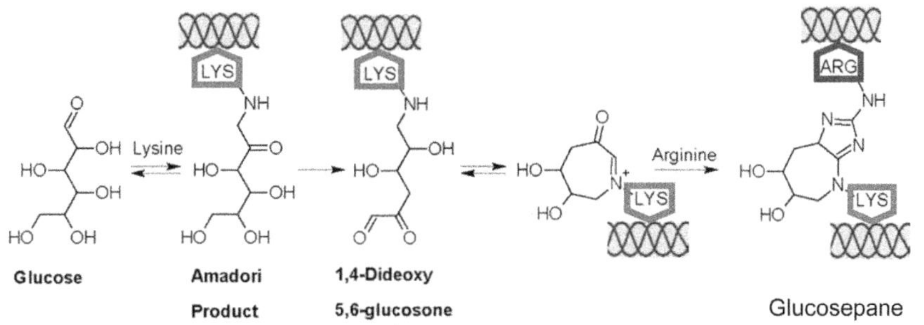

Figure 5.4. The proposed reaction for an AGE product formed from glucose. (From [35], republished with permission from the American Society for Biochemistry and Molecular Biology, through the Copyright Clearance Center Inc.)

stable end-products such as pentosidine and glucosepane, these include a range of intermediates such as fructose-lysine. There are also others formed by biological transformations of sugars (glucose) into precursors such as glyoxal and methylglyoxal that lead to AGE entities such as glyoxal-hydroimidazolone, methylglyoxal-hydroimidazolone and carboxyethyl-lysine [38]. No doubt the range of compounds to be identified is not yet complete.

Knowledge concerning AGE crosslinks and their implications for human health and associated diagnostic tests is still emerging. The mechanisms, however, are adaptable and useful for the production of crosslinked biomedical collagen products. Given the health impacts of the accumulation of AGE crosslinks, especially in diabetic patients, the development of therapeutic agents that can reverse this process is a clear research target [39, 40].

5.3 Examples of tissue structure

Tissues are the most complex elements for collagen and need to be considered in the context of the many other molecules, cell types and environmental factors (such as mechanical forces, oxygen tension) that are also components in the system. Also, the developmental stage of the tissue and its age are also important. During embryonic morphogenesis groups of cells are coordinated through families of conserved signalling mechanisms, but there is still much to understand in this area [41]. Because of the complex nature of tissue, it is necessarily the least understood aspect of collagen organisation. Understanding tissue structure, and the distribution and organisation of its components is also important for tissue engineering and regenerative medicine as in an ideal situation the new tissue should closely resemble the tissue that is being repaired, rather than being a piece of less functional tissue of the right shape and size. The recent developments in high resolution mass spectrometry have enabled proteomic analysis of tissues [42, 43] which should assist in understanding the level of similarity between natural tissues and tissue engineering constructs.

Even a tissue such as tendon, which may seem among the simpler tissues, is still very complex. Other tissues, such as a heart valve leaflet, may also appear simple in

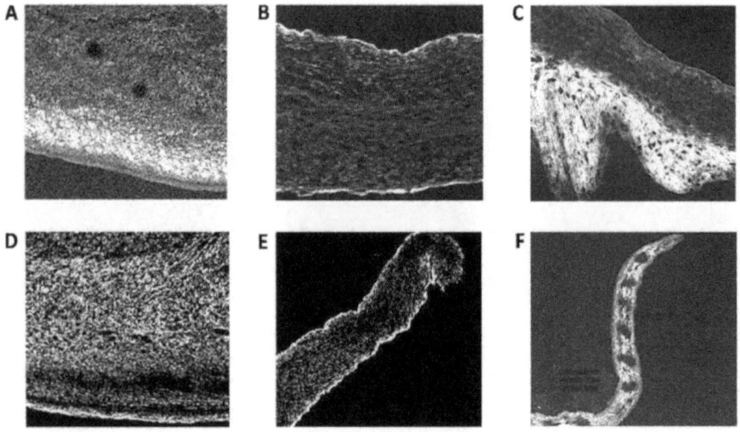

Figure 5.5. Immunofluorescence examination of heart valves. (A) Type I collagen, from the mid-region of the porcine mitral valve, (D) Type III collagen, from the mid-region of the porcine mitral valve, (B and E) Type IV collagen, from the middle and tip of the valve respectively, and (C and F) Type VI collagen, from the middle and tip of the bovine mitral valve respectively. (Modified from [44]. Reproduced by permission of ICR Publishers Ltd.)

appearance, but are complex (figure 5.5). Looking at the distributions of particular collagens in heart valves shows that type I and type III collagens are present throughout, but their distributions are different. Type IV collagen is found over the surface of the tissue, over the full length to the tip. Type VI collagen has a particularly interesting distribution; it is also present along the tissue to the tip but is present significantly in the outflow half of the leaflet [44] (figure 5.5). Can these distributions be chance or do they reflect an optimised functional requirement? Can a tissue engineered device achieve this level of sophistication? Some collagen types exist as various isotypes, including type IV collagen, and various forms can be found within a particular tissue. For example, in the testis, the seminiferous tubules and their environment show different type IV collagen isoforms (figure 5.6). Thus, the microvascular surrounding the tubules contains the more common the $(\alpha1)_2\alpha2$[IV] isoform while the tubule wall itself contains a different isoform, with α3[IV] being present while absent in the microvasculature [45].

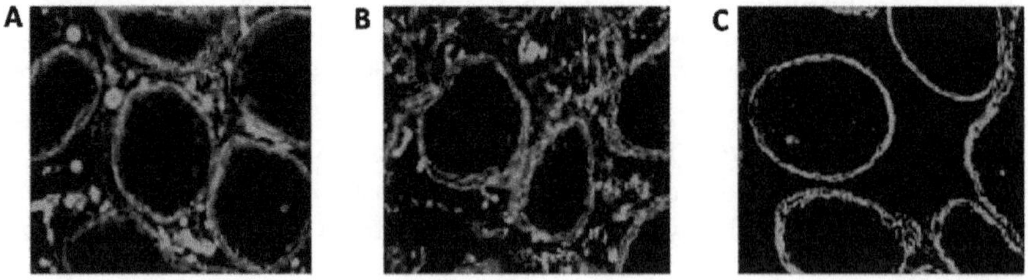

Figure 5.6. Immunohistology showing the distribution of collagen type IV isotype chains in seminiferous tubules and adjacent tissues. (A) α1[IV], (B) α2[IV] and (C) α3[IV] chains. .

5.3.1 Tissue organisation in tendon and ligament

Tendon and ligament tissues have been the object for study for many anatomists since Galen (circa 150 CE). For example, Leonardo da Vinci (1452–1519), on the basis of dissections of at least 19 corpses (see [46]), made many anatomical drawings of ligament and tendon systems (which were reported in *Quarderni d' Anatomia*), and concluded that for tendons 'they are mechanical instruments which carry out as much work as is entrusted to them' [47]. Historically, many further studies were published, especially in the last 200 years [47]. Nevertheless, fibril formation, growth and organisation are not yet fully understood.

Both tendons and ligament are well-ordered compared with many (most) tissues and have highly aligned collagen fibrils. These tissues are seen as hierarchical structures [48] and various models have been proposed for tendon structure (for example, figure 5.7) [48], although some of the smaller elements, such as possible sub-fibrils and microfibrils, are still subjects of discussion. The alignment in tendons is generally seen as greater than that found in certain ligaments [49], and tendon has been the predominant tissue that has been studied. Variations in structures and properties are observed at the attachment sites of these tissues with bone (the entheses) [50]. Tendon biology is complex, and numerous processes, including mechanical forces, and other molecules are involved in tenogenesis during development and subsequently during aging [51].

Modern studies have examined the length of fibrils in these tissues; not a simple task, but one fraught with difficulties. For example, transmission electron microscopy (TEM) studies on 1.4 μm serial sections of over 1000 fibrils from 6-month old rat-tail tendon showed no ends [52]! Modelling from these data suggested an estimate for fibril length of up to 10 mm [52]. In other exacting studies, serial sections from a 14-day chicken embryo tendon showed one complete fibril along with many incomplete segments with ends (figure 5.8) [53]. Nevertheless, the

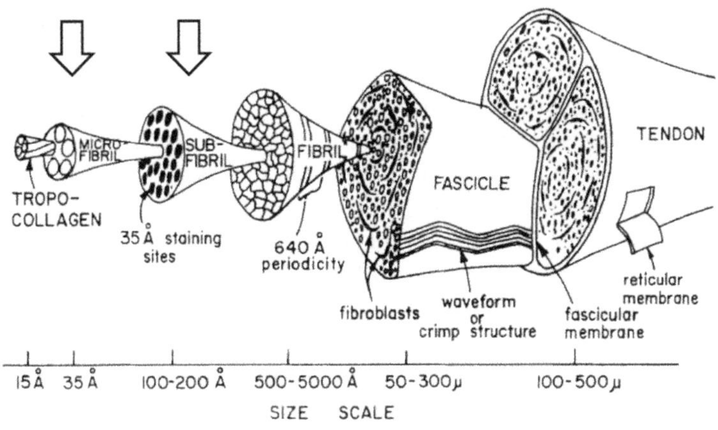

Figure 5.7. A schematic of the arrangement of tendon tissue. Arrows indicate structures that are not always uniformly accepted. (From [48], reprinted with permission of Taylor & Francis, through the Copyright Clearance Centre.)

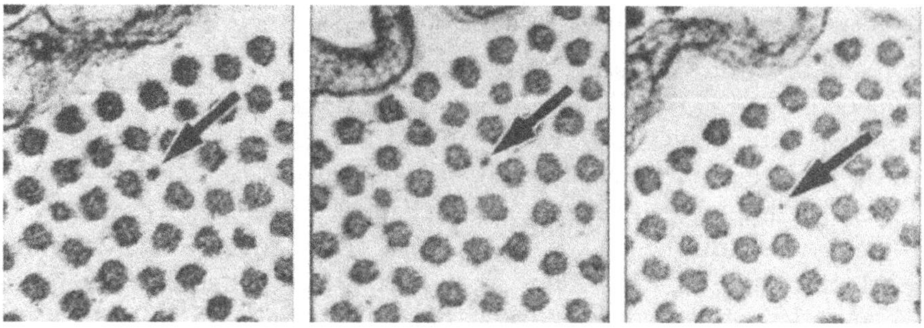

Figure 5.8. TEM images showing (left to right with arrows) the tapering end of a collagen fibril within a larger fibril assembly. (From [54], republished with permission of John Wiley & Sons Inc., through the Copyright Clearance Center Inc.)

numbers of single ends that were observed allowed a comparison between different age animals and indicated that embryonic fibrils, around 10–20 μm, were shorter than fibrils from 15-day tendons, 22–33 μm [54]. Otherwise, the diameters that were seen for each fibril were generally constant among most of their length, excepting the tapering ends [55]. In growing tissues, it has been proposed, based on studies with ligaments, that sliding between adjacent fibrils occurs during changes to ligament length [56].

Various studies have examined the distributions of collagen fibril diameters in tendon and ligament (and other tissues) and how these change with increasing age. Initially, extensive TEM studies were used [57], but more recently, using ligament, it was shown that small-angle x-ray scattering gave data that agreed with TEM data from the same samples [58]. The ready availability these days of synchrotron-based x-ray beamlines means that this approach could now be used for quick and accurate analysis of numerous samples.

In embryonic tissues the cross sections of fibrils start at a uniform small diameter until a signal stimulus leads to an increase in diameter [59]. Growing fibrils are irregular, probably as a result of aggregation and fusion of adjacent small fibrils [55]. As tissues mature the fibrils again take on a consistent round appearance, albeit with variations in size over time. The initial extensive examination of rat-tail tendons, involving over 10 thousand diameter measurements, showed a single homologous distribution of fine fibrils, about 40 nm, at birth [57]. By 2–3 weeks, the diameter of this group of fibrils has increased to 50–70 nm, while a group of larger fibrils had been formed, possibly forming a separate population with a mean diameter of about 115 nm. By 13–14 weeks these larger fibrils had become the predominant fibrils and had grown to a mean diameter of 280 nm [57]. Further studies, on rat-tail and other tissues, suggested that fibril diameter increases were multiples of an 8 nm unit [60, 61]. These data also indicated that the rates of change varied between different animals, related to their developmental stage at birth, precocious or altricious [61].

Undoubtedly, in tendon and ligament and in other tissues the collagen orientation reflects tissue function. The high tensile load for tendons and ligaments is clearly best served by the long parallel arrangement of fibrils [62]. Fibril diameter distributions may also reflect mechanical function [63]. Variations in structure could, therefore, occur with changes in loading and frequency of loading to adapt to mechanical requirements. Various ligaments and tendons of race horses, from immature tissue prior to training and through to mature tissue, about 2 years [64] and older tissues, were examined, including the effects of an extensive exercise training program [64–66]. The superficial flexor tendon showed a decrease in collagen fibril diameter (means of 132 nm to 105 nm) during training, relative to control animals, suggesting that microtrauma was happening during intense training [65]. On the other hand, neither the deep digital flexor tendon nor the suspensory ligament showed any significant changes in fibril diameters with training [66].

The other components of the ECM are also important partners in fibril development and organisation in tissues. Type III collagen is present in certain structural features of tendon, the fascicles and endotendenium, but not in the main fibril bundles (figure 5.8), while type V collagen is present at low levels throughout the tissues [67]. Possibly more important are the proteoglycans. Beyond regulation during fibril formation, especially by decorin, studies on tendon tissue show that younger fibrils have significant amounts of hyaluronic acid whereas dermatan sulfate content increases with maturation, allowing increases in fibril diameter [68]. Concomitant with fibril diameter increases, glycosaminoglycan amounts decrease [59]. They also change with mechanical changes to tendon tissue [69].

5.3.2 Tissue organisation in cornea

The cornea provides essential mechanical properties to the eye as well as contributing about two thirds of the refractive power of the eye. Critical, however, is that it is also fully transparent [70]. It was well recognised that these properties arose from the main stromal tissue [71]. Collagen is a key component of this tissue and the basic structure has been broadly understood for some time [72]. This allowed a theory for transparency to be developed [73, 74] which continues to be advanced [70]. The basic principle of the theory is that incoming light is diffracted by an ordered collagen array and that subsequent interference between scattered light waves is such that they cancel out in all but the forward direction and in the process, transfer their energy to the transmitted light [70]. A range of factors are important, one of which is the radius of the collagen fibrils. This is under a high level of control and results from a range of intermediate developmental stages [75, 76]. The detailed understanding of this tissue is important for the successful development of collagen-based artificial corneas.

The mature cornea comprises five readily identified layers [70]. Two of these are cell layers, each with an underlying structure. Externally, an epithelial layer about 50 μm thick [77] lies on the Bowman's layer [78]. At the posterior a single layer of endothelial cells lies on the Descemet's membrane, which in turn is linked to the stroma via a region rich in type VI collagen [79]. These layers are normally thin so do not disrupt the transparency. The main component, about 90% of the thickness, is the stromal tissue. The main macromolecules in this tissue are collagen, proteoglycans and other specialised glycoproteins [70, 80].

The principal collagen structures in the corneal stroma are lamellae. The presence of lamellae has been known for a considerable time, and, along with other eye tissues, were observed to swell in acetic acid [78]. A range of species variations have now been observed, for example, different thicknesses; in the human eye, there are about 240 stromal lamellae [81]. The collagen fibrils in the lamellae contain about 300–400 collagen molecules each when examined in cross section and it has been suggested that each fibril comprises around 70 microfibril structures [82]. Thus, like tendon and ligament the collagen in cornea is highly aligned, but with extra dimensionality, and also with a higher precision in fibril diameters.

In the mature eye, the fibrils are predominantly type I collagen but with a significant amount, around 20%, of type V collagen with other collagen types, such as type III being essentially absent [83, 84]. The presence of type V collagen is one of the key factors that control the precision of collagen fibril diameter [85, 86] along with proteoglycans. Indeed, in addition to the precise size and arrangement of the collagen fibrils in the lamellae, the interfibrillar gel has also long been recognised as a key aspect of transparency [87]. The gel is predominantly proteoglycans, with several distinct, different types present [70, 88].

5.3.3 Tissue organisation in skin

Skin, and most other tissues differ from tendons and corneas in that they are generally anisotropic, and contain a greater diversity of collagen types. The composition of skin varies with age and location. Initial studies on the extraction of human skin samples showed that type III collagen was more abundant in foetal tissues [89]. The use of sodium dodecyl sulfate- polyacrylamide gel electrophoresis (SDS-PAGE) allowed analysis of small samples, and also showed that there was more type III collagen in younger, particularly foetal skin [90]. Studies on human abdominal skin also showed that although type I collagen was the main component, significant amounts of type III collagen were present in equal amounts throughout the skin [91].

Further studies, using bovine skin from the rear flank region, confirmed that type III collagen was present in samples from foetal up to 10-year-old animals (figure 5.9) [92]. The use of SDS-PAGE for analysis of serial sections of frozen tissue obtained using a cryotome enabled a detailed profile to be obtained. This confirmed that type III collagen was more abundant in younger tissues. It was generally higher in the papillary dermis, associated with appendages, and its proportion declined in this region until about 18 months. However, after this time point it remained fairly constant while the proportion in the lower, reticular dermis continued to decline through to 112-month samples (figure 5.9) [92]. Other studies on rat skin, as well as other rat tissues showed that, the proportion of type III collagen decreased with age [93], while that in heart and lung increased up to 1 year [93].

In earlier studies, skin samples were generally taken from a single, reproducible region of the animal for age comparisons. However, the morphology and topology of collagen fibres is known to vary with location [94, 95]. Histology studies on porcine skin also showed these structural differences, and semi-quantitative

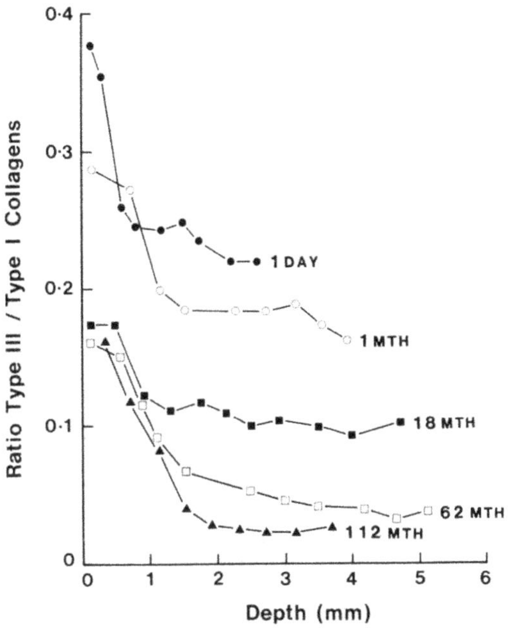

Figure 5.9. The ratio of type III to type I collagen at different depths in bovine skin from animals of various ages. The papillary dermis was between 1.0–1.3 mm deep. (From [98], reprinted with permission of Elsevier.)

histology analysis showed variation of collagen content between tissues, selected at 11 specific locations. These data showed that type III collagen was dominant in the papillary dermis and type I in the reticular dermis. However, despite the clear differences in collagen quantity, the proportion of type III to type I collagen showed limited variations between locations was seen, although there were differences between the rostral back and mid back skin regions [96]. In human skin, there is an increase in type III content in scars, particularly pathological scars, relative to adjacent normal skin [97].

The anisotropy of skin, and the resulting skin tension lines was reported by Dupuytren in 1834 [99] when he noted that a round puncture wound assumed a fusiform, ellipsoidal shape. Later, this observation was expanded by an anatomist in Vienna, Karl Langer, who punctured the skin of cadavers with a circular ice pick. The holes he produced were elliptical, as a result of the collagen orientation at that location. Lines linking the long axes defined the variations in collagen orientation over the entire body, now called Langer lines [94].

Several approaches have now been developed to provide quantitative measures of the organisation of collagen in skin. These include microscopy and several variations, including confocal microscopy [100] and second-harmonic-generation microscopy [101], as well as light scattering and x-ray diffraction [102], and in some cases combined with automated quantification programs [103]. Considerable variation in collagen organisation exists between different locations on an animal's skin and between species [104]. In more primitive species, such as shark skin [105],

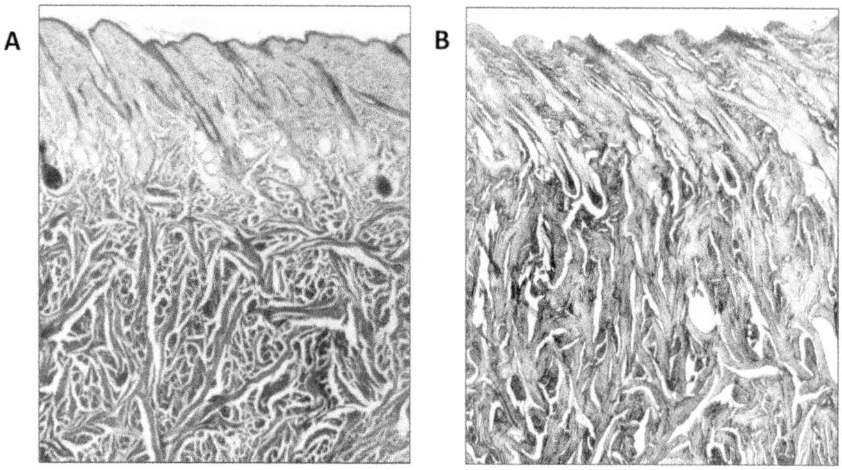

Figure 5.10. A comparison between the collagen organisation in (A) normal hide and (B) vertical fibre hide, from Hereford cattle. Photographs by the late John Bavinton of CSIRO.

teleost fish skin [106] or even some birds [107] the collagen can be well organised, with alternating sheets of fibres parallel to the surface and roughly orthogonal to an adjacent sheet. And while mammalian skin is often thought of as anisotropic, for some skins this may not be the case [100, 108].

There have also been studies on fibril diameters, although these are not as numerous as for more ordered tissues. Details of distributions on well-ordered systems with a narrow size range can potentially be obtained from measurements on just a few hundred fibrils, but in a more complex tissue like skin several thousand measurements will be needed [63].

Proteoglycan (glycosaminoglycan) content is an important component of skin, as in other tissues. In human skin the total content is greater in load bearing tissues compared with non-load bearing skin [109], consistent with the role of large proteoglycans in other tissues [110].

Compared with other tissues that are frequently used directly as tissue-based implants, where the direction of collagen orientation can be important, such as pericardium, the collagen organisation of the collagen in skin is less important for medical implants. However, in other industries, especially the leather industry, the direction of collagen fibres is particularly important as it impacts on the strength of the products [104].

Vertical fibre

A wide variety of heritable disorders in cattle have been described and have proved useful in understanding similar connective tissue disorders in humans [111]. One unusual example had been identified by the bovine leather industry [112]. Termed the 'vertical fibre defect' it is only an issue for the leather industry and does not seem to have any influence on the bovine's well-being. Examination of the butt area of hides from animals (figure 5.10) shows that in normal hides the collagen fibre bundles in the reticular dermis (corium) below the papillary dermis are aligned in a complex interwoven pattern, whereas, in vertical fibre hides, the collagen fibre bundles at the top of the reticular dermis are aligned perpendicular to the surface of the hide [113, 114]. Lower down, the fibres take on the interwoven pattern of normal hides. In the leather industry, the hide is split to give a grain (top) sheet, about 1.5 to 2.5 mm thick to provide material for furnishings, shoes, etc. The result of this splitting is that only the vertical fibres of affected hides remain so, for example, when this material is stretched over the last (shape) in shoe making it tears and splits as the strength from interwoven fibre bundles has been lost. Various studies have shown that this 'defect' is present only in horned Hereford cattle (not in Poll Herefords) [113, 115]. It probably does not reflect any changes in the collagen [116], but rather changes in the various other extracellular matrix molecules and cellular interactions and properties. This structural control of collagen fibre bundle arrangement is inherited as a recessive trait [117, 118], so the genome studies that are now possible could potentially provide a route to understanding of factors that affect collagen fibre organisation.

5.3.4 Tissue organisation in other tissues

The collagen organisation in a range of other tissues has also been examined, but generally in not as much details as the preceding tissues. Some of these are given briefly in the following examples.

Articular cartilage

Articular cartilage has evolved to withstand long periods of compression and relaxation on a daily basis. But for many individuals, it does not withstand this continuing insult, with increasing age, or trauma, leading to osteoarthritis. The high prevalence and debilitating effects of osteoarthritis has led to a large number of studies on the structure and role of collagen in normal and diseased articular cartilage. Such studies have shown that articular cartilage, especially of the knee, can be approximated as comprising three zones; a tangential zone which forms the articulating surface, a mid-zone being the main volume, and a deep zone where the major biosynthetic activity is progressing [119]. Each zone itself may comprise subzones. For example, it has long been suggested that the superficial zone has a specific layer, lamina splendens, adjacent to the joint cavity [120] and this comprises further substructures [119].

While certain substructures may have distinct compositions, the principal component of the main body of cartilage is type II collagen, in association with minor collagen types IX and XI [121]. The collagen is present as thin, tension-resisting fibrils embedded in a concentrated proteoglycan environment, such that the collagen is critical in the tissue withstanding the compression pressures [122].

The arrangements vary between zones. In the superficial layer the collagen fibrils are parallel to the surface, while at the surface a random arrangement was present. In the middle zone the fibrils were mostly perpendicular to the surface [123]. These fibrils comprise two separate populations, some around 20 nm in diameter, while the others are 40 nm or more [63, 124]. The understanding of the relationship between collagen fibril architecture and the mechanical properties of cartilage is still emerging [125, 126].

A further interesting collagenous structure that is found in cartilage is the permeable, capsule-like structure, the chondron, that surrounds the chondrocytes [127]. This structure is rich in type VI collagen as fine fibrils [128, 129]. Type VI collagen is also present, but to a lesser extent in the connections between cells in their columnar arrangement [128]. Type IX collagen is also present in chondrons, associated with type II collagen, where it may be limiting the size of the fibrils so that only very fine fibrils are present [130]. High resolution TEM shows that the chondron is attached to the chondrocyte and also to the extended collagen network [129]. The chondron has a lower proteoglycan content that the surrounding tissue. Nevertheless, this component may still be important, as it is proposed that the chondron structure is present to protect the chondrocytes from the high compressive loads occurring in cartilage [131, 132].

Pericardium

Pericardium has become a preferred tissue for use in biomedical products, especially for heart valve leaflets where there are continual mechanical changes (see chapter 13). Careful selection of tissue and its fibre orientation are therefore important for durability. Small-angle light scattering has defined the collagen organisation for bovine pericardium and showed that the anisotropic mechanical behaviour is consistent with the observed variations in fibre orientation, while suggesting that material from the left anterior side of the sac may be preferable for bioprosthesis fabrication [133].

Dura mater

Small-angle light scattering has shown that dura mater has a very variable collagen structure with fibre orientations going from highly aligned to essentially random, depending on the location, with lateral symmetry being seen [134]. The reasons for this variation are not yet fully understood.

5.4 Mineralisation

The association of collagen with calcium phosphate mineral can arise through various means. The most important, and the subject of this section, is the natural process of bone formation, termed mineralisation. Mineralisation is a complex process that involves many enzymatic steps. It is not fully understood and new insights are still likely to emerge. Others include the pathological calcification of tissues, termed calcification, of which the effects on the long-term performance of biological bioprosthetic heart valves is probably the mostly frequency discussed example (see chapter 13). More recently, there has been growing interest in the manufacture of collagen/calcium phosphate composites for bone repair (see chapter 18), and in similar materials that include cellular components for tissue engineering and regenerative medicine approaches (see chapter 16), albeit that they would likely not involve the natural process.

Bone comprises three main components. A principal one is hydroxyapatite (HAp) that provides 65%–70% by weight, ~50% by volume, and confers the fundamental hardness to bone. HAp is a distinct chemical form and is different to other forms such as tricalcium phosphate or dicalcium phosphate. Small amounts of other ions such as Mg^{++} or Cl^-, F^-, CO_3^- are typically found as well [135]. Another broad group is organic mass, comprising about 20%–25% of bone and being predominantly collagen type I which provide the framework for the mineral phase. Other protein factors are present, including osteonectin and osteocalcin, and a wide range of other very minor components, including various members of the bone morphogenic protein family. Although these are present in only very low amounts, they typically provide high levels of biological activity, acting as growth factors [136]. The third component is water, which is associated with the collagen and provides about 10% of the bone mass [135].

Composition alone does not provide functional bone with strength and toughness. These properties require an organised, hierarchical structure. In mammals this involves many enzymes and membrane transporters, comprising advanced systems that are not all present in fish [137].

The initial, primary step of mineralisation involves osteoblast derived matrix vesicles, which possess enzymes and transporters that enable accumulation of calcium and phosphate ions and which are associated with the un-mineralised organic matrix [138]. This ion accumulation leads to nucleation and crystal growth with the vesicles. The typically needle-shaped crystals form into larger stellate clusters which emerge from the vesicles to provide calcifying nodules [138]. These nodules make contact with the organic matrix, the dense crosslinked type I collagen. The collagen fibrils are synthesised by osteoblasts and are arranged in an organised staggered overlap fibril format (chapter 2). In one hypothesis, the gap zone is seen as the important region in the fibril for initial deposition during native mineralisation (and probably for calcification too) [139, 140], while one of the alternative views has been that the initial deposition takes place along the surface of the fibril [138].

TEM has provided images supporting the initial deposition of mineral in the gap region in turkey leg tendon [139, 140]. This tissue has been used as the mineral front progresses along the tendon with age and so allows isolation of the earliest stages on mineralisation. However, other TEM data has provided evidence of mineralisation proceeding from the calcifying nodules contact points on the collagen [141].

Subsequently, a secondary process follows in which the mineralisation extends throughout the collagenous tissue. The details of this process are less well understood than those of the primary steps, which in themselves are still not completely resolved and need further research.

5.5 Mechanical properties

5.5.1 Soft tissues

Historically, the mechanical properties of soft tissues have been determined using uniaxial tensile (stress/load-strain) testing [142, 143]. Despite the limitations of this early data, is was evident that 'structures containing a high percent of collagen were

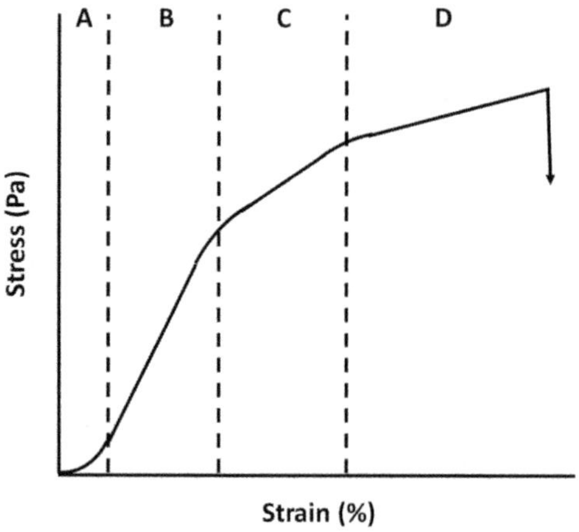

Figure 5.11. A stress–strain curve, following the profile for a rat-tail tendon, showing four behaviour regions: (A) toe region, (B) linear region, (C) failure region 1, and (D) failure region 2.

found difficult to stretch and usually broke before they elongated appreciably, while structures with a lower collagen content and a higher percent of elastic fibres were much more extensible [143]. In tissues, such as rat-tail tendon [143], a general picture of a tissue stress–strain curve emerged (figure 5.11). Qualitatively, this showed a trace that showed an initial 'toe region' (A), followed by a 'linear region' (B), that could allow a Young's modulus to be determined. Subsequently, further extension leads to regions where increasing failure (breakage) of the tissue is happening (C, D).

Understanding mechanisms at a molecular scale ought to allow significant modelling of the mechanical properties of tissues. However, such endeavours are fraught with difficulties and soft tissues have generally defied simple material modelling [144]. The range of difficulties that must be resolved include, for example, small specimens that are difficult to grip without damage changing the properties, the structural and compositional heterogeneity with a sample, even a small sample, large variations between notional replicate samples, homogeneity of behaviour within a sample, and variations arising from handling. Thus, some studies use dry samples, but for measurements of *in vivo* relevance, appropriate full hydration is needed, adding to experimental design challenges [144].

Of particular importance for planar tissues is that uniaxial experiments, even in multiple directions, are not suitable for deriving general equations to describe the material properties due to the complex anisotropy that is usually present in all tissues. The use of perpendicular biaxial testing (or preferably multiaxial testing) is an essential approach. Such protocols can go some way to resolving this issue [144], but again have difficulties including, for example, attaining uniformity in stress–strain rates, and control of edge location and maintain sufficient distance form external tethers to the target centre region to avoid deformations [144]. Hence, mechanical characterisation and modelling of tissues has become a very specialist area of research.

These different behaviours (figure 5.11) are now considered in relation to molecular and topological changes. Four different scales are generally considered. First, the 'molecular scale' which is the response of an individual molecule, and then the 'fibrillar scale' which is the response of the molecules within a single fibrillar structure. The 'microscale' represents the response of the multiple fibrils that are present in a single fibre structure, while the 'macroscale' describes the response of the full tissue [145].

At the molecular scale, the distinctly extended structure of the triple-helix, especially when compared to other motifs such as an α-helix or a β-sheet, means that it does not extend to the same extent as many other fibrous protein molecules, such as the elastomers elastin and resilin. Values for a Young's modulus have generally been derived from various models, giving values of around 5 ± 2 GPa [145], while values extrapolated from limited experimental studies [145, 146] are in the same range. The effectiveness of modelling is limited by the large data set and various approximations have been introduced, for example a coarse grain model [147].

At a fibrillar scale, modelling becomes even more difficult as the data set becomes enormous, especially as full hydration is needed for meaningful results [148]. The fibrillar scale introduces a range of additional features and constraints that must be considered. For example, the packing of individual collagen molecules into a fibril, leads to some flexibility in the 'gap zone' providing a D-period crimp [149] that is distinct from the macroscopic crimp. At the fibrillar scale, individual molecules can slip past each other, lowering the experimental modulus to approximately 0.2 to 1.0 GPa [145], with values depending on the strain rate. On one hand, modelling can give higher values [150, 151], but lower values, comparable with experimental data, can be obtained when hydration is fully considered [148]. At the fibrillar level, native crosslinking is also a factor, as it limits the extent of slippage between molecules. Modelling suggests that crosslinks, could increase the modulus several fold [151]. Modelling confirms that the various native crosslinks play an essential role in the tuneable toughness and strength of tissues [150, 152]. In collagen-based biomaterials, introduction of some types of crosslinks could have effects at the fibrillar scale, but the predominant effect is possible at the fibre or microscale where reagent penetration is more likely.

The microscale, where many fibrils form a fibre, is considerably more amenable to experimental study as individual fibres can be isolated from tissues [153]. Collagen fibres are generally present in most tissues. However, certain specialised tissues, such as the cornea, contain fibrils in an organised array, and these fibrils should not be considered as fibres [145]. Experimentally, a stress–strain curve similar to that in figure 5.11 is seen (for example [154]), although the failure mode (D) present in macroscopic samples is absent. It has been suggested that the different distributions of fibril diameters observed in tissues reflects the mechanical requirements of the particular tissue, with different loadings possible, but this is not always evident in experimental studies [155].

It is the extremely wide range of different arrangements of collagen fibres that leads to the huge diversity on collagen structures, each with distinct mechanical properties. Thus, while a human tendon ruptures at less than 10% strain at a stress of 90 MPa [156], by contrast, skin (raccoon) only takes around 10 MPa stress yet fails at a strain of around 150% [145]. In tendon, at low loads, the crimp is lost leading to the toe region of the stress–strain curve (figure 5.11). Tendons, show different properties depending on their role. For example, some tendons such as the equine superficial digital flexor tendon stores and returns energy and so must be more extensible than other tendons [157]. The molecular differences are not fully clear, but the relationship to the adjacent fascicule tissue is important [157, 158].

In tissues, the roles of other components, such as glycosaminoglycan chains will also impact on the mechanical properties of tissues [159]. This, and the variations in anisotropy of different tissues, is why detailed examination of tissues requires at least biaxial, if not multiaxial evaluation [144].

5.5.2 The effect of mineralisation

The mineralisation of collagen tissues dramatically changes their mechanical properties, especially enabling the tissues to withstand compression loads. The presence of the continuous mineral phase constrains the continuous collagen network minimising deformation while also minimising any sliding motion between collagen molecules [145]. The material is in essence a dual interpenetrating network system, and this requires modelling approaches that incorporate the synergies between the two components, especially when addressing cracking and failure modes [145]. Nevertheless, the key role of collagen is often not stressed, and it could be that the collagen changes with age are the more important factor for fracture risk [160].

References

[1] Clark R A F 1996 *The Molecular and Cellular Biology of Wound Repair* 2nd edn (New York: Plenum)

[2] Werkmeister J A, Tebb T A, White J F and Ramshaw J A M 2001 *Curr. Opin. Solid State Mat. Sci.* **5** 185–91

[3] Drubaix I, Legeais J M, Malek-Chehire N, Savoldelli M, Menasche M, Robert L, Renard G and Pouliquen Y 1996 *Exp. Eye Res.* **62** 367–76

[4] Chirila T V, Thompson-Wallis D E, Crawford G J, Constable I J and Vijayasekaran S 1996 *Graefes Arch. Clin. Exp. Ophthalmol.* **234** 193–98

[5] Werkmeister J A, Glattauer V, Tebb T A, Ramshaw J A M, Edwards G A and Roberts G 1991 *J. Long-term Effects Med. Implants* **1** 107–19

[6] Werkmeister J A and Ramshaw J A M 1991 *J. Histochem. Cytochem.* **39** 1215–20

[7] Werkmeister J A, Tebb T A, White J F and Ramshaw J A M 1993 *J. Histochem. Cytochem.* **41** 1701–06

[8] Werkmeister J A, White J F and Ramshaw J A M 1994 *Med. Prog. Technol.* **20** 231–42

[9] White J F, Werkmeister J A, Bisucci T, Darby I A and Ramshaw J A M 2014 *J. Biomed. Mater. Res. A* **102** 3550–55

[10] Kagan H M and Trackman P C 1991 *Am. J. Resp. Cell Mol. Biol.* **5** 206–10

[11] Yamauchi M and Mechanic G L 1988 *Collagen* vol 1 ed M E Nimni (Boca Raton, FL: CRC Press) pp 157–72

[12] Collier T A, Nash A, Birch H L and de Leeuw N H 2018 *J. Biomech.* **67** 55–61
[13] Naresh M D and Brodsky B 1992 *Biochim. Biophys. Acta* **1122** 161–66
[14] Verzijl N, DeGroot J, Thorpe S R, Bank R A, Shaw J N, Lyons T J, Bijlsma J W, Lafeber F P, Baynes J W and TeKoppele J M 2000 *J. Biol. Chem.* **275** 39027–31
[15] Panwar P, Butler G S, Jamroz A, Azizi P, Overall C M and Brömme D 2018 *Matrix Biol.* **65** 30–44
[16] Mott J D, Khalifah R G, Nagase H, Shield C F, Hudson J K and Hudson B G 1997 *Kidney Int.* **52** 1302–12
[17] Kohn J C, Lampi M C and Reinhart-King C A 2015 *Front. Genet.* **6** 112
[18] Henneman S, Reijers R R, Maltha J C and Von den Hoff J W 2012 *J. Periodontal Res.* **47** 383–88
[19] Henkel W and Glanville R W 1982 *Eur. J. Biochem.* **122** 205–13
[20] Yamauchi M and Sricholpech M 2012 *Essays Biochem.* **52** 113–33
[21] Tanzer M L 1968 *J. Biol. Chem.* **243** 4045–54
[22] Tanzer M L and Mechanic G 1970 *Biochem. Biophys. Res. Commun.* **39** 183–89
[23] Forrest L, Shuttleworth A, Jackson D S and Mechanic G L 1972 *Biochem. Biophys. Res. Commun.* **46** 1776–81
[24] Robins S P 1982 *Meth. Biochem. Anal.* **28** 329–79
[25] Graham L and Gallop P M 1994 *Anal. Biochem.* **217** 298–305
[26] Robins S P 2007 *Biochem. Soc. Trans.* **35** 849–52
[27] Rigby B J, Mitchell T W and Robinson M S 1977 *Biochem. Biophys. Res. Commun.* **79** 400–5
[28] Fidler A L, Vanacore R M, Chetyrkin S V, Pedchenko V K, Bhave G, Yin V P, Stothers C L, Rose K L and McDonald W H *et al* 2014 *Proc. Natl Acad. Sci. USA* **111** 331–36
[29] Bailey A J, Sims T J and Light N 1984 *Biochem. J.* **218** 713–23
[30] Benya P D and Padilla S R 1986 *J. Biol. Chem.* **261** 4160–69
[31] Reginato A M and Jimenez S A 1991 *Biochem. J.* **273** 333–38
[32] Martins S I F S, Jongen W M F and van Boekel M A J S 2000 *Trends Food Sci. Technol.* **11** 364–73
[33] Monnier V M and Cerami A 1981 *Science* **211** 491–3
[34] Sell D R and Monnier V M 1989 *J. Biol. Chem.* **264** 21597–602
[35] Sell D R, Biemel K M, Reihl O, Lederer M O, Strauch C M and Monnier V M 2005 *J. Biol. Chem.* **280** 12310–315
[36] Collier T A, Nash A, Birch H L and de Leeuw N H 2015 *Matrix Biol.* **48** 78–88
[37] Gautieri A, Redaelli A, Buehler M J and Vesentini S 2014 *Matrix Biol.* **34** 89–95
[38] Monnier V M, Genuth S and Sell D R 2016 *Glycoconj. J.* **33** 569–79
[39] Vasan S, Zhang X, Zhang X, Kapurniotu A, Bernhagen J, Teichberg S, Basgen J, Wagle D and Shih D *et al* 1996 *Nature* **382** 275–8
[40] Yang S, Litchfield J E and Baynes J W 2003 *Arch. Biochem. Biophys.* **412** 42–6
[41] Lecuit T and Lenne P F 2007 *Nat. Rev. Mol. Cell Biol.* **8** 633–44
[42] Wilson R, Belluoccio D and Bateman J F 2008 *Methods* **45** 22–31
[43] Randles M and Lennon R 2015 *Curr. Top. Membr.* **76** 171–96
[44] White J F, Werkmeister J A, Hilbert S L and Ramshaw J A M 2010 *J. Heart Valve Dis.* **19** 766–71
[45] Glattauer V, Irving-Rodgers H F, Rodgers R J, Stockwell S, Brownlee A G, Werkmeister J A and Ramshaw J A M 2007 *Reprod. Fertil. Dev.* **19** 473–81

[46] Veltman K H 1992 *Gepeinigt, begehrt, vergessen. Symbolik und Sozialbezug des Körpers im späten Mittelalter und in der frühen Neuzeit* ed K Schreiner (Bad Homburg: Werner Reimers Stiftung) pp 287–308
[47] Elliott D H 1965 *Biol. Rev.* **40** 392–421
[48] Kastelic J, Galeski A and Baer E 1978 *Connect. Tissue Res.* **6** 11–23
[49] Prantil R K, Xiu K, Kim K E, Gaitan D M, Sacks M S, Woo S L and Li Z M 2012 *Clin. Anat.* **25** 478–82
[50] Benjamin M, Toumi H, Ralphs J R, Bydder G, Best T M and Milz S 2006 *J. Anat.* **208** 471–90
[51] Gaut L and Duprez D 2016 *WIREs Dev. Biol.* **5** 5–23
[52] Craig A S, Birtles M J, Conway J F and Parry D A D 1989 *Connect. Tissue Res.* **19** 51–62
[53] Birk D E, Zycband E I, Winkelmann D A and Trelstad R L 1989 *Proc. Natl Acad. Sci. USA* **86** 4549–53
[54] Birk D E and Zycband E 1994 *J. Anat.* **184** 457–63
[55] Birk D E, Nurminskaya M V and Zycband E I 1995 *Dev. Dyn.* **202** 229–43
[56] Wood M L, Lester G E and Dahners L E 1998 *J. Orthop. Res.* **16** 438–40
[57] Parry D A D and Craig A S 1977 *Biopolymers* **16** 1015–31
[58] Goh K L, Hiller J, Haston J L, Holmes D F, Kadler K E, Murdoch A, Meakin J R and Wess T J 2005 *Biochim. Biophys. Acta* **1722** 183–88
[59] Scott J E and Parry D A D 1992 *Int. J. Biol. Macromol.* **14** 292–93
[60] Parry D A D and Craig A S 1979 *Nature* **282** 213–15
[61] Craig A S and Parry D A D 1981 *Proc. Roy. Soc. Lond. B Biol. Sci.* **212** 85–92
[62] Franchi M, Trirè A, Quaranta M, Orsini E and Ottani V 2007 *Sci. World J.* **7** 404–20
[63] Parry D A D, Barnes G R G and Craig A S 1978 *Proc. Roy. Soc. Lond. B Biol. Sci.* **203** 305–21
[64] Patterson-Kane J C, Parry D A D, Birch H L, Goodship A E and Firth E C 1997 *Connect. Tissue Res.* **36** 253–60
[65] Patterson-Kane J C, Wilson A M, Firth E C, Parry D A D and Goodship A E 1997 *Equine Vet. J.* **29** 121–25
[66] Patterson-Kane J C, Firth E C, Parry D A D, Wilson A M and Goodship A E 1998 *Am. J. Vet. Res.* **59** 64–8
[67] Birk D E and Mayne R 1997 *Eur. J. Cell Biol.* **72** 352–61
[68] Parry D A D, Flint M H, Gillard G C and Craig A S 1982 *FEBS Lett.* **149** 1–7
[69] Gillard G C, Reilly H C, Bell-Booth P G and Flint M H 1979 *Connect. Tissue Res.* **7** 37–46
[70] Meek K M and Knupp C 2015 *Prog. Retin. Eye Res.* **49** 1–16
[71] Eisler P 1930 *Kurzes Handbuch der Ophthalmologie* vol 1 ed F Schieck and A Bruckner (Berlin: Springer) pp 1–386
[72] Jakus M A 1962 *Invest. Ophthalmol. Vis. Sci.* **1** 202–25
[73] Maurice D M 1957 *J. Physiol.* **136** 263–86
[74] Benedek G B 1971 *Appl. Opt.* **10** 459–73
[75] Cai C X, Fitch J M, Svoboda K K, Birk D E and Linsenmayer T F 1994 *Dev. Dyn.* **201** 206–15
[76] Chen S, Mienaltowski M J and Birk D E 2015 *Exp. Eye Res.* **133** 69–80
[77] Reinstein D Z, Archer T J, Gobbe M, Silverman R H and Coleman D J 2008 *J. Refract. Surg.* **24** 571–81

[78] Bowman W 1849 *Lectures on the Parts Concerned in the Operations on the Eye* (London: Longman, Brown, Green, and Longmans)
[79] Dua H S, Faraj L A, Branch M J, Yeung A M, Elalfy M S, Said D G, Gray T and Lowe J 2014 *Br. J. Ophthalmol.* **98** 691–97
[80] Labat-Robert J and Robert L 2012 *Pathol. Biol. (Paris)* **60** 66–75
[81] Bergmanson J P, Horne J, Doughty M J, Garcia M and Gondo M 2005 *Eye Contact Lens* **31** 281–87
[82] Holmes D F and Kadler K E 2005 *J. Mol. Biol.* **345** 773–84
[83] Birk D E, Fitch J M and Linsenmayer T F 1986 *Invest. Ophthalmol. Vis. Sci.* **27** 1470–77
[84] White J F, Werkmeister J A, Ramshaw J A M and Birk D E 1997 *Connect. Tissue Res.* **36** 165–74
[85] Linsenmayer T F, Gibney E, Igoe F, Gordon M K, Fitch J M, Fessler L I and Birk D E 1993 *J. Cell Biol.* **121** 1181–89
[86] Birk D E 2001 *Micron* **32** 223–37
[87] Virchow H 1905 *Graefe-Saemisch Handbuch der gesamten Augenheilkunde* vol 1 (Leipzig: Engelmann)
[88] Funderburgh J L, Hevelone N D, Roth M R, Funderburgh M L, Rodrigues M R, Nirankari V S and Conrad G W 1998 *Invest. Ophthalmol. Vis. Sci.* **39** 1957–64
[89] Epstein E H 1974 *J. Biol. Chem.* **249** 3225–31
[90] Weber L, Meigel W N and Rauterberg J 1977 *Arch. Dermatol. Res.* **258** 251–7
[91] Epstein E H and Munderloh N H 1978 *J. Biol. Chem.* **253** 1336–37
[92] Ramshaw J A M 1986 *Conn. Tissue Res.* **14** 307–14
[93] Mays P K, Bishop J E and Laurent G J 1988 *Mech. Ageing Dev.* **45** 203–12
[94] Langer K 1882 *Lehrbuch der systematischen und topographischen Anatomie* (Vienna: Wilhelm Braumüller)
[95] Conabere G O 1944 *J. Int. Soc. Leather Trades Chem.* **28** 270–80
[96] Turner N J, Pezzone D and Badylak S F 2015 *Tissue Eng. Part C Methods* **21** 373–84
[97] Weber L, Meigel W N and Spier W 1978 *Arch. Dermatol. Res.* **261** 63–71
[98] Ramshaw J A M, Werkmeister J A and Peters D E 1990 *Current Perspectives on Implantable Devices* vol 2 ed D F Williams (London: JAI Press) pp 151–220
[99] Dupuytren G 1834 *Traité théorique et pratique des blessure par armes de guerre* vol 1 (Paris: Baillere)
[100] Jor J W, Nielsen P M, Nash M P and Hunter P J 2011 *Skin Res Technol.* **17** 149–59
[101] Cicchi R, Vogler N, Kapsokalyvas D, Dietzek B, Popp J and Pavone F S 2013 *J. Biophotonics* **6** 129–42
[102] Kronick P L and Buechler P R 1986 *J. Am. Leather Chem. Assoc.* **81** 221–30
[103] Liu Z, Pouli D, Sood D, Sundarakrishnan A, Hui Mingalone C K, Arendt L M, Alonzo C, Quinn K P and Kuperwasser C *et al* 2017 *Biomaterials* **116** 34–47
[104] Sizeland K H, Basil-Jones M M, Edmonds R L, Cooper S M, Kirby N, Hawley A and Haverkamp R G 2013 *J. Agric. Food Chem.* **61** 887–92
[105] Naresh M D, Arumugam V and Sanjeevi R 1997 *J. Biosci.* **22** 431–37
[106] Brown G A and Wellings S R 1970 *Z. Zellforsch.* **103** 149–69
[107] Prum R O and Torres R 2003 *J. Exper. Biol.* **206** 2409–29
[108] Sherman V R, Tang Y, Zhao S, Yang W and Meyers M A 2017 *Acta Biomater.* **53** 460–69
[109] Gillard G C, Reilly H C, Bell-Booth P G and Flint M H 1977 *J. Invest. Dermatol.* **69** 257–61

[110] Flint M H, Craig A S, Reilly H C, Gillard G C and Parry D A D 1984 *Connect. Tissue Res.* **13** 69–81

[111] Jolly R D and Windsor P A 2010 *Diseases of cattle in Australasia: a Comprehensive Textbook* ed T J Parkinson, J J Vermunt and J Malmo (Wellington, NZ: New Zealand Veterinary Association Foundation for Continuing Education) pp 759–76

[112] Amos J 1958 *J. Soc. Leather Trades Chem.* **42** 79–90

[113] Dufty J H, Peters D E and Bavinton J H 1983 *J. Soc. Leather Technol. Chem.* **67** 70–3

[114] Kronick P L and Sacks M S 1991 *Connect. Tissue Res.* **27** 1–13

[115] Peters D E and Dufty J H 1985 *J. Am. Leather Chem. Assoc.* **80** 42–6

[116] Ramshaw J A M 1985 *J. Am. Leather Chem. Assoc.* **80** 219–25

[117] Dufty J H and Peters D E 1983 *J. Soc. Leather Technol. Chem.* **68** 35

[118] Cundiff L V, Buechler P R, Hannigan M V, Everett A L and Dahms M P 1987 *J. Hered.* **78** 24–8

[119] Fujioka R, Aoyama T and Takakuwa T 2013 *Osteoarth. Cart.* **21** 1092–98

[120] MacConaill M A 1951 *J. Bone Joint Surg.* B **33** 251–57

[121] Holmes D F and Kadler K E 2006 *Proc. Natl Acad. Sci. USA* **103** 17249–254

[122] Bilton M A, Thambyah A and Clarke R J 2018 *Biomech. Model. Mechanobiol.* **17** 1297–315

[123] Minns R J and Steven F S 1977 *J. Anat.* **123** 437–57

[124] Dahmen G 1973 *Connective Tissue and Ageing* ed H G Vogel (Amsterdam: Excerpta Medica) pp 109–13

[125] Hosseini S M, Wu Y, Ito K and van Donkelaar C C 2014 *Biomech. Model. Mechanobiol.* **13** 41–51

[126] Meng Q, An S, Damion R A, Jin Z, Wilcox R, Fisher J and Jones A 2017 *J. Mech. Behav. Biomed. Mater.* **65** 439–53

[127] Poole C A, Flint M H and Beaumont B W 1987 *J. Orthop. Res.* **5** 509–22

[128] Poole C A, Ayad S and Schofield J R 1988 *J. Cell Sci.* **90** 635–43

[129] Poole C A, Ayad S and Gilbert R T 1992 *J. Cell Sci.* **103** 1101–110

[130] Poole C A, Wotton S F and Duance V C 1988 *Histochem. J.* **20** 567–74

[131] Poole C A, Honda T, Skinner S J, Schofield J R, Hyde K F and Shinkai H 1990 *Connect. Tissue Res.* **24** 319–30

[132] Adams M A 2006 *Biorheology* **43** 537–45

[133] Sacks M S, Chuong C J and More R 1994 *ASAIO J.* **40** M632–7

[134] Hamann M C, Sacks M S and Malinin T I 1998 *J. Anat.* **192** 99–106

[135] Tomoaia G and Pasca R D 2015 *Clujul Med.* **88** 15–22

[136] Ampuja M and Kallioniemi A 2018 *Genes Chromosomes Cancer* **57** 3–11

[137] Blair H C, Larrouture Q C, Li Y, Lin H, Beer-Stoltz D, Liu L, Tuan R S, Robinson L J, Schlesinger P H and Nelson D J 2017 *Tissue Eng. Part B Rev.* **23** 268–80

[138] Hasegawa T 2018 *Histochem. Cell Biol.* **149** 289–304

[139] Landis W J 1986 *J. Ultrastruct. Mol. Struct. Res.* **94** 217–38

[140] Traub W, Arad T and Weiner S 1989 *Proc. Natl Acad. Sci. USA* **86** 9822–26

[141] Hasegawa T, Li M, Hara K, Sasaki M, Tabata C, de Freitas P H, Hongo H, Suzuki R and Kobayashi M *et al* 2011 *Biomed. Res.* **32** 259–69

[142] Gratz C M 1931 *J. Bone Joint Surg.* **13** 334–40

[143] Rigby B J, Hirai N, Spikes J D and Eyring H 1959 *J. Gen. Physiol.* **43** 265–83

[144] Sacks M S and Sun W 2003 *Annu. Rev. Biomed. Eng.* **5** 251–84

[145] Sherman V R, Yang W and Meyers M A 2015 *J. Mech. Behav. Biomed. Mater.* **52** 22–50

[146] Cusack S and Miller A 1979 *J. Mol. Biol.* **135** 39–51

[147] Gautieri A, Vesentini S and Redaelli A 2010 *J. Mol. Model.* **16** 1845–51

[148] Gautieri A, Vesentini S, Redaelli A and Buehler M J 2011 *Nano Lett.* **11** 757–66

[149] Parry D A D 1988 *Biophys. Chem.* **29** 195–209

[150] Buehler M J 2006 *Proc. Natl Acad. Sci. USA* **103** 12285–90

[151] Buehler M J 2008 *J. Mech. Behav. Biomed. Mater.* **1** 59–67

[152] Depalle B, Qin Z, Shefelbine S J and Buehler M J 2015 *J. Mech. Behav. Biomed. Mater.* **52** 1–13

[153] White J F, Werkmeister J A, Darby I A, Bisucci T, Birk D E and Ramshaw J A M 2002 *J. Struct. Biol.* **137** 23–30

[154] Gentleman E, Lay A N, Dickerson D A, Nauman E A, Livesay G A and Dee K C 2003 *Biomaterials* **24** 3805–13

[155] Rigozzi S, Stemmer A, Müller R and Snedeker J G 2011 *J. Struct. Biol.* **176** 9–15

[156] Benedict J V, Walker L B and Harris E H 1968 *J. Biomech.* **1** 53–63

[157] Thorpe C T, Udeze C P, Birch H L, Clegg P D and Screen H R C 2012 *J. Roy. Soc. Interface.* **9** 3108–17

[158] Shearer T, Thorpe C T and Screen H R C 2017 *J. Roy. Soc. Interface.* **14** 20170261

[159] Lavagnino M, Arnoczky S P, Frank K and Tian T 2005 *J. Biomech.* **38** 69–75

[160] Burr D B 2002 *Bone* **31** 8–11

John A M Ramshaw and Veronica Glattauer

Chapter 6

Collagen stability

Maintaining the collagen triple-helical structure is an important aspect in the manufacturing of collagen-based biomedical products. Loss of this tertiary structure leads to gelatin, which exhibits different properties and can be rapidly resorbed. Gelatin is much cheaper than collagen so it is important to retain the investment in collagen use. This chapter examines the stability of collagens under a wide variety of conditions, including changed pH and changed solvent compositions. The use of peptides to provide a more detailed understanding of collagen stability is included.

6.1 Molecular stability

6.1.1 Individual soluble collagen molecule stability

Various methods, including circular dichroism spectroscopy (CD) and optical rotary dispersion (ORD) [1, 2], calorimetry [3], viscosity [4], protease susceptibility [5] and hydrogen exchange [6], have been used to determine the denaturation temperature (T_d), otherwise known as the melting temperature (T_m), of soluble collagens (see [7]). While T_m is generally equated with stability, from a thermodynamic standpoint, the correct parameter determining stability is the free energy, $\Delta G°$ [8].

CD spectroscopy, followed by calorimetry, have emerged as the methods of choice for examining the stability of soluble collagens. The CD spectrum for collagens shows a maximum around 222–244 nm, (figure 6.1) and this signal is lost with increasing temperature as the collagen unfolds (figure 6.1). The T_m value is taken generally as the temperature of the midpoint of the transition, where the fraction of molecules folded is half, i.e. $F(T_m) = 1/2$. In measuring this transition, care is required as the short wavelength, UV light used in CD experiments will lead to protein degradation (chain scission) if long, continuous irradiation is used [2].

Hydrogen exchange data indicated two processes in unfolding, one being non-denaturing associated with micro-unfolding, possibly limited to only one or two

doi:10.1088/978-0-7503-2096-2ch6

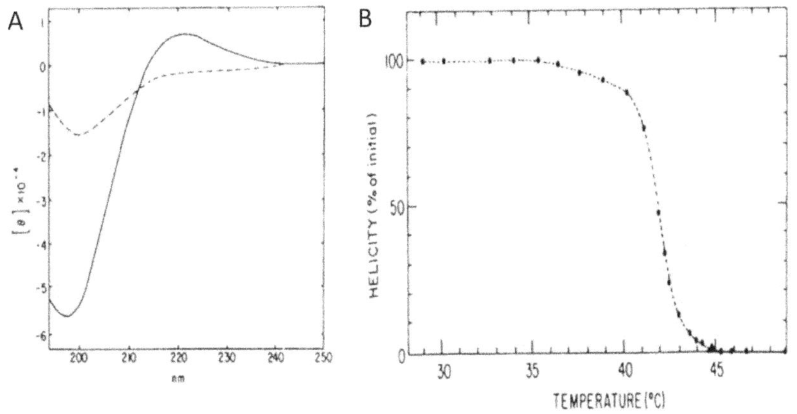

Figure 6.1. CD spectra of chick embryo type I collagen at pH 4.3, (A) at 20 °C (—) and after denaturation by heating to 50 °C (- - -) and (B) intensity changes at 221 nm during heating from 20 °C to 50 °C, using a strategy that minimised UV irradiation. (Reprinted with permission from [2]. Copyright 1979, American Chemical Society.)

triplets, while the other was complete unfolding of the triple-helix [9]. Micro-unfolding can also be detected by protease susceptibility at temperatures lower than the T_m [10].

Calorimetry provides data on the extent of cooperativity during unfolding. Interstitial collagens types I, II and III all show single, sharp, highly cooperative melting transitions, at about 41 °C, with no evidence for any intermediate part-folded segments [7]. Types V and XI collagens show different and distinct melting transitions, with evidence of two major and two minor transitions [7]. The specific T_m values obtained by calorimetry are dependent of the heating rate [11], which is typically performed at a much higher rate than CD melting experiments. For example, at a heating rate of 0.004 °C min^{-1} a T_m of about 38.1 °C was obtained, increasing to around 40.7 °C at a heating rate of 1.0 °C min^{-1} [11]; some data are reported at heating rates up to 10 °C min^{-1}.

For animal collagens, the Hyp content more than any other sequence feature is key to the thermal stability of the triple-helix [12] (see chapter 2). The extent of Hyp formation varies across the animal kingdom, with species that live in colder environmental conditions having less Hyp as less stability is needed [13, 14]. When a species is taken from its low temperature conditions and slowly acclimatised to higher temperatures, in some cases the extent of Hyp can also increase [15].

Effects of pH on stability

One of the early topics for study was the effect of pH on collagen and tissues, particularly as significant industrial processes in the production of gelatin and leather put the tissues at extremes of pH. However, from a biomaterials perspective, although there are occasions when stability has been reported under acid conditions that can relate to production strategies, those reported at physiological conditions will generally be more relevant for the properties of products and how they behave after implant. For collagen solutions, greatest stability is found at neutral pH while lower stability is found at acidic and basic conditions [4, 16], reflecting the pKa values of the side chain carboxyl and amino groups.

Effects of salts on stability
A wide range of salts have been examined for their effects on the stability of soluble collagen. Several, such as KSCN and LiBr, which are known to be particularly destabilising for globular proteins, are equally destabilising for the collagen triple-helix [17], while others, such as Na_2SO_4 and KCl, while still destabilising are less so [16, 17]. Certain salts, particularly NaCl and $(NH_4)_2SO_4$, are used without issues for collagen precipitation during preparations (see chapter 8).

Effects of chemical modifications on stability
A selection of single point chemical modifications has been reported for soluble collagens. The stability of the collagen triple-helix depends on the amino acid sequence (chapter 2) so any single point modification is likely to disrupt this stabilisation, although reactions involved in crosslinking will usually add stability. The amount of destabilisation will depend also on the extent of the modification of the collagen.

Treatment of collagen with alkali, which is common in the bovine gelatin and in the leather industry, leads to deamidation, where the amide residues are lost from Asn and Gln residues giving Asp and Glu residues, increasing the negative charge along the molecule. The result of deamidation is that the T_m drops with the increasing extent of change such that with full deamidation the T_m is about 3 °C lower [18].

Further examples of single point modifications include those that add functions, such as methylation and succinylation [18]. With full methylation, which reduces the number of acidic charges, the T_m is reduced by about 6 °C, although at around 60% methylation this change is only around 1.5 °C [18]. In succinylation the Lys amino groups are modified and become negatively charged, so that ion pair stabilisation could be reduced as well as extra repulsion introduced. Thus, with full succinylation the T_m drops by about 4 °C, as well as changing the solubility properties [18].

More recently, there has been an interest in adding methacrylate groups or similar vinyl functions to collagen, to enable photo-crosslinked scaffolds to be assembled. As expected from other studies, these chemical modifications also led to destabilisation of the collagen triple-helix, up to around 5 °C depending on the extent of modification. Conversely, once photo-crosslinking is initiated the thermal stability of the aggregate increases [19].

6.1.2 Tissue collagen stability

In mammalian tissues the individual collagen molecules form higher order structures (see chapter 4) that are stabilised by intermolecular crosslinking, especially for the interstitial collagens (see chapter 5). A consequence of the crosslinking is that the tissues, in an aqueous environment, show an enhanced stability compared to individual molecules. Frequently, the stability of tissues is quoted as a shrinkage temperature (strictly a hydrothermal shrinkage temperature), reflecting the dimensional change that occurs with temperature increase and denaturation.

Examination of shrinkage and denaturation temperatures for a wide range of species showed that frequently the difference in these values was around 24 °C–28 °C, irrespective of the denaturation temperature of the individual molecule [14].

In biomaterials studies, the additional crosslinking that is used for either tissue-based or soluble purified collagens (see chapter 12) also leads to an increase in shrinkage temperature. This tends to increase with the extent of crosslinking for a given reagent, up to a maximum value. Comparisons between the shrinkage temperatures when different reagents are used should be used with caution as the chemistry involved differs (see chapter 12). Typically, for example, treatment of fresh tissue with glutaraldehyde (GA) increase the shrinkage temperature from a low 60 °C range to a mid-80 °C range, depending on the tissue type and the specific reaction conditions [20]. Conversely, single point modifications, such as acetylation, can lower the stability of collagen fibres [21].

6.2 Solvent effects on stability

A wide variety of potential solvent variations on the stability of collagens are possible, including organic solvents, salts and pH. Early studies on the effects on stability of additions of organic solvents and salts at neutral pH initially involved tissue samples [22, 23]. Various compounds, including KSCN, $NaClO_4$, KI, salicylate, guanidine and urea, all led to a significant reduction in tissue shrinkage temperature (stability). Addition at 1 M of various primary amines all showed a decrease in stability, with the effect increasing with the number of carbons in the compounds. The effects of numerous other compounds, more than 70, were also tested. These included a limited range of 1 M solutions of non-ionised organic solvents, mostly alcohols, which showed little effect on stability. Studies on the effect of pH showed that between pH 4 and pH 10 changes in pH did not alter stability, but below pH 4 and above pH 10 stability decreased substantially, with swelling occurring in acidic conditions [22, 23].

Subsequently, when preparations of soluble collagen became readily available, these were also examined in a large number of studies, particularly using various organic solvents over a range of concentrations. Monohydric alcohols in solutions up to around 2 M reduce the collagen stability, with larger compounds, with more methylene groups, having the greater effect [24, 25]. On the other hand, certain diols, including ethylene glycol, showed a stabilising effect [24] as did propane 1,3-diol, while propane 1,2-diol, was destabilising [26]. Various polyols as sugars in many cases also show collagen stabilising effects, although some such as xylose show little effect [27]. In some cases, diols and certain polyols may inhibit fibril reformation [28]. The stabilising effect for ethylene glycol was also seen with glycerol [1, 29] and for tissues, ethylene glycol and higher molecular weight polyethylene glycols also all showed a similar effect [30].

In peptide studies [31] trimethylamine N-oxide (TMAO) solutions were shown to be effective in stabilising the collagen triple-helix. TMAO is found in various marine species, including elasmobranchs, where it counteracts the destabilising effects on proteins of urea that is also present [32] and had been shown to increase the melting temperature of proteins *in vitro* [33].

A range of other studies have more recently been reported, extending and adding more detailed data to describe the effects of addition of alcohols. Most recently, fluorinated solvents have been suggested for use with collagens, especially hexa-fluoro-iso-propanol in electrospinning of collagen, but these were found to denature the collagen and produce a gelatin product [34]. Studies on other similar solvents, including 2,2,2-trifluoroethanol [35], have confirmed the unsuitability of these as solvents for collagens.

6.3 Peptide models to study stability

6.3.1 Polypeptide models

The initial use of polyamino acid structures, notably $(Gly)_n$ and $(Pro)_n$, was important for elucidating the triple-helical structure of collagen (see chapter 2). Subsequently, the development of various polytripeptide and polyhexapeptide structures (see [36]) gave opportunities for examining the role of certain amino acids in providing triple-helical stability [37, 38]. This approach, however, was limited as synthetic collagen-peptides of defined length were becoming more readily available for study [39, 40].

6.3.2 Defined peptide models

Several studies have looked at how variation in triplet composition can change triple-helical stability. The concept of measuring the changes arising from individual triplets was examined using several polytripeptides or block polytripeptides, generally incorporating the more frequently occurring triplets [41]. With these shorter defined peptides, a C-terminal crosslinking site, based on peptide synthesis from the three amino groups of a Lys-Lys entity, was included to ensure triple-helix formation [42]. These peptides formed stable triple-helices when there were eight or more triplets, and were readily soluble [42]. Studies on the stability of a large number of peptides showed that different triplet sequences could affect stability, with the transition temperatures being higher for peptides containing the Gly-Xaa-Hyp sequences followed by those with the Gly-Pro-Yaa sequences. Incorporation of imino acid free tripeptides resulted in a weakening of the triple-helical structure [42–44]. However, the design of this initial system meant that the amount of change from a single Gly-Xaa-Yaa unit was difficult to measure because stability differences were often very small, there were variations in structures, and the system was affected by difficulties in peptide purification.

Theoretical approaches to collagen stability were also suggested to overcome the experimental issues. Initially, it was assumed that Gly-Xaa-Yaa sequences found in clusters with frequently occurring triplets, such as Gly-Pro-Hyp, had a high stability, while those with less frequent triplets have a low stability [43]. Subsequently, this approach was extended to include some experimental data, for example from polypeptides, and led to a proposed stability scale in which each Gly-Xaa-Yaa triplet was allocated to one of four stability categories [45].

A large study based on collecting numerical data on the effect of individual amino acids stability was developed by Brodsky and Ramshaw and was completed between 1996 and 2005. The basis of this study was the use of host–guest peptides [46–50]. These host–guest peptides followed the following structures, with stability determined by slow melting measurement by CD spectroscopy,

(1) acetyl-(Gly-Pro-Hyp)$_3$-**Gly-Pro-Hyp**-Gly-Pro-Hyp-(Gly-Pro-Hyp)$_3$-Gly-Gly$_{amide}$
(2) acetyl-(Gly-Pro-Hyp)$_3$-**Gly-Xaa-Hyp**-Gly-Pro-Hyp-(Gly-Pro-Hyp)$_3$-Gly-Gly$_{amide}$
(3) acetyl-(Gly-Pro-Hyp)$_3$-**Gly-Pro-Yaa**-Gly-Pro-Hyp-(Gly-Pro-Hyp)$_3$-Gly-Gly$_{amide}$
(4) acetyl-(Gly-Pro-Hyp)$_3$-**Gly-Xaa-Yaa**-Gly-Pro-Hyp-(Gly-Pro-Hyp)$_3$-Gly-Gly$_{amide}$

The host–guest peptides were designed to be of the same length, eight triplets, and uniform sequence (1), with variations inserted in a central triplet that allows substitution to any residue in either the Xaa position (2), or the Yaa position (3) or both of these positions (4). Acetylation of the N-terminus and amidation of the C-terminus were used to eliminate destabilising terminal charge repulsion [51] and a C-terminal Gly-Gly sequence to eliminate the likelihood of diketopiperazine formation during synthesis. The length of the peptide, eight triplets, was based on the stability of various length (Gly-Pro-Hyp)$_n$ peptides, such that the peptide was short enough so that the effect of a single guest triplet would not be masked, while not being too short so that some substitutions would not form stable entities. Experimentally, it was found over various studies that all peptides formed stable triple-helix structures, with stabilities between 20 °C and 47 °C depending on the specific guest residues.

Early studies on 12 host–guest peptides, representing about a third of triplets in interstitial collagen chains, showed that the host–guest peptides gave CD spectra that were characteristic of the collagen triple-helical fold (figure 6.2) and that melting curves were different for specific substitutions (figure 6.2) [46]. Further, initial studies using various Ala, Leu and Phe substitutions, showed that the ranking of the calculated relative free energy values for 12 peptides was consistent with the order of their observed thermal stabilities; hence, in further studies T_m values were frequently preferred. However, no simple relationship was found between enthalpy (or entropy) rankings and thermal stability, showing that the mechanism of stabilisation varies between peptides.

Further studies led to the completion of all 20 normal amino acid substitutions in both the Xaa position and the Yaa positions, but with Pro restricted to the Xaa position and Hyp restricted to the Yaa position (figure 6.3). These data allowed a propensity scale of triple-helical stability effects to be established [48]. These data also indicated that many residues show quite different propensities between the Xaa and Yaa positions, which arises from the structural non-equivalence of these positions. It was interesting that that Gly was consistently destabilising in both the Xaa and Yaa positions [52], while aromatic amino acids, Tyr and Phe, were also

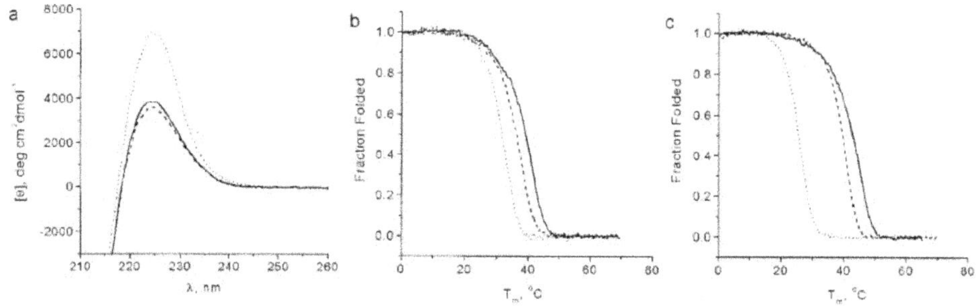

Figure 6.2. (a) CD spectra at 5 °C in PBS (pH 7) of host–guest peptides with guest triplets GPM (—), GPT (- - -) and GPW (· · · ·), (b) Thermal transition curves for host–guest peptides, with GMO (—), GTO (- - -), and GWO (· · · ·) guest triplets, showing the fraction folded versus temperature, and (c) Thermal transitions for host–guest peptides, with GPM (—), GPT (- - -), and GPW (· · · ·) guest triplets, showing the fraction folded versus temperature. (Reprinted with permission from [48]. Copyright 2000, American Chemical Society.)

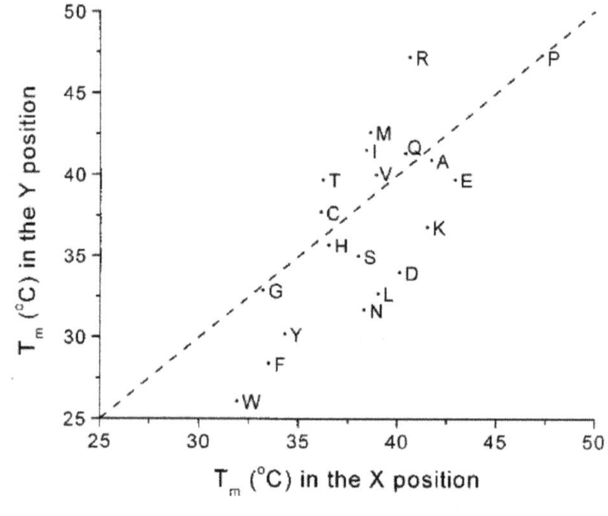

Figure 6.3. The T_m values of all 20 amino acids as guests in the Yaa position of Gly-Pro-Yaa triplets are plotted against the T_m values of the same amino acid in the Xaa position of Gly-Xaa-Hyp guest triplets. The dashed line shows same values of T_m in the Xaa and Yaa positions. (Reprinted with permission from [48]. Copyright 2000, American Chemical Society.)

destabilising. An unexpected feature was that Arg residues in the Yaa position, when not adjacent to a similar triplet, are particularly stabilising, as much as a Hyp residue [53]. The different triplets showed a good correlation between stability and occurrence [48].

Unfortunately, the number of potential triplets, 400, makes it a considerable task to examine all possible triplets. However, about one third of these have not been observed as yet in a collagen, and many others have been observed only once or very few times [54]. On the other hand, only a few triplets, often of the form Gly-Xaa-

Hyp or Gly-Pro-Yaa, else Gly-Glu-Arg and Gly-Glu-Lys, are found at high frequency [54].

Further studies were reported (see [50, 55]) that covered over 75% of observed triplets. These data allowed predictions for other triplets based on a simple 'additive model'. This approach was also extended to hexapeptide, two triplet substitutions to the common host–guest framework.

acetyl-$(\text{Gly-Pro-Hyp})_3$-**Gly**-**<u>Xaa</u>**-**<u>Yaa</u>**-**Gly**-**<u>Xaa′</u>**-**<u>Yaa′</u>**-$(\text{Gly-Pro-Hyp})_3$-Gly-$\text{Gly}_{\text{amide}}$

The 'additive model' worked well for small uncharged residues. For example, substitutions of Ala into the Xaa and Yaa positions led to destabilisation by ~5 °C and ~6 °C respectively, while the double substitution Gly-Ala-Ala, is destabilised by around the sum of these. Similarly, a single Gly-Ala-Hyp triplet is destabilising by 5 °C while two adjacent such triplets, Gly-Ala-Hyp-Gly-Ala-Hyp, are destabilising by the sum, 10 °C [50, 55]. This additive approach is not true for all adjacent residues and triplets. Thus, the peptide Gly-Leu-Hyp-Gly-Leu-Hyp was found to be 7 °C more stable that predicted by the additive model, as a result of hydrophobic interactions along the triple-helix [50]. The effects of charged residues can be more substantial still. In this case sequences involving oppositely charged residues, Glu or Asp and Lys (but not Arg) on either side of a Gly residue, such as Gly-Xaa-Lys-Gly-Asp-Yaa, the additional stability above the predicted, additive model value was as much as 17 °C, giving a stability equivalent to a peptide with all Pro and Hyp residues [50].

Together, all these data enabled an algorithm to be developed to allow prediction of the stability of collagen-like homotrimeric peptides of 6 to 30 triplets and variations in stability along a collagen chain [55]. This algorithm, (http://compbio.cs.princeton.edu/csc/),

$$T_{\text{m}} = T_{\text{m}}^{o} + \Delta T_{\text{m}}^{cap} - \sum \Delta T_{\text{m}}^{GXY} + \sum \Delta T_{\text{m}}^{\text{int}}$$

for the T_{m} of the peptide, is based on the stability of a (Gly-Pro-Hyp) of the same length, T_{m}^{o}, plus addition of stability if the N- and/or C-terminal is blocked, $\Delta T_{\text{m}}^{cap}$, less the stability that is lost from substitutions of Xaa and Yaa residues, $\sum \Delta T_{\text{m}}^{GXY}$, plus the extra stability that arises from certain proximal residues, $\sum \Delta T_{\text{m}}^{\text{int}}$, such as Lys-Gly-Asp charge pairs. Overall, testing has suggested that this algorithm works well, except for highly charged sequences, where further refinements are needed.

6.3.3 Effects of solvents on peptide models

Compared with native collagens, relatively few studies have been made on peptide models of collagens to examine the effects of solvents. Most frequently, the effects of pH can be seen for peptides examined at neutral conditions in PBS and also under acidic conditions in dilute acetic acid, typically at around pH 2.5 (for example, [47]). In these cases, the stability at acid conditions may be less than at neutral conditions, depending on the amino acid sequence, while in some cases the peptide may not be stable at all under acid conditions [56].

Table 6.1. Examples of strategies that have been use to form triple-helical collagen peptide models.

Strategy	Type	Structure	
Di-Lysine link	Crosslinked	$(GXY)_n(GPO)_4GPPSGG\text{-}Ahx$ $(GXY)_n(GPO)_4GPPSGG\text{-}Ahx\text{-}Lys\text{-}Lys$ $(GXY)_n(GPO)_4GPPSGG\text{-}Ahx$	[44]
Trimeric template	Crosslinked	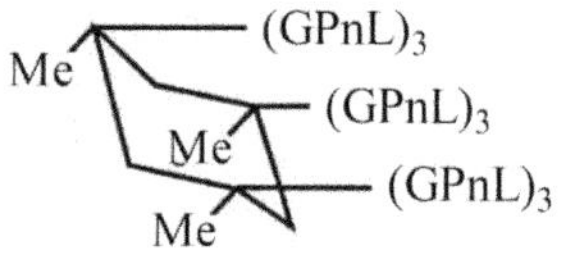	[59]
Di-disulfide links	Crosslinked	$Ac\text{-}(GPO)_5(GXY)_4\,GCG$ $Ac\text{-}(GPO)_5(GXY)_4\,GCCG$ $H\text{-}(GPO_5(GXY)_4\,GCG$	[61]
host–guest	Associating	$Ac\text{-}(GPO)_3(GXY)(GPO)_4GG$	[46]
Mono-end capped	Associating	$Ac\text{-}(GXY)_6(GPO)_4GV$	[64]
Bis-end capped	Associating	$GPC(GPO)_3(GXY)_{10}(GPO)_3GPC$	[65]
Trimer domains	Recombinant	$(GPP)_{10}$ -foldon	[67]

nL: norLeucine (N-isobutylglycine); Ahx: ε aminohexanoic acid

Additives to the buffer solution have been used, for example, ethylene glycol [57] and trimethylamine N-oxide [31] to increase peptide stability so as to allow measurement and comparison of transitions when one or more peptide lacks stability under normal PBS conditions. Of interest, the stability of peptide (Pro-Pro-Gly)$_{10}$ has been compared between H_2O and D_2O solutions [58]. In D_2O, the stability is greater, possibly due to enthalpic considerations, although the extent reported varies between studies. Stability increases steadily in graded increases in composition to pure D_2O [58].

6.4 Other uses for peptide models

6.4.1 Synthetic peptide model designs

In addition to providing a way for studying the stability of the collagen triple-helix, synthetic peptide models have been used to study a wide variety of other features of collagen structure and function, as well as being proposed as a source of collagen-like biomaterials in their own right. Various strategies have been developed to meet the needs of particular scientific questions (some examples are in the following subsection) and to ensure that a sufficiently stable triple-helical conformation is obtained (table 6.1)

Initially a di-lysine sequence was used at the C-terminal and the three polypeptide chains synthesised simultaneously from the α and two ε amino groups [44]. The difficulty with this method is that an extremely high coupling efficiency is needed for each of the many amino acid additions (90 for three 30-residue chains). If less than 100%, then a range of by-products lacking amino acids builds up across the chains, making purification of a single accurate structure exceedingly hard; any structure lacking a single amino acid is almost identical to the accurate structure. Similar issues exist for other trimeric template structures such as Kemp's triacid [59] or Tris [60]. A resolution of this difficulty was the development of structures linked by disulfide bonds [61] as each of the individual chains can be purified prior to covalent trimer formation. A particular advantage of this strategy is that it allows hetero-trimer formation.

Other approaches have relied of the self-assembly of trimer structures from defined, purified monomer peptides, as had been demonstrated using $(\text{Pro-Pro-Gly})_{10}$ and $(\text{Pro-Hyp-Gly})_{10}$ structures [62, 63]. The individual triplets Pro-Pro-Gly and Pro-Hyp-Gly, are particularly stabilising to the triple-helix and so were used as the basis for structures where a single triplet was varied in a set of host–guest peptides used to study helical stability [55]. They were also used for structures that incorporated real collagen sequence segments through use as stabilising caps at the ends of the peptides (table 6.1). In one case a single end cap was used to stabilise a 6-triplet native sequence [64], while in another example caps at both ends were used to stabilise a 10-triplet native sequence [65]. This latter approach has been significantly extended to allow full sets of overlapping peptides that cover an entire collagen chain to be established and used to probe for functional domains [66] (see chapter 7).

Most recently defined recombinant structures have emerged for studying collagen functions, and in these cases a trimerizing domain is used to stabilise the initial structure. Initially, a 29-residue foldon motif, from a phage protein, was used [67], while more recently the trimerizing domains from bacterial collagens have been used [68] (see chapter 2).

Defining the triple-helix structure and the role of hydroxyproline

More than 60 model peptides of defined collagen sequences have now been used for crystal structure analyses that have been used to define the structure of the collagen triple-helix at an atomic scale [69] (see chapter 2). In addition, a variety of peptide models have been used in NMR studies of the structure of collagen [70]. The various crystal structures have provided more detailed information on the role of Hyp than any other method (see chapter 2).

Effects of sequence interruptions

Peptide models have been used to study the effects of sequence interruptions on collagen structure, for example those found naturally in type IV collagen, by using crystallography [71], NMR and other biophysical methods [72, 73].

Mapping functional sites

Collagens are now clearly recognised as having many functional domains embedded in the triple-helical sequence. Some are present on individual molecules while others are found as fibrils and as cryptic sites that may emerge during catabolic or disease changes. In many cases the binding sites are poorly defined, for example, to within a large segment such as a CNBr fragment. Peptide models have shown their utility in allowing binding sites to be defined to a few residues in two or three triplets. These have included mapping of a monoclonal antibody binding site [64], the site of interaction of the α2β1 integrin [65], and more recently a range of sites have been defined using a 'toolkit' of numerous sequential peptides [66] (see chapter 7).

Consideration of disease severity

Mutations in various collagen genes have now been identified as leading to a wide variety of diseases [74]. Of these, the best characterised are the mutations of Gly residues in the triple-helix that are a major cause of *Osteogenesis imperfecta.* These mutations have been examined in peptide models by crystallography [75], NMR and other biophysical methods [76–78]. A single mutation at a Gly codon can lead to eight potential amino acid substitutions. A study of seven of these (not Trp) using single Gly substitutions in host–guest peptides allowed a ranking of the destabilising effects for each substitution, Ala $\leqslant$ Ser $<$ Cys_{red} $<$ Arg $<$ Val $<$ Glu $\leqslant$ Asp [31] that broadly correlated with disease severity. Thus, small amino acids Ala and Ser tend to cause mild forms of the disease while the acidic residues, Glu and Asp, leading to the severe forms of the disease, with Asp typically leading to perinatal lethal forms [31]. The individual effects are further modulated by the surrounding residues in adjacent triplets [77, 78].

Effects of chemical changes

Chemical modifications to collagen, typically during processing, can occur to collagens, including deamidation, the loss of amide groups from Gln and Asn residues by alkaline treatment, and racemisation. Studies of these using host–guest peptides showed that deamidation of Gln residues in the Yaa position are the most destabilising [79] while racemisation of Asp residues was found to be particularly destabilising [57]. The effects of a wide variety of substitutions, for example variations to proline structures can be studied. Also, a variety of novel molecules can be produced by 'click' chemistry [80].

Self-assembling systems

A small number of self-assembling fibre systems based on short collagen-like peptides have been described. It has been suggested that such structures could form the basis of synthetic, animal-free collagen-like biomaterials, although comparative costs have not been presented. Axial growth of fibres has been achieved through in-line N-to C-terminal bonding between Lys and Asp residues [81] and by cation-π interactions [82]. Lateral growth can be achieved through adjustment of side chain electrostatic interactions [81]. In one case a fibre with short periodic structure visible by TEM was established using a simple polypeptide [83]. A particularly versatile concept was the formation of disulfide crosslinked trimer precursors where two chains extended in the opposite direction to the third chain. The interaction of the third chain of one molecule with the two chains of a separate molecule to form a stable triple-helix allows a fibre to form by sequential additions of precursors [84]. Lateral growth of these fibres could be achieved through use of complementarity of adjacent molecules, and the ready adjustment of the sequence enables variation in the stability of the fibres [85], while also allowing the incorporation of binding domains modelled from animal collagens that show comparable characteristics [86].

New approaches
Recently, recombinant bacterial collagens (see chapter 9) have been used to explore the structure and properties of functional domains from animal collagens, providing an additional approach to the previous studies using peptides. The recombinant systems have been used to further explore the properties of functional domains identified using peptide models [87, 88] and to discover the detailed nature of other sites [89, 90]. This system is of value when longer segments of native sequence are required so that the influence of adjacent triplets can be examined [68], although suffer from the disadvantage of lacking Hyp that may be important in some functional domains. The new bacterial constructs have frequently been made using the Scl2 gene of *S. pyogenes* which behaves as a 'blank slate' with no observed binding functions [87]. The new molecules featuring designed, specifically added functions have been proposed as targeted biomedical materials [87, 91] (see chapter 9).

References

[1] Russell A E 1973 *Biochem. J.* **131** 335–42
[2] Hayashi T, Curranpatel S and Prockop D J 1979 *Biochemistry* **18** 4182–87
[3] McClain P E and Wiley E R 1972 *J. Biol. Chem.* **247** 692–97
[4] Hayashi T and Nagai Y 1973 *J. Biochem.* **73** 999–1006
[5] Bruckner P and Eikenberry E F 1984 *Eur. J. Biochem.* **140** 397–99
[6] Heidemann E and Bernhardt H W 1967 *Nature* **216** 263–64
[7] Bächinger H-P, Morris N P and Davis J M 1993 *Am. J. Med. Genet.* **45** 152–62
[8] Cantor C R and Schimmel P R 1980 *Biophysical Chemistry: Part III: The Behavior of Biological Macromolecules* (Oxford: WH Freeman)
[9] Privalov P L 1982 *Adv. Protein Chem.* **35** 1–104
[10] Ryhänen L, Zaragoza E J and Uitto J 1983 *Arch. Biochem. Biophys.* **223** 562–71
[11] Leikina E, Mertts M V, Kuznetsova N and Leikin S 2002 *Proc. Natl Acad. Sci. USA* **99** 1314–18
[12] Rosenbloom J, Harsch M and Jimenez S 1973 *Arch. Biochem. Biophys.* **158** 478–84
[13] Rigby B J 1968 *Nature* **219** 166–7
[14] Mathews M B 1975 *Connective Tissue* (Berlin: Springer)
[15] Cohen-Solal L, Castanet J, Meunier F J and Glimcher M J 1986 *Comp. Biochem. Physiol.* B **83** 483–86
[16] Bianchi E, Conio G, Ciferri A, Puett D and Rajagh L 1967 *J. Biol. Chem.* **242** 1361–69
[17] Woodlock A F and Harrap B S 1968 *Aust. J. Biol. Sci.* **21** 821–26
[18] Rauterberg J and Kühn K 1968 *Hoppe Seylers Z. Physiol. Chem.* **349** 611–22
[19] Tronci G, Russell S J and Wood D J 2013 *J. Mater. Chem.* B **1** 3705–15
[20] Ruijgrok J M, de Wijn J R and Boon M E 1994 *Clin. Mater.* **17** 23–7
[21] Fujimoto D 1970 *Biochim. Biophys. Acta* **200** 172–4

[22] Lennox F G 1947 *Biochem. J.* **41** xlvii
[23] Lennox F G 1949 *Biochim. Biophys. Acta* **3** 170–87
[24] Harrap B S 1969 *Int. J. Protein Res.* **1** 245–52
[25] Bianchi E, Rampone R, Tealdi A and Ciferri A 1970 *J. Biol. Chem.* **245** 3341–45
[26] Hart G J, Russell A E and Cooper D R 1971 *Biochem. J.* **125** 599–604
[27] Gekko K and Koga S 1983 *J. Biochem.* **94** 199–205
[28] Kuznetsova N, Chi S L and Leikin S 1998 *Biochemistry* **37** 11888–895
[29] Na G C 1986 *Biochemistry* **25** 967–73
[30] Snowden J M 1982 *Biochim. Biophys. Acta* **707** 142–46
[31] Beck K, Chan V C, Shenoy N, Kirkpatrick A, Ramshaw J A M and Brodsky B 2000 *Proc. Natl Acad. Sci. USA* **97** 4273–78
[32] Yancey P H, Clark M E, Hand S C, Bowlus R D and Somero G N 1982 *Science* **217** 1214–22
[33] Yancey P H and Somero G N 1979 *Biochem. J.* **183** 317–23
[34] Zeugolis D I, Khew S T, Yew E S, Ekaputra A K, Tong Y W, Yung L Y, Hutmacher D W, Sheppard C and Raghunath M 2008 *Biomaterials* **29** 2293–305
[35] Shanmugam G, Reddy S M, Natarajan V and Madhan B 2013 *Int. J. Biol. Macromol.* **54** 155–59
[36] Fraser R D B and MacRae T P 1973 *Conformation in Fibrous Proteins* (New York: Academic)
[37] Heidemann E and Bernhardt H W 1968 *Nature* **220** 1326–27
[38] Heidemann E and Roth W 1982 *Adv. Polym. Sci.* **43** 143–203
[39] Merrifield R B 1965 *Science* **150** 178–85
[40] Merrifield R B 1973 *Harvey Lect.* **67** 143–73
[41] Dölz R and Heidemann E 1986 *Biopolymers* **25** 1069–80
[42] Roth W and Heidemann E 1980 *Biopolymers* **19** 1909–17
[43] Thakur S, Vadolas D, Germann H P and Heidemann E 1986 *Biopolymers* **25** 1081–86
[44] Germann H P and Heidemann E 1988 *Biopolymers* **27** 157–63
[45] Bächinger H-P and Davis J M 1991 *Int. J. Biol. Macromol.* **13** 152–56
[46] Shah N K, Ramshaw J A M, Kirkpatrick A, Shah C and Brodsky B 1996 *Biochemistry* **35** 10262–268
[47] Chan V C, Ramshaw J A M, Kirkpatrick A, Beck K and Brodsky B 1997 *J. Biol. Chem.* **272** 31441–446
[48] Persikov A V, Ramshaw J A M, Kirkpatrick A and Brodsky B 2000 *Biochemistry* **39** 14960–967
[49] Persikov A V, Ramshaw J A M, Kirkpatrick A and Brodsky B 2002 *J. Mol. Biol.* **316** 385–94
[50] Persikov A V, Ramshaw J A M, Kirkpatrick A and Brodsky B 2005 *Biochemistry* **44** 1414–22
[51] Venugopal M G, Ramshaw J A M, Braswell E, Zhu D and Brodsky B 1994 *Biochemistry* **33** 7948–56
[52] Shah N K, Sharma M, Kirkpatrick A, Ramshaw J A M and Brodsky B 1997 *Biochemistry* **36** 5878–83
[53] Yang W, Chan V C, Kirkpatrick A, Ramshaw J A M and Brodsky B 1997 *J. Biol. Chem.* **272** 28837–40
[54] Ramshaw J A M, Shah N K and Brodsky B 1998 *J. Struct. Biol.* **122** 86–91
[55] Persikov A V, Ramshaw J A M and Brodsky B 2005 *J. Biol. Chem.* **280** 19343–349
[56] Anachi R B, Siegel D L, Baum J and Brodsky B 1995 *FEBS Lett.* **368** 551–5

[57] Shah N K, Brodsky B, Kirkpatrick A and Ramshaw J A M 1999 *Biopolymers* **49** 297–302
[58] Gough C A and Bhatnagar R S 1999 *J. Biomol. Struct. Dyn.* **17** 481–91
[59] Feng Y, Melacini G, Taulane J P and Goodman M 1996 *Biopolymers* **39** 859–72
[60] Cai W, Wong D, Kinberger G A, Kwok S W, Taulane J P and Goodman M 2007 *Bioorg. Chem.* **35** 327–37
[61] Fiori S, Saccà B and Moroder L 2002 *J. Mol. Biol.* **319** 1235–42
[62] Olsen B R, Berg R A, Sakakibara S, Kishida Y and Prockop D J 1971 *J. Mol. Biol.* **57** 589–95
[63] Sakakibara S, Inouye K, Shudo K, Kishida Y, Kobayashi Y and Prockop D J 1973 *Biochim. Biophys. Acta* **303** 198–202
[64] Glattauer V, Werkmeister J A, Kirkpatrick A and Ramshaw J A M 1997 *Biochem. J.* **323** 45–9
[65] Knight C G, Morton L F, Onley D J, Peachey A R, Messent A J, Smethurst P A, Tuckwell D S, Farndale R W and Barnes M J 1998 *J. Biol. Chem.* **273** 33287–294
[66] Farndale R W, Lisman T, Bihan D, Hamaia S, Smerling C S, Pugh N, Konitsiotis A, Leitinger B and de Groot P G *et al* 2008 *Biochem. Soc. Trans.* **36** 241–50
[67] Stetefeld J, Frank S, Jenny M, Schulthess T, Kammerer R A, Boudko S, Landwehr R, Okuyama K and Engel J 2003 *Structure* **11** 339–46
[68] Yu Z, Visse R, Inouye M, Nagase H and Brodsky B 2012 *J. Biol. Chem.* **287** 22988–997
[69] Bella J 2016 *Biochem. J.* **473** 1001–25
[70] Baum J and Brodsky B 1999 *Curr. Opin. Struct. Biol.* **9** 122–28
[71] Bella J, Liu J, Kramer R, Brodsky B and Berman H M 2006 *J. Mol. Biol.* **362** 298–311
[72] Mohs A, Popiel M, Li Y, Baum J and Brodsky B 2006 *J. Biol. Chem.* **281** 17197–202
[73] Thiagarajan G, Li Y, Mohs A, Strafaci C, Popiel M, Baum J and Brodsky B 2008 *J. Mol. Biol.* **376** 736–48
[74] Royce P M and Steinmann B 2003 *Connective Tissue and Its Heritable Disorders* 2nd edn (New York: Wiley-Liss)
[75] Bella J, Eaton M, Brodsky B and Berman H M 1994 *Science* **266** 75–81
[76] Bhate M, Wang X, Baum J and Brodsky B 2002 *Biochemistry* **41** 6539–47
[77] Hyde T J, Bryan M A, Brodsky B and Baum J 2006 *J. Biol. Chem.* **281** 36937–43
[78] Xiao J, Madhan B, Li Y, Brodsky B and Baum J 2011 *Biophys. J.* **101** 449–58
[79] Silva T, Kirkpatrick A, Brodsky B and Ramshaw J A M 2005 *J. Agric. Food Chem.* **53** 7802–06
[80] Erdmann R S and Wennemers H 2012 *Org. Biomol. Chem.* **10** 1982–86
[81] O'Leary L E, Fallas J A, Bakota E L, Kang M K and Hartgerink J D 2011 *Nat. Chem.* **3** 821–8
[82] Chen C C, Hsu W, Kao T C and Horng J C 2011 *Biochemistry* **50** 2381–83
[83] Rele S, Song Y, Apkarian R P, Qu Z, Conticello V P and Chaikof E L 2007 *J. Am. Chem. Soc.* **129** 14780–87
[84] Koide T, Homma D L, Asada S and Kitagawa K 2005 *Bioorg. Med. Chem. Lett.* **15** 5230–33
[85] Yamazaki C M, Asada S, Kitagawa K and Koide T 2008 *Biopolymers* **90** 816–23
[86] Yamazaki C M, Kadoya Y, Hozumi K, Okano-Kosugi H, Asada S, Kitagawa K, Nomizu M and Koide T 2010 *Biomaterials* **31** 1925–34
[87] Cosgriff-Hernandez E, Hahn M S, Russell B, Wilems T, Munoz-Pinto D, Browning M B, Rivera J and Höök M 2010 *Acta Biomater.* **6** 3969–77

[88] Peng Y Y, Stoichevska V, Schacht K, Werkmeister J A and Ramshaw J A M 2014 *J. Biomed. Mater. Res.* A **102** 2189–96

[89] An B, Abbonante V, Yigit S, Balduini A, Kaplan D L and Brodsky B 2014 *J. Biol. Chem.* **289** 4941–51

[90] An B, Abbonante V, Xu H, Gavriilidou D, Yoshizumi A, Bihan D, Farndale R W, Kaplan D L and Balduini A *et al* 2016 *J. Biol. Chem.* **291** 4343–55

[91] Peng Y Y, Yoshizumi A, Danon S J, Glattauer V, Prokopenko O, Mirochnitchenko O, Yu Z, Inouye M and Werkmeister J A *et al* 2010 *Biomaterials* **31** 2755–61

IOP Publishing

John A M Ramshaw and Veronica Glattauer

Chapter 7

Interactions

As well as providing essential structural stability to tissues, collagens also demonstrate a wide range of critical molecular interactions that play a key role in tissue functions. In many cases the interaction has been localised to a specific site on a collagen molecule, or on a fibril, and in some cases detailed structural data for an interaction have been determined. The detailed data allow the design of novel recombinant collagens. For biomaterials based on xenogeneic collagens, a key interaction is that with the host's immune system. An understanding of this process can assist with ensuring a safe biomedical product.

7.1 Describing interactions with other molecules

Key interactions between collagens molecules are essential in the formation of functional tissues (see chapters 4 and 5) and in the response to a medical implant. Further to the physical contribution to tissue organisation and structure, collagens play a key role in functional interactions with many other components of the extracellular matrix (ECM). In examining the range of potential interactions, readers should pay attention to the methods used in establishing the interaction. The molecules need to be proximal in tissue for interaction. Some molecules, such as heparin, seem to be particularly 'sticky' and so could be reported as binding in *in vitro* studies while the interaction may not be present in native tissues. The presence of molecules at specific locations in native tissue is important and not always represented faithfully in an *in vitro* test.

7.1.1 Collagen network plots

Various interaction networks have been established. The network plot could include molecule to molecule interactions with little molecular details, yet for design of novel molecules for recombinant production more detail is needed. An initial example was

doi:10.1088/978-0-7503-2096-2ch7

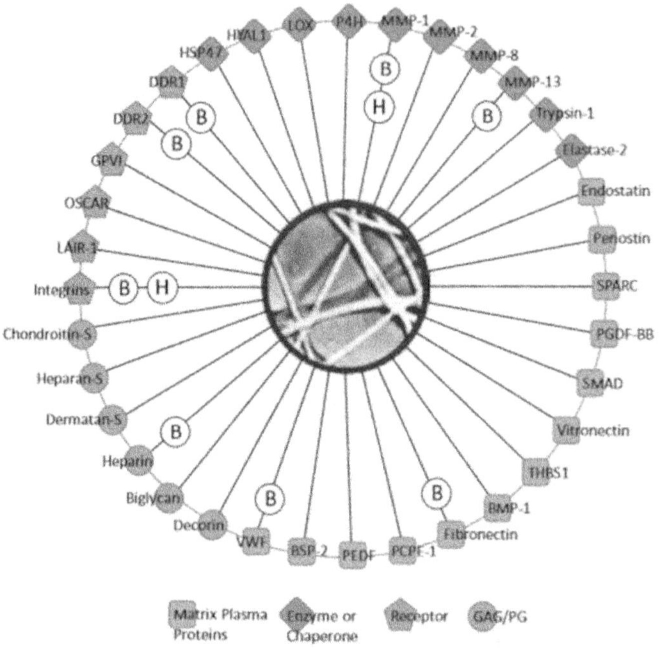

Figure 7.1. A schematic representation of the interaction network of collagen with some of the molecules of different classes that interact with collagen. Interactions that have been further studied in recombinant human (H) and bacterial (B) collagen systems are indicated. (From [3], reprinted by permission from Springer Nature, Copyright 2017).

the network plot for the C-terminal fragment of type XVIII collagen (endostatin) [1]. Subsequent plots have been reported for type I collagen [2, 3] (figure 7.1) and the combination of type I, II and III collagens [4]. In circumstances where the detailed structure of the binding domain has been established, these data have been used in recombinant triple-helical model constructs to allow further study of the nature of the binding sites [3] (figure 7.1).

7.1.2 Collagen interactome maps

Specific domains within the collagen triple-helical molecule and higher order fibrillar structures have been identified as recognition motifs for cell surface receptors, enzymes, chaperones and plasma proteins. Many of these interactions only occur with the native triple-helix and not on single chains after denaturation. Methodologies applied to native and fragments of collagens have elucidated ligand binding regions along the collagen molecule. Peptide models and recombinant systems (see chapter 6) have facilitated the process of defining recognition amino acid sequences and have been used to explore structural and functional properties. Molecular modelling, NMR or crystallographic data have revealed in some cases direct evidence of structural inter-relationships between collagen and ligands.

From the vast array of experimental data arising from studies on diseases related to collagen sequence mutations and the identification of interacting partners to collagens,

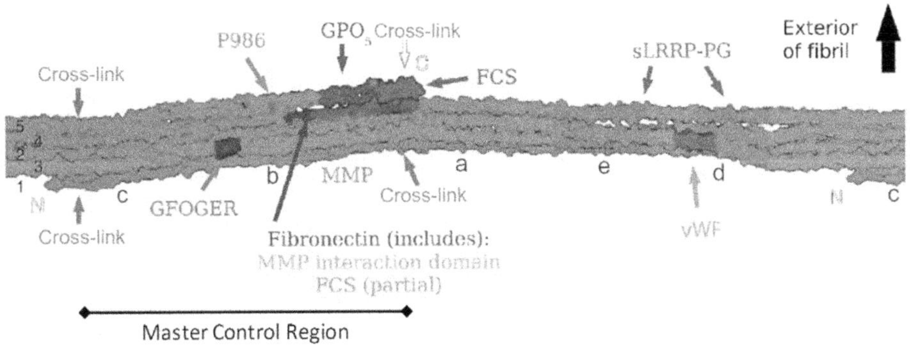

Figure 7.2. Collagen molecules arranged to form a microfibril, with a 234 residue D-period stagger. The locations of well-defined binding sites are indicated, along with the active portion termed the master control region. The proposed orientation of the microfibril relative to the exterior fibril surface is indicated (black arrow). See [10] for further details. (Reprinted from [10], by permission of Taylor & Francis Inc., through the Copyright Clearance Center Inc.)

collagen maps depicting interactions sites have been produced [5–8]. For the initial map describing type I collagen, the assembly was based on a D-period repeat of the collagen fibril, 67 nm long, comprising five collagen monomers in a staggered manner, encompassing an overlap and gap zone (see chapter 4) depicting a 2D model of a collagen fibril. These maps are referred to as 'interactomes' and the first of these was assembled for collagen type I heterotrimer α1α1α2 [5]. Mapping of ligand binding and functional sites revealed three major ligand binding regions 1 to 3 (MLBR-1/3) with most interaction sites found on the C-terminal half of the collagen molecule. A 'cell interaction domain' close to MLBR-2 contains at its core the integrin binding sequence GFOGER site with no co-localised ligands on the monomer and well away from proteoglycan binding, mineralisation and crosslinking sites (known as the 'matrix interaction domain') on the fibril ensuring clear cellular access.

Using fibre x-ray diffraction structure for collagen type I, previously identified functional sites have been mapped onto the fibrils [9–11]. Most of the crucial functional sites are zoned fairly close together on the fibrils, this is described as the 'master control region' (figure 7.2) [10]. As these microfibrils then form higher order structures, contiguous repeat zones are present.

Interactome maps have now been described in detail for other collagens, types III and IV. Others are emerging, especially where detailed overviews have been prepared, such as for type VI collagen [12]. Nevertheless, for many collagens the number of interactions that have been specifically established is typically low, or zero, so in these cases there is little value in initiating a map.

Collagen type IV is the major collagen found in basement membranes, typically separating epithelial and or endothelial cells from an underlying stroma. Three heterotrimeric collagen type IV interactome maps have been constructed for α1α1α2, α3α4α5 and α5α5α6 isoforms [7]. These maps indicate major structural zones, ligand binding sites, both known and predicted, as well as missense mutations. These maps have enabled predictions of functional domains for angiogenesis, haemostasis, and disease domains for autoimmunity, tumour growth, infection and glycation to be made.

The interactome map for type III collagen [8], a homotrimer, reveals similarities to collagen type I, with nearly identical structural and functional sites. Type III collagen has 2 major sites involved in haemostasis, comprising binding motifs for von Willebrand factor, integrin $\alpha 2\beta 1$, platelet binding octapeptides and glycoprotein VI. Sequence variants in these sites are responsible for bleeding tendency of some COL3A1 associated diseases [8].

The interactome maps are more specific than the network plots in that they indicate the sequence regions that are involved in interactions, while network plots may contain information that is only at the molecule to molecules level. On the other hand, the interactome maps contain proposed sites that are based on the extension of sequence motifs determined at other locations. In general, there are relatively few sites for interactions that have been explored and located at a molecular level; a selection of these are discussed in the following sections.

7.2 Interactions with other collagens

The interactions with other collagens include the predominant like-to-like interactions, such as in fibril or network formation, as well as the various interactions between different collagen types, such as types I, III and V, within a collagen fibril (see chapter 4). These interactions also include those between the fibril-associated collagens with interrupted triple helices (FACIT collagens) [13] and interstitial collagen fibrils, where the FACIT collagens attach to the surface of a fibril in a D-periodic arrangement [14]. This interaction can be visualised by rotary shadowing [13, 15]. Despite the importance of these interactions, the extent of molecular knowledge on the specific interactions involved in these higher order structures that are formed is still limited.

7.3 Interactions with proteoglycans

7.3.1 Heparin and heparan sulfate glycan chains

There have been a range of studies on the binding of heparin to collagens, and less often binding of heparan sulfate chains, with heparin being a convenient functional and structural analogue for heparan sulfate proteoglycan chains. The heparan sulfate proteoglycans are one of the core components of basement membranes. In some cases, the binding is to the retained globular domains of the collagens, while in others it is to the triple-helical domain. In some cases, binding to both globular and triple-helical domains has been observed. For globular proteins, binding is seen at clusters of basic amino acids, Arg and/or Lys. These could be present as a linear sequence, such as 'basic-basic-Xaa-basic' or also in a cluster arising from the secondary and tertiary structures of the protein to bring basic amino acids from different, non-linear chain locations into close proximity [16–18]. For the collagen triple-helix, a site would be linear, but would also need topological input from the other chains to present a suitable clustering of basic residues in a single molecule [19] that could be augmented in fibrillar assemblies [20].

Examples of the binding of heparin/heparan sulphate chains to collagens are given in table 7.1, along with the regions of the sequences that are involved in these interactions, where they have been determined. It is suggested that binding to fibrils was better than to individual molecules, indicating the value of the additional clustering of basic residues [21]. It was found that type V collagen showed the best binding affinity of the fibril forming collagens, type I, II, III and V, as well as types IV and VI collagens [22]. This may reflect a role in early stages of fibril formation since, if type V collagen is a core filament within a collagen fibril in certain tissues (see chapter 4) [23, 24], then it may not be available for binding in more mature tissues. Similarly, a sequence identified for binding by type I collagen, KGHRGF at positions 87–92 in the α1-chain [25], is also the site for crosslink formation (see chapter 5), so also may not be as accessible in more mature tissues. In some cases, although the binding domain is located within the main collagen triple-helical zone, the binding site may be associated with an interruption in the Gly-Xaa-Yaa repeat [26, 27].

Two heparin binding sites have also been reported in the collagen-like tail of acetylcholine esterase [28]. It was later shown that these sites had different affinities [29] and they required the triple-helical structure for optimal binding [30]. Spectroscopy using model peptides suggested that the two binding sequences, GRKGR *vs* GKRGK, had distinctive, different manners of interaction [31]. Inclusion of the high affinity sequence into the 'blank slate' collagen domain of the *S. pyogenes* Scl2 collagen structure leads to increased heparin binding [32].

Table 7.1. Examples of heparin/heparan sulphate chain binding to various collagen types.

Collagen type	Globular domain		Triple-helical domain	
	Binding zone		Binding zone	
Type I	Near N-propeptide terminus	[20]	Fibril better than monomer	[21]
			'a-band' region in fibril	[20]
			α1, 87–92, KGHRGF	[25]
Type IV	TAGSCLRKFSTM (α1 NCI)	[26]	206 nm from terminus	[47]
	LAGSCLARFSTM (α2 NC1)			
			GEFYFDLRLKGDK (α1)	[26]
Type V	N-terminal domain, α4-chain	[48]	30 kDa CNBR fragment	[49]
			Ile824 to Pro950 (α1)	[50]
			Arg at 912, 918, 921	
			Lys905 and Arg909	[51, 52]
Type XI	N-propeptide, residues 147–152	[53]	GKPGPRGQRGPTGPRGSRGAR (α1)	[54]
Type XIII			Ectodomain (pepsin site(s))	[27]
Type XIV	NC1 domain, residues 94–111	[55]	Location not defined	[56]
Type XIX	N-terminal globular domain	[57]		

7.3.2 Interactions with other proteoglycans

Proteoglycans are key components of the ECM and are known to contribute to nucleation, growth and calcification of the collagen fibrils through interactions with collagens [33]. Early studies in solution utilising light scattering, turbidimetric and precipitation techniques, demonstrated that proteoglycans and single glycosaminoglycan chains could interact with collagens [34], with binding observed to chondroitin 4-sulphates, dermatan sulphate, heparin sulphate and heparin.

Interactions in solution were thought not to be necessarily representative of what occurs when proteoglycans are associated with collagen fibrils. The first tissues to be investigated ultrastructurally, with electron microscopy, were rat-tail tendon and rabbit Achilles tendon [35, 36]. New, electron dense dyes, especially Cupromeronic and Cuprolinic blue, were developed for these studies [36]. These studies showed a regular array of electron dense proteoglycan filaments arranged orthogonally to the collagen fibrils separated by the D-period repeat distance. The staining protocol caused the condensation of the proteoglycans allowing a more specific location to be identified. Initially, small dermatan rich proteoglycan was identified as associated with the d-band of the gap region (figure 7.3) [37], while subsequently keratan sulfate proteoglycan was identified binding to corneal fibrils at the a- and c-bands [38]. Studies utilising the same methods have been applied to other tissues such as skin, sclera, cornea, and bone, [38–40] and cartilage [41]. While the main proteoglycan in soft tissues is dermatan sulphate, with orthogonal staining at the surface of the fibrils, in decalcified bone chondroitin sulphate rich proteoglycans are found parallel to the collagen fibrils.

Decorin is a small proteoglycan, so called from its decoration along the collagen fibril [42], is a leucine rich repeat proteoglycan with a core protein called decoron. Decorin plays a role in fibrillogenesis [43] and organisation of collagen fibrils. It was identified as binding near to the C-terminal of type I collagen [44]. With decoron's

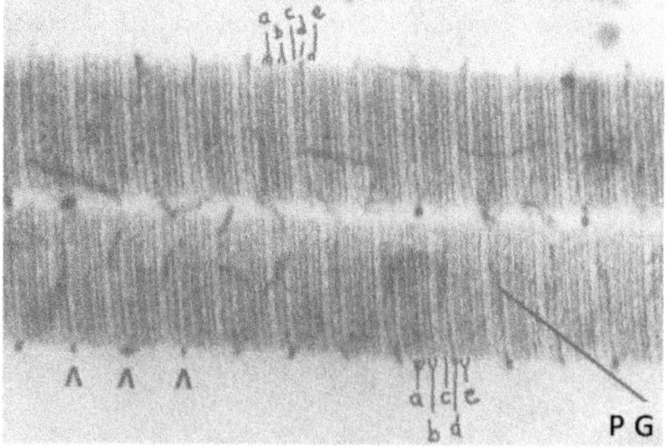

Figure 7.3. Rat-tail tendons, with Cupromeronic blue staining, showing the location of small, dermatan sulfate proteoglycans along banded collagen fibrils. The individual bands, a–e, within a collagen D-period are indicated along with the proteoglycan locations Λ and PG. One D-period is 67 nm. (From [37], reprinted with permission from Portland Press ©1981.)

crystal structure determined [45] along with already known fibrillar structure of collagen type I [9], interactions were investigated through molecular modelling. It was proposed that one molecule of decoron interacts with four to six collagen molecules through hydrogen bonding and electrostatic interactions. Design of peptides based on decorin binding, or other binding systems, allows the development of systems for attachment of biological entities to collagen products or for cross-linking collagen-based materials [46].

7.4 Interaction with globular proteins

7.4.1 Signalling molecules

Integrin binding to collagen
Collagen interactions with other cells present in the tissue environment occur through binding to cell surface receptors, one group of which are integrins. These are metal ion-dependent adhesion molecules, with interaction leading to integrin activation and subsequent cellular responses, adhesion, migration and proliferation. Integrin receptors are heterodimers composed of an α and a β subunit which in combination give rise to 24 different integrin types [58–60] of which collagen interacts with only a subset, $\alpha_1\beta_1$, $\alpha_2\beta_1$, $\alpha_{10}\beta_1$ and $\alpha_{11}\beta_1$.

Integrin binding to collagen types has been determined with microscopic methods such as rotary shadowing and electron microscopy, surface plasmon resonance, SPR [61], cell adhesion in cell culture and solid phase protein binding assays. Studies have employed intact collagen proteins, cyanogen bromide fragments and synthetic peptide models with purified or recombinant integrin or its domains.

The $\alpha_1\beta_1$ integrin is found on fibroblasts, chondrocytes, hepatocytes and mesenchymal stem cells as well as on circulating white blood cells. In addition, it is localised on vascular, visceral smooth muscle, pericytes and endothelial cells closely associated with the basement membrane. Collagen ligands include collagens I, II, III, IV, VI, IX XIII, XVI [62]. The affinity for collagen IV is higher than that for collagen I [63]. $\alpha_2\beta_1$ is found on cells associated with interstitial collagen matrices, including on fibroblasts, myeloid, T cells and megakaryocytes and platelets. These bind with collagen types I, II, III, IV V, XI, XVI and XXIII.

Unlike the extensively studied $\alpha_1\beta_1$ and $\alpha_2\beta_1$ receptors, there is less known on $\alpha_{10}\beta_1$ and $\alpha_{11}\beta_1$ interactions. However, the $\alpha_{10}\beta_1$ has homology to the $\alpha_1\beta_1$ and preference to collagen IV over the fibrillar collagens. It is found on chondrocytes, fibroblasts, cells associated with the endosteum and periosteum [64]. The $\alpha_{11}\beta_1$ integrin, found on fibroblasts and subsets of mesenchymal stem cells, resembles the $\alpha_2\beta_1$ integrin with a higher affinity to fibrillar collagens.

Initial studies, in particular to the $\alpha_2\beta_1$ receptor in platelets localised binding to the cyanogen bromide α_1(I) CB3 fragment in collagen I [65]. Specific sequence domains were further explored with the synthesis of collagen peptides, leading to the identification of the first binding sequence, GFOGER [66]. These peptide studies identified the importance of the triple-helical conformation, while identifying the way to establish other binding sites. The development of large synthetic sets of overlapping triple-helical peptide models in the collagen 'Toolkits' [67] has enabled

Table 7.2. Examples of integrin binding sequences that have been identified in triple-helical collagen.

Collagen type	Interacting molecule	Residues/sequence	
I	Integrin $\alpha_1\beta_1$	GFOGER (residues 502–507)	[66]
		GLOGER and GASGER (weak)	[68]
		GROGER, GAOGER (weak)	[61]
	Integrin $\alpha_{11}\beta_1$	GFOGER	[69]
II	Integrin $\alpha_1\beta_1$	GVOGEA	[70]
III	Integrin $\alpha_1\beta_1$	GROGER, GAOGER (weak)	[61]
		GROGER, GLOGEN, GLKGEN (moderate)	[71]
		GLOGEN	[70]
	Integrin $\alpha_2\beta_1$	GROGER, GAOGER (weak)	[61]
		GROGER, GLOGER, GLKGEN	[71]
	Integrin $\alpha_{10}\beta_1$	GLOGEN	[72]
IV	Integrin $\alpha_1\beta_1$	Asp α1(IV)461 and Arg α2(IV)461	[63, 73]
	Integrin $\alpha_2\beta_1$	GFOGER	[66]

many recognition sites to be elucidated on the collagen molecule of different collagen types (table 7.2). Generally, the sequence binding motifs reveal a hexapeptide Gxx′GEx″ sequence, where x is a hydrophobic residue such as phenylalanine (F), y′ is hydroxyproline (O) or sometimes a serine (S) residue and y″ is usually arginine (R), alanine (A) or glutamine (N).

Co-crystallisation of the I-domain of integrin $\alpha_2\beta_1$ with triple-helical peptide containing the GFOGER sequence, $(GPO)_2GFOGER(GPO)_3$, led to the understanding of the structural basis for this metal ion-dependent adhesion site (MIDAS) interaction. It was determined that the binding was with the upper edge of the I-domain, through two of the strands, with the majority of the interactions with the middle strand [74]. Since this study, 2 further molecular structures have been determined for the α_1 and α_2 I-domains with peptide sequences containing GLOGEN and GFOGER. This has provided further insight into the involvement of the glutamic acid in the coordination with the divalent cation, magnesium ion, involved in the integrin ligand interaction [75, 76]. One of these studies used NMR spectroscopy and a triple-helical consensus peptide GLOGEN, which is a high affinity ligand [75] (figure 7.4). These data indicated that the nature of integrin binding may vary with collagen type, but all still use a common transduction mechanism.

Discoidin domain receptors

Triple-helical collagen molecules bind and activate another class of cell surface receptors, the discoidin domain receptors 1 and 2 (DDR1 and DDR2) [77, 78]. These have been named as such, as they have homology in the extracellular domains to the *Dictyostelium discoideum* protein discoidin-1. These are tyrosine kinase receptors that regulate cell adhesion, proliferation and migration. In addition, DDRs are associated with ECM remodelling during developmental processes, in

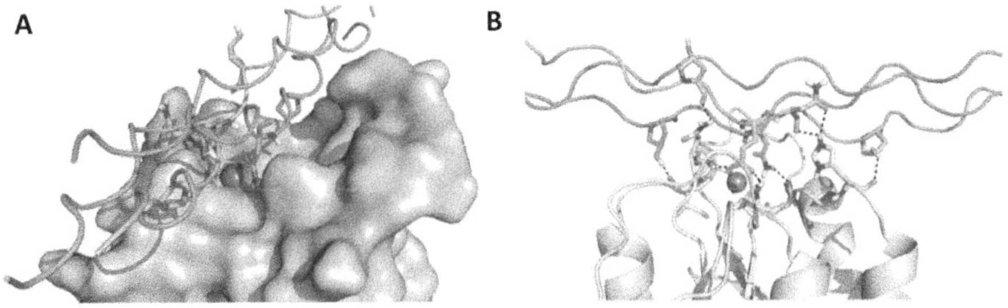

Figure 7.4. The calculated structure of the $\alpha 1$I-GLOGEN complex, with the peptide bound along the edge of a surface trench. A and B show different representations of the lowest energy conformer for the complex. The middle (orange) strand contains the glutamate residue that coordinates with the Mg^{2+}. (Republished from [75], with permission from the American Society for Biochemistry and Molecular Biology, through the Copyright Clearance Center Inc.).

particular mammary glands [79] and growth of long bones [80]. They have also been attributed to human diseases such as organ fibrosis, atherosclerosis, osteoarthritis and cancers [81].

Through the use of Collagen Toolkits II and III both DDR1 and DDR2 were determined to bind to the GVMGFO sequence found in collagen types I, II and III, with DDR1 mainly binding to this sequence whilst DDR2 also binding to two other sites on collagen type II. DDR1 has been found to have a preference for non-fibrillar type IV collagen that is found in basement membranes [82, 83]. The molecular basis for the interaction between a GVMGFO containing triple-helical peptide and DDR2 was determined through crystallographic studies. This identified that the apolar GVMGFO motifs of two of the collagen chains are recognised by the amphiphilic pocket in DDR2.

Immunoglobulin superfamily receptors

Other cellular plasma membrane receptors which are members of the immunoglobulin superfamily of receptors have also been identified to bind to collagen, the glycoprotein immune receptor GPVI on platelets, leukocyte-associated immunoglobulin-like receptor-1, LAIR-1 on immune cells and the osteoclast-associated receptor, OSCAR.

Glycoprotein immune receptor (GPVI)

GPVI receptor is involved in signalling and activation of platelets. Through the use of cyanogen bromide fragments and synthetic peptides, binding regions and recognition sequences have been identified [65, 84, 85]. This receptor binds to collagens on various sites on the collagen molecule with sequences frequently containing GPO triplets with higher affinity binding and aggregation associated with higher order structures.

Leukocyte-associated immunoglobulin-like receptor-1 (LAIR-1)
Collagens I-III and IV are ligands for LAIR-1 [86] and like GPVI binding studies with synthetic peptides from Collagen Toolkits II and III identified recognition sequences with GPO content [87].

Both GPVI and LAIR-1 have similarities in their overlapping collagen binding specificities and high-resolution structures have been determined for both [88, 89]. Computational docking simulations, mutational analysis and studies with nuclear magnetic resonance have elucidated the binding in the immunoglobulin domain with the collagen triple-helical molecule. Both have conserved residues in their binding domains, suggesting common mechanism for collagen recognition.

Osteoclast-associated immunoglobulin-like receptor (OSCAR)
OSCAR is an activating immune receptor expressed in osteoclasts, monocytes and dendritic cells. The crystal structure of the ectodomain was determined both in free form and complexed with a triple-helical peptide, GPOGPAGFO. This direct determination of its molecular interaction of binding revealed differences in the D1 and D2 ectodomain binding to the predicted binding preference of GPVI and LAIR- 1 D1 domains [90].

Von Willebrand factor (vWF)
Von Willebrand factor (vWF) is a plasma protein which mediates binding of the GpIb receptor on platelets with types I, II and III collagens. It plays a crucial role in recruiting and adhering to platelets during blood vessel wall damage. vWF is a multimeric glycoprotein with up to 80 monomers linked by disulphide bonds. Each monomer consists of four types of domains, A, B, C and D with the A domains having the binding sites for collagens. The main collagen binding site is found on the A3 domain [91, 92] with further supporting data from its crystal structure [93]. A high affinity binding site on collagen type III was identified utilising the Collagen Toolkit III approach, identifying RGQOGVMGF as the minimal binding sequence [94] found in both types II and III collagens. Using the identified collagen sequence with flanked GPO triplets, a stable triple-helical peptide was co-crystallised with a recombinant A3 domain to determine its molecular interaction [95]. This crystal structure reveals binding to all 3 α-chains of types I and II collagens and provides further insight into the binding interaction with the heterotrimeric collagen type I molecule. Interestingly, this collagen binding site is overlapping with the binding of two other structurally and distinct proteins, DDR2 and secreted protein acidic and rich in cysteine (SPARC) which is also called osteonectin or Basement Membrane Protein 40 (BM-40) [96, 97]. Other clusters are found and determinates for binding are clustered into four regions along the collagen molecule [4].

7.4.2 Protease sites

In addition to various interactions including those with proteases during biosynthesis of collagen (see chapter 3), there are interaction sites that are specific to the turnover and degradation of collagens. Some of these target the telopeptides and gelatin-like fragments [98, 99] but those involved in the cleavage of an intact triple-helix have perhaps generated the greatest interest.

Matrix metalloproteases (MMP, collagenases)
The degradation of collagens, in particular interstitial types I, II and III collagens, is important in biological events such as embryo development, morphogenesis, tissue repair and remodelling, and in diseases such as cancer, atherosclerosis, fibrosis and arthritis. A group of metalloproteinases referred to as collagenases, MMP-1, MMP-8, MMP-13 and MMP-14 have the ability to cleave the native collagen triple-helix normally resistant to most proteolytic enzymes [100].

The first collagenolytic activity characterised was in the culture medium of a pathogenic anaerobic bacterium, *Clostridium histolyticum* [101, 102]. Sometime later, in 1962, was the discovery of vertebrate collagenase activity in tissues of tadpoles undergoing metamorphosis [103].

The vertebrate collagenases cleave the triple-helix at a single site of the collagen molecule, three quarters from the N terminus [103, 104]. Hydrolysis occurs following a Gly residue in a consensus sequence Gly-(Ile or Ala)-(Ala or Leu) of the mature triple-helix [105, 106]. MMP-1 consists of a catalytic domain with an active zinc ion and with a linker region connected to a C-terminal hemopexin domain. It was through the crystallographic data obtained from MMP-1 [107, 108] and molecular docking to an existing determined structure of a collagen peptide [109] that it became apparent that the cleft of the catalytic site of MMP-1 was too narrow to accommodate the triple-helical collagen rod [110]. They reported that there is likely a partial unwinding of the triple-helix before the peptide bond is hydrolysed. Others have also reported structural instability around this cleavage site due to lower imino acid and low charge residue content making this region more susceptible to proteolysis at 37 °C [106, 111, 112]. Further insights into the triple-helix binding and cleavage mechanism by MMP-1 was provided by structural studies conducted with co-crystallisation of a MMP-1 mutant (E200A) and a triple-helical peptide [113]. This study showed that there are multiple exosite interactions, binding sites distant to the catalytic site which position the scissile bond near the active site. In addition, a binding mode is proposed suggesting an axial rotation of the collagen triple-helix facilitated by the enzyme which allows thermal loosening to enable it to interact with the catalytic cleft. A wide range of recombinant and peptide model systems have been used to study the MMP sites (see chapters 6 and 9).

7.4.3 Other interaction sites

As indicated by the collagen interactome maps [5–8], there are a wide variety of further interaction sites. A small selection of the other sites are discussed briefly below. Expanding the knowledge of binding domains will be valuable in producing new materials through designed, recombinant collagens.

Fibronectin

It has been known that fibronectin (Fn) binds strongly to gelatin, with no difference between collagen types, but it also bonds less strongly to triple-helical collagen [114]. A strongly binding peptide was isolated that came from the sequence region close to the MMP-1 cleavage sequence and the structure of the complex with a Fn fragment was established [115]. Later studies, using peptide sequences inserted into a triple-helical collagen structures [116, 117], allowed studies on the less well understood binding to triple-helical collagen. These studies showed using a type II collagen sequence that a minimum 6-triplet sequence was needed, ^{775}GLAGQRGIVGLPGQRGER^{792}, that is highly homologous to the equivalent type I collagen sequence [116], and that fibronectin bound to the triple-helical form, albeit not as strongly as when the collagen was denatured to give a gelatin [116]. Further studies, using Gly to Ser mutations of the inserted sequences, along with modelling studies [117], showed that certain mutations reduced binding efficiency for the gelatin chain, while the triple-helical form showed a significant increase in proteolytic sensitivity to trypsin [117].

Thrombospondins

Thrombospondins are calcium-binding glycoproteins that bind with collagens and other ECM molecules [118]. The binding to collagens is particularly to fibrillar collagens [119]. The binding to type V collagen resides in a single chain 19-kDa fragment, and is inhibited by heparin [119, 120].

Pathogens

A wide variety of pathogens are able to bind to collagens, generally to assist in the mechanism of infection. The molecules involved in this binding, bacterial adhesins [121, 122], have been the subject of structural and functional studies, but the application of these binding properties to biomaterials development is limited to date.

7.5 Interactions with the immune system

A potentially quite significant interaction of collagen is when a collagen-based biomedical device is introduced into a xenogeneic host. The interaction proceeds with a complex system of different types of T cells, B cells and antigen presenting cells. The T cells are obligatory, as shown through experiments using athymic nude mice [123]. Subsequently, additional supporting evidence for T cell involvement has been obtained [124]. In addition to the humoral, or antibody response, cell-mediated immunity (CMI) or delayed type hypersensitivity (DTH) responses may occur [124]. For example, injection into animals of collagen in complete Freund's adjuvant can produce a long lasting DTH response independent of antibody generation [125]. The antigenic determinants involved in cellular type responses were different to those described for the humoral response to collagens [124]. Clearly, autologous collagen would be advantageous, whether derived from tissue or by recombinant production.

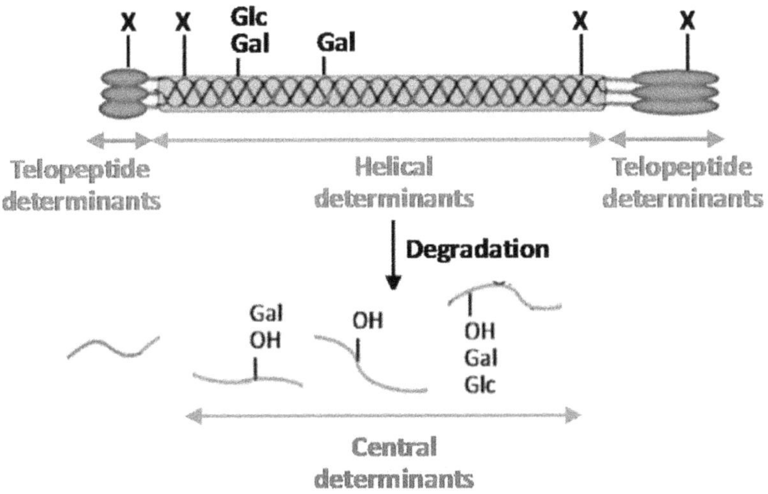

Figure 7.5. Structure and antigenic domains of interstitial collagens. X represents the locations of potential crosslink sites, and Gal (galactose) and Glc (glucose) represent potential glycosylation of hydroxylysine residues.

Notwithstanding, any immunological response will determine the efficacy of the device—it has to be 'immunologically inert'.

It has long been known that collagens, especially the most abundant type I that is generally the main component of collagen-based devices, are poorly immunogenic [126], particularly when compared to most globular proteins. This possibly reflects some of the unique structural features of collagen, including the tightly wound polypeptide chains, side chain accessibility and the low aromatic amino acid content. Data initially arose from animal studies, but more recently data on implants in humans has been accumulating, especially on gels for tissue augmentation. However, although collagen-based devices have proved successful in many clinical applications, issues are still raised about their immunological safety. Therefore, understanding the nature of any response so that it can be minimised and ideally eliminated is important.

Antigenic determinants in globular proteins had been classified as either 'sequential' or 'conformational' [127]. Sequential determinants are linear segments of amino acid sequence, independent of any secondary or tertiary structural features of the protein. In contrast, conformational determinants are dependent on higher order structures, such as secondary and tertiary structures, that bring amino acids from disparate, non-continuous parts of the sequence into defined local features.

For collagens, both types of determinants have been identified. Sequential determinants, based on sections of unwound, non-native collagen chains have been observed, infrequently, in certain species, such as chicken [128]. For collagens, these are termed 'central' determinants (figure 7.5). Conformational determinants have also been found, and predominate in other species such as mouse, and depend on having an intact triple-helix [129]. These determinants are termed 'helical determinants' (figure 7.5) and are lost when the collagen is denatured. The rigid rod-like triple-helical structure of collagen means that different regions of the structure are unable to come together and fold to give a conformational determinant like those found for globular proteins. However, it is the close proximity of the individual collagen chains in the triple-helix that forms the conformation dependent domain. A structural study of a collagen binding site for an integrin shows that most interactions arise from a single chain but that the proper folded conformation is essential [74]. Hypothetically, it is possible that collagen fibrils in tissues could provide conformation determinants formed from adjacent collagen molecules, but to date there is no evidence for this type of determinant.

For the processed, soluble collagens that are used in certain biomedical applications, the processing steps themselves can have immunological implications. In particular, the use of enzymes to produce soluble collagen (see chapter 8) removes much of the telopeptides, which, with their aromatic content and greater chain mobility, are the more immunogenic parts of the collagen structure [128].

On the other hand, the removal of the telopeptides will generate neo-epitopes at the N- or C- terminal of the collagen [130] and the residues of crosslinked telopeptides attached to the helix may increase the response [131]. Similarly, denaturation of collagen also exposes neo-epitopes, and could occur during degradation of collagen-based biomaterials, but it is unlikely to lead to antibodies that will cross react with native host collagens and lead to an autoimmune response. Nevertheless, the most likely immunological response to a collagen-based biomaterial is not from the collagen itself, but rather impurities, such as serum proteins, that are present [132].

For the vast majority of collagen-based devices, chemical stabilisation is a necessary step, and this has the potential benefit of reducing and potentially eliminating immunogenicity. In particular, processing of tissue-based materials should include steps to remove immunogenic non-collagenous proteins prior to stabilisation, but this may not always happen in proprietary processes.

7.5.1 Responses to collagen-based products

A range of collagen-based devices, using collagen from bovines and from other species, have been tested in animals and reported as non-immunogenic. These include, for example, a bovine collagen sponge, Haemostagen (Bioetica, Lyon, France) [133], an equine collagen haemostatic and wound dressing (Tachotop) [134] and a porcine collagen membrane (Bio-Gide) [135]. Collagens from other, non-mammalian, more primitive animal species have been suggested for use in biomaterials, but with the exception of avian collagen [131] few immunological studies have been reported.

The most extensively used collagen is bovine type I collagen, especially as a gel/fibrous dispersion for soft tissue augmentation (see chapter 14). Although initially examined in animal systems, these materials provide the vast majority of the data that is available on the human response to an animal collagen.

The initial animal studies in guinea pigs showed that an injectable bovine collagen, Zyderm, was minimally immunogenic [132]. However, comparisons to other bovine collagen-based products, which were haemostats, (Avitene, Collastat, Gelfoam) showed a much greater response, with Avitene showing the greatest response through reactions to several components of the material, including bovine serum albumin [132].

In humans injected with bovine collagen, no single immunodominant region on the collagen was detected and it appeared that there were immuno-reactive sites along the entire length of the collagen triple-helix, with potential hot spots in α1(I)-CB6 and α2(I)-CB4 zones [136]. A library of monoclonal antibodies (MAbs) to type III collagen is a good example that shows this spread of helical determinants along the full length of the molecule (figure 7.6), as well as showing that type III collagen, which can also be present in skin derived preparations, is considerably more immunogenic than type I collagen [137]. The epitope recognised by one of these antibodies, #7 (2G8-B1/Col3), has been determined using peptide studies; this MAb has the interesting effect of inhibiting platelet adhesion to type III collagen [138] and is in a similar, equivalent location to the GFOGER site seen in type I collagen [66].

Significant studies using Zyplast and Zyderm I or II products (then made by Collagen Corporation) and a comparable product made by Koken have been reported; these products are pepsin-soluble collagen from bovine dermis [140–143]. Prior to clinical treatment, patients were subjected to a small dose skin test. About 3% of patients showed hypersensitivity as an underlying reaction [144, 145] and indicated that they are not suited to full treatment. A strong association was found between the presence of the anti-bovine collagen antibodies and positive skin testing. Despite this test, some 1% to 2% of non-responding patients subsequently show clinical symptoms of hypersensitivity [142, 145]. The incidence of immunoreactivity was reasonably similar for the different manufacturers [142, 143], but higher incidences were found when a similar but GA crosslinked material was used as a bulking agent for treating urinary stress incontinence [146].

Autoimmune issues

An important question regarding the immunogenicity of collagen-based biomaterials is whether there is cross-reactivity in any immune response to the implanted collagen to allow the development of autoimmune diseases. However, there are many autoimmune and heritable diseases where anti-collagen antibodies have been detected and could be associated with the disease in the absence of a collagen implant. For these clinical issues, it is not clear whether the anti-collagen type

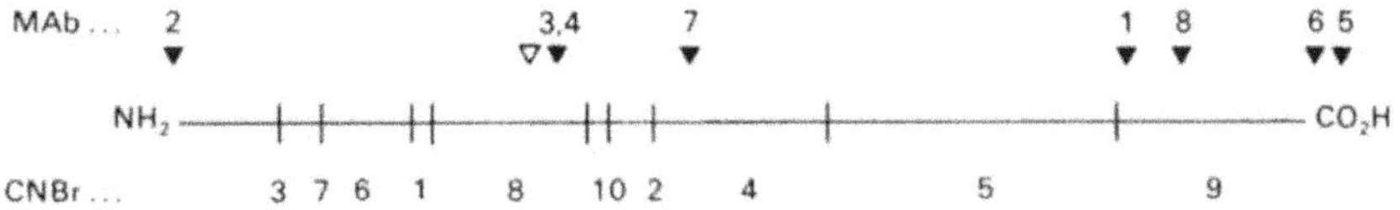

Figure 7.6. The location of epitopes on type III collagen showing the possible locations of the epitopes recognised by the eight MAbs from one study [137] along with that from a previous study (∇) [139]. (From [137], reprinted with permission from Portland Press ©1991).

antibodies are responsible for the emergence of the disease or a consequence of the progression of the disease [124]. The autoimmune diseases that have attracted some attention are two very rare conditions that involve the destruction of muscle tissue, polymyositis and dermatomyositis (PM/DM). Do collagen injections increase their frequency? Epidemiological analyses of large numbers of patients receiving collagen implants have broadly found no scientific data to support the alleged link between PM/DM and collagen injections [147–149].

Certainly, a patient immune response may already exist to collagen for unknown reasons or may develop merely as a result of disease state or due to the surgery performed for implantation. So, patients with pre-existing autoimmune type disorders would probably have an elevated immunity and stringent testing prior to further deliberate collagen priming would be strongly advised, along with on-going vigilance of data that emerges [124].

7.6 Antibodies to collagens as biochemical reagents

Collagens are typically very poor immunogens for production of polyclonal and monoclonal antibodies (MAbs). Production of reagents has always required use of adjuvants to enhance the persistence and presentation of the immunogen. This is usually Freund's complete adjuvant, along with high dose, repetitive injections. For other proteins this can lead in many cases to high titre antibodies or monoclonal antibodies with good affinity. For collagens, this approach can help, but this is not universally the case.

7.6.1 Polyclonal antibodies

Various researchers have developed polyclonal antibodies as reagents. However, in their initial form they tend to lack specificity. Hence to make them useful, for example specific to a particular collagen type, various purification approaches must be used, including affinity chromatography on the corresponding collagen type followed by assessments using immunoblotting and by radioimmunoassay [150, 151]. These reagents remain very valuable, but in some cases have been superseded by MAbs, where particular features are required, such as species specificity.

7.6.2 Monoclonal antibodies

Monoclonal antibody technology [152, 153] emerged as a method that could generate antibody reagents that were highly specific to both the collagen type and its species of origin. Initially, MAbs were derived against chicken collagens, principally for studies on the eye [154–157]. Subsequently, MAbs were developed against human and mammalian collagens, which are more relevant for the study of collagen-based biomaterials; in some cases, these MAbs were specifically developed for studying biomaterial performance [130, 137, 158, 159]. These MAbs were not only specific to individual types, but also exhibited different species specificities so that both the persistence or the augmentation of collagen-based implanted devices could be examined [160].

7.6.3 Defining epitopes

In certain cases, the location of the immunological reactive site, the epitope, on the collagen molecule has been defined, with a number of different methods being useful for this process [161]. Knowledge of epitope locations can be useful, for example in defining regions of importance in diseases such as in rheumatoid arthritis [162], and explaining the biological properties of the antibodies [138].

For sequential, central determinants, a peptide scan [163] provides a reliable method to map epitopes, and this approach has been used, for example, for studies on type II collagen [164]. For the conformation dependent, helical determinants, epitope location and identification can be more difficult, especially for heterotrimeric collagens.

Imaging of antibody–collagen complexes using transmission electron microscopy has been used for many collagen types to identify the region of binding using samples prepared by using rotary shadowing [161, 165]. Atomic force microscopy could provide and alternative approach [166]. The main difficulty for interstitial collagens is in identifying the N- from the C-terminal of the collagen in the images. If precursor forms of collagen are available, they can resolve this issue, else the imaging data needs to be combined with evidence from other methods. The resolution which can be obtained by rotary shadowing potentially allows an estimate of the position of the epitope to within 50 residues. For homotrimer collagens, and possibly heterotrimers if there is a dominant chain, examination of reaction to refolded fragments can be useful. The fragments could be the two that can be formed by specific digestion with MMP-1 or MMP-8 or else by CNBr fragments, either separated by chromatography or on SDS-PAGE gels followed by Western blotting. Both these approaches identify only the broad regions of the interaction but can be valuable in conjunction with rotary shadowing data. Type III collagen is unusual in that the three chains are covalently linked at the C-terminal by disulphide bonds, so that when the collagen is denatured the three chains remain linked and refold readily on cooling. If proteolysis is used during the refolding, a series of refolded fragments that increase in size from the C-terminal of the molecule are formed, which can then be analysed for binding, for example, using an ELISA assay, giving the relative positions of epitopes. These data can be used along with comparisons of sequence data. In some cases, a single change relative to the mouse sequence may be the only difference in structure that originally led to the generation of a MAb. Although not definitive, these comparisons should be consistent with observed species specificities and can allow selection of peptides for synthesis that define an accurate location [138].

References

[1] Faye C, Chautard E, Olsen B R and Ricard-Blum S 2009 *J. Biol. Chem.* **284** 22041–7

[2] An B and Brodsky B 2016 *Blood* **127** 521–2

[3] Brodsky B and Ramshaw J A M 2017 *Subcell. Biochem.* **82** 601–29

[4] An B, Lin Y S and Brodsky B 2016 *Adv. Drug Deliv. Rev.* **97** 69–84

[5] Di Lullo G A, Sweeney S M, Korkko J, Ala-Kokko L and San Antonio J D 2002 *J. Biol. Chem.* **277** 4223–31

[6] Sweeney S M, Orgel J P, Fertala A, McAuliffe J D, Turner K R, Di Lullo G A, Chen S, Antipova O and Perumal S *et al* 2008 *J. Biol. Chem.* **283** 21187–97

[7] Parkin J D, San Antonio J D, Pedchenko V, Hudson B, Jensen S T and Savige J 2011 *Hum. Mutat.* **32** 127–43

[8] Parkin J D, San Antonio J D, Persikov A V, Dagher H, Dalgleish R, Jensen S T, Jeunemaitre X and Savige J 2017 *PLoS One* **12** e0175582

[9] Orgel J P R O, Irving T, Miller A and Wess T 2006 *Proc. Natl Acad. Sci. USA* **103** 9001–5

[10] Orgel J P, San Antonio J D and Antipova O 2011 *Connect. Tissue Res.* **52** 2–17

[11] Orgel J P, Antipova O, Sagi I, Bitler A, Qiu D, Wang R, Xu Y and San Antonio J D 2011 *Connect. Tissue Res.* **52** 18–24

[12] Cescon M, Gattazzo F, Chen P and Bonaldo P 2015 *J. Cell Sci.* **128** 3525–31

[13] Ricard-Blum S, Dublet B and Van der Rest M 2000 *Unconventional collagens: Types VI, VII, VIII, IX, X, XII, XIV, XVI and XIX* (Oxford: Oxford University Press)

[14] Bruckner P, Mendler M, Steinmann B, Huber S and Winterhalter K H 1988 *J. Biol. Chem.* **263** 16911–7

[15] Ruggiero F, Petit B, Ronziere M C, Farjanel J, Hartmann D J and Herbage D 1993 *J. Histochem. Cytochem.* **41** 867–75

[16] Margalit H, Fischer N and Ben-Sasson S A 1993 *J. Biol. Chem.* **268** 19228–31

[17] Fromm J R, Hileman R E, Caldwell E E, Weiler J and Linhardt R J 1997 *Arch. Biochem. Biophys.* **343** 92–100

[18] Rullo A and Nitz M 2010 *Biopolymers* **93** 290–8

[19] Doss-Pepe E, Deprez P, Silva T, Inestrosa N C, Kirkpatrick A, Ramshaw J A M and Brodsky B 2004 *Biochim. Biophys. Acta* **1698** 187–95

[20] San Antonio J D, Lander A D, Karnovsky M J and Slayter H S 1994 *J. Cell Biol.* **125** 1179–88

[21] Koda J E and Bernfield M 1984 *J. Biol. Chem.* **259** 11763–70

[22] LeBaron R G, Höök A, Esko J D, Gay S and Höök M 1989 *J. Biol. Chem.* **264** 7950–6

[23] Wenstrup R J, Florer J B, Brunskill E W, Bell S M, Chervoneva I and Birk D E 2004 *J. Biol. Chem.* **279** 53331–7

[24] White J F, Werkmeister J A, Bisucci T, Darby I A and Ramshaw J A M 2014 *J. Biomed. Mater. Res.* A **102** 3550–5

[25] Sweeney S M, Guy C A, Fields G B and San Antonio J D 1998 *Proc. Natl Acad. Sci. USA* **95** 7275–80

[26] Koliakos G G, Kouzi-Koliakos K, Furcht L T, Reger L A and Tsilibary E C 1989 *J. Biol. Chem.* **264** 2313–23

[27] Tu H, Sasaki T, Snellman A, Göhring W, Pirilä P, Timpl R and Pihlajaniemi T 2002 *J. Biol. Chem.* **277** 23092–9

[28] Deprez P N and Inestrosa N C 1995 *J. Biol. Chem.* **270** 11043–6

[29] Deprez P, Inestrosa N C and Krejci E 2003 *J. Biol. Chem.* **278** 23233–42

[30] Deprez P, Doss-Pepe E, Brodsky B and Inestrosa N C 2000 *Biochem. J.* **350** 283–90

[31] Doss-Pepe E, Deprez P, Inestrosa N C and Brodsky B 2000 *Biochemistry* **39** 14884–92

[32] Peng Y Y, Stoichevska V, Schacht K, Werkmeister J A and Ramshaw J A M 2014 *J. Biomed. Mater. Res.* A **102** 2189–96

[33] Scott J E 1988 *Biochem. J.* **252** 313–23

[34] Obrink B 1973 *Eur. J. Biochem.* **33** 387–400
[35] Scott J E 1980 *Biochem. J.* **187** 887–91
[36] Scott J E 1980 *J. Microsc.* **119** 373–81
[37] Scott J E and Orford C R 1981 *Biochem. J.* **197** 213–6
[38] Scott J E and Haigh M 1985 *Biosci. Rep.* **5** 765–74
[39] Scott J E and Haigh M 1985 *Biosci. Rep.* **5** 71–81
[40] Young R D 1985 *J. Cell Sci.* **74** 95–104
[41] Orford C R and Gardner D 1984 *Connect. Tissue Res.* **12** 345–8
[42] Ruoslahti E 1988 *Annu. Rev. Cell Biol.* **4** 229–55
[43] Pogány G and Vogel K G 1992 *Biochem. Biophys. Res. Commun.* **189** 165–72
[44] Keene D R, San Antonio J D, Mayne R, McQuillan D J, Sarris G, Santoro S A and Iozzo R V 2000 *J. Biol. Chem.* **275** 21801–4
[45] Scott P G, McEwan P A, Dodd C M, Bergmann E M, Bishop P N and Bella J 2004 *Proc. Natl Acad. Sci. USA* **101** 15633–8
[46] Federico S, Pierce B F, Piluso S, Wischke C, Lendlein A and Neffe A T 2015 *Angew. Chem., Int. Ed. Engl.* **54** 10980–4
[47] Laurie G W, Bing J T, Kleinman H K, Hassell J R, Aumailley M, Martin G R and Feldmann R J 1986 *J. Mol. Biol.* **189** 205–16
[48] Rothblum K, Stahl R C and Carey D J 2004 *J. Biol. Chem.* **279** 51282–8
[49] Yaoi Y, Hashimoto K, Koitabashi H, Takahara K, Ito M and Kato I 1990 *Biochim. Biophys. Acta* **1035** 139–45
[50] Delacoux F, Fichard A, Geourjon C, Garrone R and Ruggiero F 1998 *J. Biol. Chem.* **273** 15069–76
[51] Delacoux F, Fichard A, Cogne S, Garrone R and Ruggiero F 2000 *J. Biol. Chem.* **275** 29377–82
[52] Ricard-Blum S, Beraud M, Raynal N, Farndale R W and Ruggiero F 2006 *J. Biol. Chem.* **281** 25195–204
[53] Warner L R, Brown R J, Yingst S M and Oxford J T 2006 *J. Biol. Chem.* **281** 39507–16
[54] Vaughan-Thomas A, Young R D, Phillips A C and Duance V C 2001 *J. Biol. Chem.* **276** 5303–9
[55] Montserret R, Aubert-Foucher E, McLeish M J, Hill J M, Ficheux D, Jaquinod M, van der Rest M, Deléage G and Penin F 1999 *Biochemistry* **38** 6479–88
[56] Brown J C, Mann K, Wiedemann H and Timpl R 1993 *J. Cell Biol.* **120** 557–67
[57] Myers J C, Li D, Amenta P S, Clark C C, Nagaswami C and Weisel J W 2003 *J. Biol. Chem.* **278** 32047–57
[58] Hynes R O 2002 *Cell* **110** 673–87
[59] Barczyk M, Carracedo S and Gullberg D 2010 *Cell Tissue Res.* **339** 269–80
[60] Zeltz C and Gullberg D 2016 *J. Cell Sci.* **129** 653–64
[61] Kim J K, Xu Y, Xu X, Keene D R, Gurusiddappa S, Liang X, Wary K K and Höök M 2005 *J. Biol. Chem.* **280** 32512–20
[62] Gardner H 2014 *Adv. Exp. Med. Biol.* **819** 21–39
[63] Eble J A, Golbik R, Mann K and Kühn K 1993 *EMBO J.* **12** 4795–802
[64] Varas L, Ohlsson L B, Honeth G, Olsson A, Bengtsson T, Wiberg C, Bockermann R, Järnum S and Richter J *et al* 2007 *Stem. Cells Dev.* **16** 965–78
[65] Morton L F, Zijenah L S, McCulloch I Y, Knight C G, Humphries M J and Barnes M J 1991 *Biochem. Soc. Trans.* **19** 439S

[66] Knight C G, Morton L F, Peachey A R, Tuckwell D S, Farndale R W and Barnes M J 2000 *J. Biol. Chem.* **275** 35–40
[67] Farndale R W, Lisman T, Bihan D, Hamaia S, Smerling C S, Pugh N, Konitsiotis A, Leitinger B and de Groot P G *et al* 2008 *Biochem. Soc. Trans.* **36** 241–50
[68] Xu Y, Gurusiddappa S, Rich R L, Owens R T, Keene D R, Mayne R, Höök A and Höök M 2000 *J. Biol. Chem.* **275** 38981–9
[69] Zhang W M, Kapyla J, Puranen J S, Knight C G, Tiger C F, Pentikainen O T, Johnson M S, Farndale R W, Heino J and Gullberg D 2003 *J. Biol. Chem.* **278** 7270–7
[70] Hamaia S W, Pugh N, Raynal N, Némoz B, Stone R, Gullberg D, Bihan D and Farndale R W 2012 *J. Biol. Chem.* **287** 26019–28
[71] Raynal N, Hamaia S W, Siljander P R, Maddox B, Peachey A R, Fernandez R, Foley L J, Slatter D A, Jarvis G E and Farndale R W 2006 *J. Biol. Chem.* **281** 3821–31
[72] Hamaia S W, Luff D, Hunter E J, Malcor J D, Bihan D, Gullberg D and Farndale R W 2017 *Matrix Biol.* **59** 80–94
[73] Golbik R, Eble J A, Ries A and Kühn K 2000 *J. Mol. Biol.* **297** 501–9
[74] Emsley J, Knight C G, Farndale R W, Barnes M J and Liddington R C 2000 *Cell* **101** 47–56
[75] Chin Y K, Headey S J, Mohanty B, Patil R, McEwan P A, Swarbrick J D, Mulhern T D, Emsley J, Simpson J S and Scanlon M J 2013 *J. Biol. Chem.* **288** 36796–809
[76] Carafoli F, Hamaia S W, Bihan D, Hohenester E and Farndale R W 2013 *PLoS One* **8** e69833
[77] Vogel W, Gish G D, Alves F and Pawson T 1997 *Mol. Cell* **1** 13–23
[78] Shrivastava A, Radziejewski C, Campbell E, Kovac L, McGlynn M, Ryan T E, Davis S, Goldfarb M P and Glass D J *et al* 1997 *Mol. Cell* **1** 25–34
[79] Vogel W F, Aszódi A, Alves F and Pawson T 2001 *Mol. Cell. Biol.* **21** 2906–17
[80] Labrador J P, Azcoitia V, Tuckermann J, Lin C, Olaso E, Mañes S, Brückner K, Goergen J L, Lemke G and Yancopoulos G 2001 *EMBO Rep.* **2** 446–52
[81] Vogel W F, Abdulhussein R and Ford C E 2006 *Cell Signal.* **18** 1108–16
[82] Konitsiotis A D, Raynal N, Bihan D, Hohenester E, Farndale R W and Leitinger B 2008 *J. Biol. Chem.* **283** 6861–8
[83] Xu H, Raynal N, Stathopoulos S, Myllyharju J, Farndale R W and Leitinger B 2011 *Matrix Biol.* **30** 16–26
[84] Kehrel B 1995 *Semin. Thromb. Hemost.* **21** 123–9
[85] Jarvis G E, Raynal N, Langford J P, Onley D J, Andrews A, Smethurst P A and Farndale R W 2008 *Blood* **111** 4986–96
[86] Lebbink R J, De Ruiter T, Adelmeijer J, Brenkman A B, van Helvoort J M, Koch M, Farndale R W, Lisman T and Sonnenberg A *et al* 2006 *J. Exp. Med.* **203** 1419–25
[87] Lebbink R J, Raynal N, de Ruiter T, Bihan D G, Farndale R W and Meyaard L 2009 *Matrix Biol.* **28** 202–10
[88] Horii K, Kahn M L and Herr A B 2006 *Blood* **108** 936–42
[89] Brondijk T H, de Ruiter T, Ballering J, Wienk H, Lebbink R J, van Ingen H, Boelens R, Farndale R W, Meyaard L and Huizinga E G 2010 *Blood* **115** 1364–73
[90] Zhou L, Hinerman J M, Blaszczyk M, Miller J L, Conrady D G, Barrow A D, Chirgadze D Y, Bihan D, Farndale R W and Herr A B 2016 *Blood* **127** 529–37
[91] Cruz M A, Yuan H, Lee J R, Wise R J and Handin R I 1995 *J. Biol. Chem.* **270** 10822–7

[92] Lankhof H, van Hoeij M, Schiphorst M E, Bracke M, Wu Y P, Ijsseldijk M J, Vink T, de Groot P G and Sixma J J 1996 *Thromb. Haemost.* **75** 950–8
[93] Huizinga E G, Martijn van der Plas R, Kroon J, Sixma J J and Gros P 1997 *Structure* **5** 1147–56
[94] Lisman T, Raynal N, Groeneveld D, Maddox B, Peachey A R, Huizinga E G, de Groot P G and Farndale R W 2006 *Blood* **108** 3753–6
[95] Brondijk T H, Bihan D, Farndale R W and Huizinga E 2012 *Proc. Natl Acad. Sci. USA* **109** 5253–8
[96] Giudici C, Raynal N, Wiedemann H, Cabral W A, Marini J C, Timpl R, Bächinger H-P, Farndale R W, Sasaki T and Tenni R 2008 *J. Biol. Chem.* **283** 19551–60
[97] Hohenester E, Sasaki T, Giudici C, Farndale R W and Bächinger H-P 2008 *Proc. Natl Acad. Sci. USA* **105** 18273–7
[98] Murphy G and Nagase H 2011 *FEBS J.* **278** 2–15
[99] Yamamoto K, Murphy G and Troeberg L 2015 *Matrix Biol.* **44–46** 255–63
[100] Visse R and Nagase H 2003 *Circ. Res.* **92** 827–39
[101] Jennison M W 1947 *J. Bacteriol.* **54** 55
[102] Mandl I, Maclennan J D and Howes E L 1953 *J. Clin. Invest.* **32** 1323–9
[103] Gross J and Lapiere C M 1962 *Proc. Natl Acad. Sci. USA* **48** 1014–22
[104] Miller E J, Harris E D, Chung E, Finch J E, McCroskery P A and Butler W T 1976 *Biochemistry* **15** 787–92
[105] Gross J, Harper E, Harris E D, McCroskery P A, Highberger J H, Corbett C and Kang A H 1974 *Biochem. Biophys. Res. Commun.* **61** 605–12
[106] Fields G B 1991 *J. Theor. Biol.* **153** 585–602
[107] Li J, Brick P, O'Hare M C, Skarzynski T, Lloyd L F, Curry V A, Clark I M, Bigg H F and Hazleman B L *et al* 1995 *Structure* **3** 541–9
[108] Iyer S, Visse R, Nagase H and Acharya K R 2006 *J. Mol. Biol.* **362** 78–88
[109] Kramer R Z, Bella J, Brodsky B and Berman H M 2001 *J. Mol. Biol.* **311** 131–47
[110] Chung L, Dinakarpandian D, Yoshida N, Lauer-Fields J L, Fields G B, Visse R and Nagase H 2004 *EMBO J.* **23** 3020–30
[111] Gross J and Nagai Y 1965 *Proc. Natl Acad. Sci. USA* **54** 1197–204
[112] Brown R A, Hukins D W, Weiss J B and Twose T M 1977 *Biochem. Biophys. Res. Commun.* **74** 1102–8
[113] Manka S W, Carafoli F, Visse R, Bihan D, Raynal N, Farndale R W, Murphy G, Enghild J J, Hohenester E and Nagase H 2012 *Proc. Natl Acad. Sci. USA* **109** 12461–16
[114] Dessau W, Adelmann B C, Timpl R and Martin G R 1978 *Biochem. J.* **169** 55–9
[115] Erat M C, Slatter D A, Lowe E D, Millard C J, Farndale R W, Campbell I D and Vakonakis I 2009 *Proc. Natl. Acad. Sci. USA* **106** 4195–200
[116] An B, Abbonante V, Yigit S, Balduini A, Kaplan D L and Brodsky B 2014 *J. Biol. Chem.* **289** 4941–51
[117] Chhum P, Yu H, An B, Doyon B R, Lin Y S and Brodsky B 2016 *J. Biol. Chem.* **291** 27073–86
[118] Adams J C and Lawler J 2011 *Cold Spring Harb. Perspect. Biol.* **3** a009712
[119] Galvin N J, Vance P M, Dixit V M, Fink B and Frazier W A 1987 *J. Cell Biol.* **104** 1413–22
[120] Takagi J, Fujisawa T, Usui T, Aoyama T and Saito Y 1993 *J. Biol. Chem.* **268** 15544–9
[121] Berisio R and Vitagliano L 2012 *Curr. Protein Pept. Sci.* **13** 855–65

[122] Avilés-Reyes A, Miller J H, Lemos J A and Abranches J 2017 *Mol. Oral Microbiol.* **32** 89–106

[123] Nowack H, Hahn E and Timpl R 1976 *Immunology* **30** 29–32

[124] Werkmeister J A and Ramshaw J A M 2000 *Handbook of Biomaterials Engineering* ed D L Wise (New York: Marcel Dekker) pp 739–59

[125] Adelmann B C, Kirrane J and Glynn L E 1972 *Immunology* **23** 723–77

[126] Furthmayr H and Timpl R 1976 *Int. Rev. Connect. Tissue Res.* **7** 61–99

[127] Sela M, Schechter B, Schetcher M and Borek F 1967 *Cold Spring Harbor Symp. Quant. Biol.* **32** 537–45

[128] Furthmayr H, Stoltz M, Becker U, Beil W and Timpl R 1972 *Immunochemistry* **9** 789–98

[129] Beil W, Timpl R and Furthmayr H 1973 *Immunology* **24** 13–24

[130] Werkmeister J A, Ramshaw J A M and Ellender G 1990 *Eur. J. Biochem.* **187** 439–43

[131] Peng Y Y, Glattauer V, Ramshaw J A M and Werkmeister J A 2010 *J. Biomed. Mater. Res.* A **93** 1235–44

[132] DeLustro F, Condell R A, Nguyen M A and McPherson J M 1986 *J. Biomed. Mater. Res.* **20** 109–20

[133] Anselme K, Bacques C, Charriere G, Hartmann D J, Herbage D and Garrone R 1990 *J. Biomed. Mater. Res.* **24** 689–703

[134] Adelmann-Grill B C and Otto K 1987 *Arzneimittel-Forsch.* **37** 802–5

[135] Schlegal A K, Möhler H, Busch F and Mehl A 1997 *Biomaterials* **18** 535–8

[136] Ellingsworth L R, DeLustro F, Brennan J E, Sawamura S and McPherson J M 1986 *J. Immunol.* **136** 877–82

[137] Werkmeister J A and Ramshaw J A M 1991 *Biochem.* J **274** 895–8

[138] Glattauer V, Werkmeister J A, Kirkpatrick A and Ramshaw J A M 1997 *Biochem. J.* **323** 45–9

[139] Keene D R, Sakai L Y, Burgeson R E and Bächinger H-P 1987 *J. Histochem. Cytochem.* **35** 311–8

[140] Cooperman L and Michaeli D 1984 *J. Am. Acad. Dermat.* **10** 647–51

[141] McCoy J P, Schade W, Siegle R J, Waldinger T P, Vanderveen E E and Swanson N A 1985 *Arch. Dermatol.* **121** 990–4

[142] Keefe J, Wauk L, Chu S and DeLustro F 1992 *Clin. Mat.* **6** 155–62

[143] Charriere G, Bejot M, Schnitzler L, Ville G and Hartmann D J 1989 *J. Am. Acad. Dermatol.* **21** 1203–8

[144] DeLustro F, Mackinnon V and Swanson N A 1988 *J. Dermatol. Surg. Oncol.* **14** 49–56

[145] Cooperman L, Mackinnon V, Bechler G and Pharris B B 1985 *Aesth. Plastic Surg.* **9** 145–52

[146] McClelland M and DeLustro F 1996 *J. Urol.* **155** 2068–73

[147] Elson M L 1993 *J. Dermatol. Surg. Oncol.* **19** 165–8

[148] Rosenberg M J and Reichlin M 1994 *Arth. Rheum.* **37** 747–53

[149] Hanke C W, Thomas J A, Lee W T, Jolivette D M and Rosenberg M J 1996 *J. Am. Acad. Dermatol.* **34** 450–4

[150] Magloire H, Callé A, Hartmann D J, Joffre A, Serre B, Grimaud J A and Schué F 1986 *Cell Tissue Res.* **244** 133–40

[151] Magloire H, Joffre A and Hartmann D J 1988 *Histochem.* **88** 141–9

[152] Köhler G and Milstein C 1975 *Nature* **256** 495–7

[153] Galfrè G and Milstein C 1981 *Meth. Enzymol.* **73** 3–46

[154] Linsenmayer T F, Hendrix M J and Little C D 1979 *Proc. Natl Acad. Sci. USA* **76** 3703–7
[155] Linsenmayer T F and Hendrix M J 1980 *Biochem. Biophys. Res. Commun.* **92** 440–6
[156] Linsenmayer T F, Fitch J M, Schmid T M, Zak N B, Gibney E, Sanderson R D and Mayne R 1983 *J. Cell Biol.* **96** 124–32
[157] Fitch J M, Gibney E, Sanderson R D, Mayne R and Linsenmayer T F 1982 *J. Cell Biol.* **95** 641–7
[158] Werkmeister J A, Tebb T A, White J F and Ramshaw J A M 1993 *J. Histochem. Cytochem.* **41** 1701–6
[159] Werkmeister J A and Ramshaw J A M 1991 *J. Histochem. Cytochem.* **39** 1215–20
[160] Werkmeister J A, Glattauer V, Tebb T A, Ramshaw J A M, Edwards G A and Roberts G 1991 *J. Long Term Eff. Med. Implants* **1** 107–19
[161] Glattauer V, Ramshaw J A M, Tebb T A and Werkmeister J A 1991 *Int. J. Biol. Macromol.* **13** 140–6
[162] Cook A D, Gray R, Ramshaw J A M, Mackay I R and Rowley M J 2004 *Arthritis Res. Ther.* **6** R477–83
[163] Geysen H M, Rodda S J, Mason T J, Tribbick G and Schoofs P G 1987 *J. Immunol. Meth.* **102** 259–74
[164] Morgan K, Turner S L, Reynolds I, Hajeer A H, Brass A and Worthington J 1992 *Immunology* **77** 609–16
[165] Mayne R, Wiedemann H, Irwin M H, Sanderson R D, Fitch J M, Linsenmayer T F and Kühn K 1984 *J. Cell Biol.* **98** 1637–44
[166] Avci R, Schweitzer M, Boyd R D, Wittmeyer J, Steele A, Toporski J, Beech I, Arce F T and Spangler B *et al* 2004 *Langmuir* **20** 11053–63

Part III

The production of collagen-based biomaterials

John A M Ramshaw and Veronica Glattauer

Chapter 8

Production of tissue-derived collagens

Purification is still highly relevant, especially because of the commercial use of collagens by the biomedical industry. New approaches are still emerging, due to the novel collagens being discovered and the increasing scale needed for certain applications. Development of acellular matrix collagen-based materials is providing a promising scaffold material for tissue engineering.

8.1 Tissue and fibrous collagen

For some applications of native collagen tissue for devices, such as the use of heart valves or pericardium, the principle step appears to be the selection of suitable undamaged tissue of appropriate size or thickness. In many cases, this tissue may be collected directly into a fixative solution, generally glutaraldehyde (GA), so that residual antigens such as serum albumin may be fixed in place, which could be a disadvantage. On the other hand, such an approach may maintain the presence of hyaluronic acid and proteoglycans which may be beneficial for tissue that will be used in mechanically challenging situations. In some cases, the native tissue has been treated, mildly, with an enzyme such as ficin to partly eliminate unwanted antigens while minimising any degradation of collagen [1]. In many instances the exact processes are not generally available and are commercial in confidence, proprietary processes. In some, perhaps many cases, the processes used were well established prior to the current, more rigorous regulatory processes. Hence for these effective, 'grandfathered' processes the process remains unchanged as any improvements would cause significant regulatory costs. This 'grandfathering' is certainly the case for 'catgut' sutures, where the chromic stabilisation of collagen 'catgut' sutures [2] would undoubted create regulatory hurdles?

For sutures, purified bovine serosa or ovine submucosa have been the principal collagen source, twisted to provide a filament that is then accurately ground to

doi:10.1088/978-0-7503-2096-2ch8

provide a defined single filament. Porcine small intestine submucosa (SIS) has become a valuable commercial biomedical material, in various forms, and in applications other than for sutures, where it can be used as is, or as a perforated material [3]. Alternatively, comminuted material can be reformed into sponges [4] or a SIS gel can be made [5], or used as multilayer constructs [6]. A protocol for producing SIS [7], for example, includes mechanical treatment to obtain clean tissue, washing in MeOH/$CHCl_3$ to remove fats, rinsing with water, digestion with trypsin/EDTA, and saline washing. It was then washed in SDS/PBS followed by saline. It had been noted that full removal of porcine cells and DNA was important for good clinical performance [8], while the method chosen for sterilisation can also affect performance [9]. On the other hand, the presence of residual growth factors, such as FGF, can have positive effects [10].

A comminuted hide product was developed (see [11]), initially as a dry haemostatic material, but also has been developed as a gel, cream, film and porous mat [12]. The process initiated in 1955, but for producing a rayon paste, and led to experiments on other materials including cellulose, that led to successful commercial materials, and also collagen (see [11]). Some ten years later, the 'microcrystalline' collagen emerged [13], with an initial application as a haemostat [12, 14]. Microcrystalline collagen has been prepared from ground, freeze-dried cleaned bovine hide that was blended first with 0.01 M HCl followed by further blending after addition of 2-propanol (or an alternative solvent) and freeze drying [15]. It has been pointed out, however, from studies on these and a range of related bovine products, that unless adequate washing or another step to remove potential non-collagenous immunogens is part of the production process, an immunological response to impurities is possible in clinical use [16].

Tendon provides a tissue for production of microfibrous collagen for biomedical products that have been extensively used. Tendons, such as bovine or ovine, are an ideal source. They can be thoroughly cleaned and then the clean collagen released by trimming both exposed ends while removing the protective external membrane, without any risk of contamination. Tendons were then sliced at 0.5 mm, and these tissue disks treated briefly with <1% w/w ficin for 1 h, followed by enzyme inactivation. After thorough washing the tendon tissue was dispersed in 0.5% lactic acid and then this mixture blended to give a uniform dispersion [17].

Amnion provides a human tissue that can be readily obtained and used, that does not have complex ethical questions. Initially it was proposed for skin transplantation [18], for scalded, burned and ulcerated skin [19, 20], more than a century ago. The extent of preparation was limited to multiple washes in saline solution [19]. The initial enthusiasm was restrained and the material was rediscovered in the 1930s for surgical use [21], whence occasional use was discussed for various surgical indications (see [22]). During this time other preparation procedures emerged, including dehydration in 70% EtOH and drying [23]. Significant interest again re-emerged in the 1970s (see [22, 24]). Rigorous washing was still used, but also treatment with Daikin's solution (which contains hypochlorite) [24]. Most recently, the processing of amnion has advanced significantly, so that it is now a commercially available processed material. Various preparation approaches have emerged

[25], of which preparation as an acellular matrix material (see the following section), typically using detergent processing, has proved of most interest [25–27]. In some cases, the acellular amnion has been further processed to a soluble product for use in coatings [28].

Bone provides a material from which collagen, generally as demineralised bone matrix [29], can be extracted, normally as segments, fragments or particles. Two methods have been developed that are normally used for demineralisation, although other strategies are possible [29, 30]. In one, treatment of bone with HCl was used, typically using a 0.5 to 0.6 M solution [30, 31]. Demineralisation of bone by HCl has a history going back to at least the late 19th century [32] when demineralised bone segments were used in clinical applications. The other laboratory method that has been used is 0.3 to 0.4 M EDTA extraction [31, 33]. Demineralised bone can be used to prepare soluble collagen (see below) but is generally used as a powder or paste as a filler material for bone repair (see chapter 14). Its function in this role is particularly effective due to the residual presence of a variety of osteogenic factors [34–36], although the composition of these factors varies with the method for demineralisation that has been used [37, 38].

8.2 Acellular matrix

While intact tissue and comminuted collagen tissue have been used in developing a range of materials and products develop for several decades, the more recent innovation has been the development of acellular matrix materials. These have been of particular interest for researchers interested in tissue engineering and regenerative medicine (see chapter 16). The extracellular matrix is, of course, the natural biologic scaffold and material [39]. The main aim is to remove, cells, cell contents and debris, as these can be significant immunogens in clinical use. The processes may also remove other immunogens, such as serum proteins, and generally the aim is to retain as much as possible the three-dimensional structure and characteristics of the original tissue, including the integrity of biologically active sites. A very wide range of possible approaches have been described (table 8.1), with many used in combination strategies. Further details on these and other processes are included in various reviews [40, 41]. Taken together the number of individual methods along with the various combinations and orders of use (table 8.2) means that the number of compete strategies described has become very large, and that no single best choice approach has yet emerged, particularly as efficiencies will vary between different tissue sources, each needing a specific approach.

The production of a collagen-based acellular matrix typically minimises any immunological response when implanted, while allowing the potential for xenogeneic implants. It is capable of being stored, so that development of off-the-shelf products is possible. However, the difficulties vary with tissues, and full characterisation is necessary. Further, care has to be taken that the sterilisation used does not significantly alter the structure and properties of the material.

For xenogeneic materials, a significant understanding on residual antigens has emerged from studies on potential organ transplantation, including the importance

Table 8.1. An overview of some of the methods that have been proposed for production of acellular matrix.

Physical methods		
Snap freezing	(and freeze drying)	
Freeze/thaw cycles		
Mechanical force	Pressure	
	Delamination	
Mechanical flow	Agitation	
	Perfusion	
Chemical methods		
Alkaline and/or acid treatments		
Solvent extraction	Organic solvents	e.g., EtOH, Acetone
	Liquid CO_2	
	Tri(*n*-butyl)phosphate	
Detergent action	Non-ionic detergents	e.g., Triton X-100
	Anionic detergents	e.g., SDS
	Cationic detergents	e.g., CTAB
	Zwitterionic detergents	e.g., CHAPS
Hypo- and hyper-tonic treatment		
Chelating agents		e.g., EDTA
Enzymatic methods		
Proteases		e.g., Trypsin
Endonucleases		e.g., DNase, RNase
Exonucleases		

of α-Gal and MHC 1 antigens [42, 43]. Opportunities for α-Gal reduced porcine tissues have been explored [44].

A wide variety of tissues have now been examined for production of acellular matrix (table 8.2), with many other additional examples from various species also being considered [41]. A wide variety of commercial materials are also now available (see [41, 45]).

Evaluation of the acellular product is important as certain residues could cause an adverse response on implantation. In some cases, the residues could be introduced during the processing, such as enzymes and detergents. Residual SDS, for example, may cause issues, but it is possible that matrix topology changes may be equally important [46]. Various approaches have been described, and a full study would include a range of different methods targeting different potential residues. Histology is a widely used approach. Stains such as Haematoxylin and Eosin will show residual nucleated cells and cell fragments, while methyl green-pyronin will show DNA and RNA residues. Various stains, such as Masson's trichrome, Picrosirius red and Miller's Elastin stain can be used to examine the integrity of the collagen/elastin matrix, while immunostaining can allow discrimination of individual components [41]. Other imaging methods, such as SEM and TEM can be used to examine the acellular matrix structure and integrity. Various biochemical tests are available for

Table 8.2. Examples of tissue that have been converted to acellular matrix with retention of tissue structure

Tissue	Species	Method summary	
Adipose	Human	Repeated freeze–thaw cycles, polar solvent extraction, enzymatic digestion	[55]
Amnion	Human	SDS, hypotonic buffer and protease inhibitors, nuclease treatment	[26]
Artery	Canine	Hypotonic solution plus protease inhibitors and antibiotics, high NaCl with non-ionic detergent, DNase and RNase, anionic detergent, PBS	[56]
Bladder	Porcine	Hypotonic buffer with protease inhibitors, Triton X-100, rinsing with buffer, DNase and RNase	[57]
Cardiac muscle	Porcine	Trypsin/Triton-based perfusion	[58]
Cardiac valve	Porcine	PBS with Triton X-100, EDTA, DNase and RNAse	[59]
Cartilage	Porcine	Freeze/thaw cycles, cycles of hypotonic buffer - SDS in buffer plus protease inhibitors. RNase and DNase	[60]
Colon	Ovine	Buffer, SDS, Triton X, Trypsin, multiple PBS washes	[61]
Dermis	Human	1 M NaCl, SDS in Hanks solution, pre-freezing solution, freeze drying	[62, 63]
Heart (entire)	Porcine	Perfused by trypsin/EDTA, Triton X-100/EDTA, deoxycholic acid, peracetic acid/EtOH	[64]
Intestine	Rat	Sodium deoxycholate, PBS, DNase-I in 1 M NaCl, PBS	[65]
Liver	Porcine	Distilled water, ammonium hydroxide with Triton X-100, PBS, freeze drying	[66]
Lung	Primate (non-human)	Distilled water, Triton X-100, repeated distilled water, 1 M NaCl, DNase solution, PBS solution	[67]
Pericardium	Bovine	Hypotonic tris buffer with PMSF, Triton X-100, rinse in Hanks' solution, DNase and RNase, Triton X-100, wash in Hanks' solution	[68]
Testis	Human	Triton X-100 or SDS, washing	[69]

quantitative analysis of both residues such as DNA, while also examining the extracellular matrix components [41].

While many, probably most studies to produce ACM have focussed on retaining the structural as well as the compositional integrity of the original native tissue, an alternative strategy has also emerged where the tissue structure is lost, but the composition maintained, giving an acellular paste that can be used as coatings [47] or in potential clinical applications [48]. Various tissue sources have been examined, typically using an SDS-based approach for decellularisation, including, for example, adipose tissue [49], porcine brain [50] skeletal muscle [51] and pancreas [52]. The

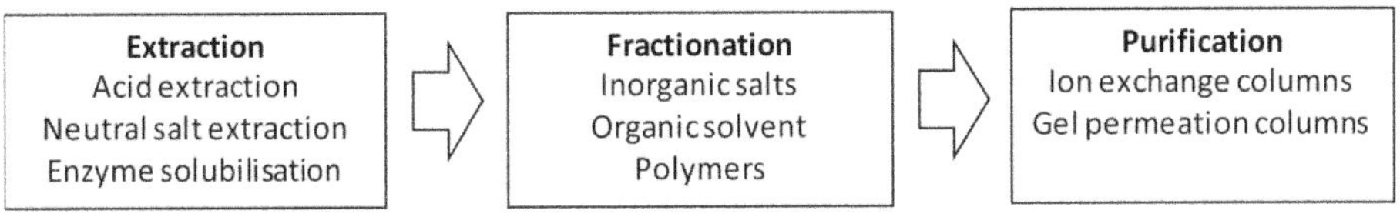

Figure 8.1. Examples of purification approaches for soluble collagens.

focus, however, has been on development of collagenous materials for repair of myocardial tissues [48, 53, 54].

A significant issue that has emerged is the moral and ethical questions that arise when human tissues are harvested for use. This present text is not the ideal vehicle for discussion of these complex issues, as it is difficult to present in depth the variety of views, some of which contrast greatly. It is highly recommended that readers further explore this important issue. Some of the issues, but there are others too, that have emerged include, for example, whether if a person bequeaths their body to science, should parts be used for commercial applications and profit. Another issue is the use of donor tissue from prisoners, in exchange for benefits, and whether the use of tissue for executed prisoners is acceptable. For views on various issues, see, for example, these and other authors [70–73].

8.3 Soluble collagens

Purification of soluble collagen (figure 8.1) is important, particularly the ability to produce and purify collagens in large quantities for biomedical applications, where it is now a main reason for collagen purification, whether from tissues or as recombinant products (chapter 9). It has also been key for the study of the biochemistry of collagens and their role in natural tissues as well as in diseased or damaged tissues.

Early histologists, when treating sections with acid (e.g., [74]), would have been extracting soluble collagen, but possibly saw this as an inconvenience rather than as being useful. For example, tendons were observed to swell and dissolve in dilute acetic acid [75]. Later, a soluble collagen fraction from rat was reported by Nageotte [76], who also showed that a reconstituted fibrillar gel could be reformed from this material [76, 77]. Subsequently, a limited number of broad strategies for purification of collagen emerged. After initial extraction of soluble fractions (figure 8.1) various other steps have been used to further purify collagens (figure 8.1). Usually, fractionation using salts [78] is a key step, although reagents other than salts can also be used for this fractionation step. Alternatively, or subsequent to the fractionation step, a variety of chromatographic approaches for purification have been used, including ion exchange chromatography and in some circumstances gel permeation chromatography (figure 8.1). Preparation of tissue prior to extraction, for example by dissection, can assist in purification. Even then, the presence of several different collagen types in a single tissue adds to the complexity of the purification required to provide pure samples of the different collagen types. Preparation of tissue prior to extraction also includes the demineralisation of bone, for example, using either EDTA or HCl solutions. Fat may also need removal, either by appropriate solvents or by supercritical carbon dioxide [79].

While the successful methods have been developed in most cases using mammalian tissues, there has recently been an increased interest in the extraction and purification of collagens from more primitive species. This interest has been driven, in part, by the medical manufacturing sector, where concerns had emerged about the transmission of various zoonoses from bovines, especially spongiform encephalopathy (mad cow disease) [80].

The other source of collagen that has emerged is production of collagen in recombinant systems (chapter 9). This has allowed production of collagens that would not normally be available readily through extraction of tissues. It also allows production and characterisation of collagens from novel sources [81]. Once extracted from these systems, the subsequent purification steps are generally based on prior animal collagen studies.

8.3.1 Extraction of soluble collagen

Early workers found that dilute acetic acid could extract a functional soluble collagen from rat tendons [75–77]. Subsequently, various protocols were described that enabled extraction of readily soluble collagen, by acid solutions such as citric acid [82, 83], or by neutral NaCl solutions [84, 85] (see below). These classical methods are still among the most useful of approaches and continue to be used extensively, especially in the purification of biosynthetic precursor molecules [86]. However, in most tissues the collagens are often crosslinked to give an extensive insoluble matrix requiring additional strategies for collagen extraction.

In many cases a key step prior to extraction is cleaning the source material. This may be through washing to remove blood and other extraneous materials, or by hair removal, or by demineralising of mineralised tissues. Hair removal can be achieved readily for small samples by immersing in liquid nitrogen and then brushing off the hair, while for larger samples a wet splitting machine (used in the leather industry) is suitable. For some tissues, such as calf skin, soaking in acetic acid allows the hair layer to be removed by scraping. For some species fat can be an issue and for small preparations can be removed by solvent extraction. However, for larger extractions this may be impractical and the temperature must be kept low at all times. In many cases careful dissection to select the appropriate tissue, such as a cornea or a lens capsule, may be necessary to allow a focus on a particular collagen type.

8.3.2 Extraction of acid soluble collagen

Acid can solubilise collagen from various tissues [76]. It is probable that two different factors are present. In one case, recently synthesised collagen molecules that have not yet been incorporated or crosslinked into the stable extracellular matrix are available to dissolve in an appropriate solvent. In the other, the initial crosslinking of the collagen (chapter 5) is in a state that can be reversed, so that the released collagens can be dissolved.

Acetic acid was the first recognised as useful in solubilising collagen [76], and has continued to be valuable ever since. The best-known example is probably the solubilisation of collagens from rat-tail tendon. A detailed account of a full process has been given [87]. Rat-tail collagen is still regularly purified for small scale laboratory studies, as the tissue is readily available, the process is easy and the material can be made sterile for cell culture studies. The retention of the telopeptides on the collagen significantly enhances the rate of collagen gel formation [88], especially when compared to atelocollagen where the telopeptides have been removed, for example by proteolysis [89]. The soft collagen gels that form can be used 3-dimensional scaffolds for cell growth [90]. When certain cells are used, for example fibroblasts, the soft collagen gel contracts in volume over several days to give a denser, more tissue-like material [91].

The other tissue where acetic acid extraction is frequently used is with fish skin, where a substantial proportion of the collagen can often be solubilised (e.g., [92]). It has been suggested that a limitation in oxygen availability may result in the slowing down of crosslink maturation, allowing enhanced solubility to be retained longer [93]. Overall, collagen has been extracted and examined now from several dozens of species of fish.

An alternative acid system, using citric acid, was introduced by Russian scientists [82, 83]. They used 100 mM citrate buffers, noting that the yield of collagen from rat skin was best at pH 4.1, while the yield from rabbit skin was best at pH 5.7 [82]. Subsequently, skin tissue pastes were extracted using citrate buffer at pH 4.0, giving filtrates that were a viscous transparent solution of soluble collagen from which an amorphous precipitate could be collected after addition of NaCl solution to 5% final concentration [82].

This approach has been used significantly for collagen preparations, particularly including precursor molecules, where inclusion of protease inhibitors is strongly preferred. The approach has also been used for extracting collagen from tendons (e.g., bovine), which has gained popularity in the biomedical industry. The tendons are covered in a fascia membrane that protects the isolated inner fibrillar tissue from contamination. The external surfaces can be washed in a suitable sanitising/sterilising reagent and then the ends and the membrane removed under appropriate clean conditions to give a high-quality material for extraction. Although the yields are not good, the collagen from a citric acid extraction retains its telopeptides and so can be transformed into a gel in a controlled and reproducible manner. Tendons are frequently a preferred source for collagen extraction by acetic acid for biomedical and tissue engineering applications [94].

8.3.3 Neutral salt soluble collagen

Neutral salt (NaCl) treatment of tissues will extract newly synthesised collagen, as it is unlikely that these conditions will have any effect on the crosslinking system. The approach was introduced subsequent to the use of acetic acid and citric acid extraction approaches [84, 85]. In one example, young rabbit skin was extracted with M/15 (as given in the paper) sodium phosphate buffer and the resulting soluble collagen collected by addition of NaCl to 16% (w/v). Purification was by repeated dissolution and precipitation; on standing afterwards, the collagen formed into banded fibrils that were examined by electron microscopy [85]. In another example, minced new-born calf skin was extracted with a range of different phosphate buffers and un-buffered salt solutions, with extracted soluble collagens collected by precipitation through dialysis against distilled water. The best neutral salt yield was at pH 7.0 with sodium phosphate buffer [84]. Heating to 37 °C to induce collagen gel formation was used to isolate banded collagen fibrils visible in the electron microscope [84]. The authors continued to refine the neutral salt extraction approach, for example for studies on skin composition and growth [95, 96]. It was also adopted by others (e.g., [97, 98]) and not just for mammalian tissues but also for fish tissues [99]. Subsequently, it has remained a recommended method with a well-defined protocol for soluble collagen extraction [100]. However, the method is not frequently used as it appears that, unlike the citrate extraction method, it has not gained favour with the biomedical industry for larger scale processing, possibly because the neutral pH conditions favour proteolysis and microbial growth. The inclusion of a broad range of protease inhibitors is an essential improvement to the method.

8.3.4 Increasing soluble collagen yield

Historically, the extraction of collagens has been enhanced by feeding animals a lathyritic agent, but such approaches would possibly not meet today's animal ethics protocols. Lathyrism is a neurological disease in humans and domestic animals, characterised by lack of strength in, or inability to move, the lower limbs, and has been known for millennia. Indeed, Hippocrates (460 to 377 BC) is thought to have been aware that a toxic pea caused this paralysis of the legs [101]. It was later recognised that eating certain pea seeds of the genus *Lathyrus*, particularly *L. sativus*, in moderate to large quantities led to the condition and that it became more prevalent in certain parts of the world where it was consumed when normal grain crops were in low supply [101]. Subsequently, it was noticed that seeds of the sweet pea, *L. odoratus*, gave a similar, but distinct outcome (sometimes called Odoratism). The toxic compound in sweet peas was identified as β-(*N*-γ-glutamyl)-aminopropionitrile, and it was shown that the β-aminopropionitrile (BAPN) entity was the toxic component in this form of lathyrism [102]. It was also observed that BAPN ingestion led to aneurysms in various animals including rats, turkeys and chicks [103–105]. Studies on the biochemistry of lathyritic chicks recognised that the high tissue fragility found in this form of lathyrism was accompanied by a pronounced increase in the amount of soluble collagen that could be extracted by neutral NaCl [105]. Subsequently, it was recognised that the BAPN was interfering with cross-linking of collagen in fibrils, due to irreversible inhibition of the lysyl oxidase activity (chapter 3) [106]. So, while this approach was historically very valuable in understanding the biosynthesis of collagen, today it's use could be limited to cell culture studies where stopping collagen crosslinking may be necessary (see [107]).

Another system that could be considered for cell culture is addition of various α-amino thiol compounds, such as cysteine, cysteamine and penicillamine, which also lower crosslink formation, by via a different mechanism [108]. Nevertheless, D-penicillamine was successfully introduced clinically to control Wilson's disease (hepatolenticular degeneration) [109]. The thiol compounds lead to an accumulation of neutral salt soluble collagen in certain tissues principally as a result of their binding to the free aldehydes from the LOX activity, although at high concentrations inhibition of LOX may occur [108].

8.3.5 Extraction of soluble collagen by tissue digestion

The resolution of low collagen yields was through the use of proteolysis. It had been known for a considerable time that tissue collagen was resistant to short trypsin digestion [110]. A small amount of digestion could occur on the surface of hide collagen particles if they were pre-treated in acid and had a longer digestion period [111]. Subsequently, it was also shown that soluble collagen was essentially resistant to trypsin digestion [99], and this was extended to show that the collagen triple-helix was also resistant to chymotrypsin and pepsin [112]. For soluble collagen preparations that contain molecular dimers, trimers and small polymers, proteolysis led to a depolymerisation of these components [112]. Prior to and around this period, in the different discipline of meat and food technology, there was also interest in using proteolysis, particularly using plant-derived proteases such as papain and bromelain, to digest collagen to tenderise meat [113, 114]. Collagen was typically determined on gelatinised samples by Hyp analysis so the integrity of the extracted collagen was not established.

From these studies, it was recognised that digestion of the short terminal non-triple-helical telopeptide domains (chapter 2), that contain the tyrosine content, was associated with these changes. The effect of protease treatment of insoluble tissues led to development of a method for production of soluble collagen extracts [115, 116]. A wide variety of enzymes were examined by various researchers, including trypsin, chymotrypsin, papain, ficin and pepsin. Subsequently, other enzymes have been examined for specific tissues, including actinidain [117]. The proteolytic methods, especially using pepsin, generally gave excellent yields of soluble collagen from juvenile tissues [100, 115, 116]. The anticipated utility and commercial value of this approach was indicated, for example, by patent applications [118], with priority dates from 1960.

The effectiveness of the protease approach is due to it targeting the short non-helical peptides (telopeptides) at the N- and C-terminal of the triple-helical domain (chapter 2) while usually not degrading the triple-helix. Cleavage of these telopeptides breaks the crosslink network and allows solubilisation of the collagen. Pepsin, in conjunction with acetic acid at 4 °C, which also provides a chaotropic benefit, has proved the most commonly used method [100]. This can be regarded as a standard method, and has been applied to a wide range of species, from primitive species such as jellyfish [119] through the animal kingdom to a variety of mammalian species [120] (figure 8.2).

Proteolytic digestion to solubilise collagens also has some problems. In some cases, the target protein is susceptible to the protease activity. For example, this has been observed with type III collagen, with various groups noting its partial susceptibility to certain enzymes. For example, to trypsin [121] and to thermolysin [122], neither of which is a primary choice of enzyme to use in extraction and purification. Susceptibility of type III collagens to trypsin was further described for a

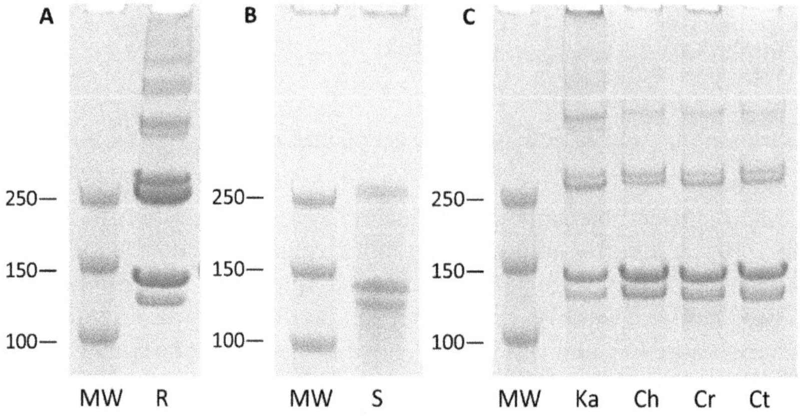

Figure 8.2. SDS-PAGE of collagens prepared by different methods, shown by SDS-polyacrylamide gel electrophoresis, where MW indicates molecular weight standards (kDa). A: Extraction of rat-tail collagen (R) by acetic acid. B: Extraction of fish (salmon) collagen (S) using citric acid. C: Examples of type I collagen extracted from skin samples by pepsin digestion; Kangaroo (Ka), Chicken (Ch), Crocodile (Cr) and Cane Toad (Ct). (Gel photographs courtesy of Yong Peng, CSIRO.)

range of species, where it mirrored the susceptibility to collagenase digestion [123]. These studies showed that human, dog and cat type III collagens were cleaved more rapidly than that from guinea pig, while type III collagen from chicken was particularly resistant to trypsin [123], in this case possibly reflecting its greater thermal stability [124]. In all these cases the cleavage was proximal to the collagenase cleavage site. Complete susceptibility to pepsin has been shown for kangaroo type III collagen, where the digestion to solubilise type I collagen from mature skin was sufficient to entirely digest the type III collagen [125]. The presence of type III collagen was shown by immunohistology, and it could be purified by enzyme free extraction of juvenile tissue [126]. Also, proteins that contain intervening, non-triple-helical segments, such as type IV collagen [127], are particularly susceptible to fragmentation.

8.3.6 Sources for specific collagen types

The preceding sections dealt mainly with the preparation of the most abundant collagens, particularly type I collagen. However, most of the 28 different collagen types have been purified often from human tissues or cell cultures, albeit in many cases in small quantities, for example for structural characterisation or for production of antibodies for tissue distribution and developmental studies.

Interstitial collagens

As well as from human, type I has been purified, mainly using the skin, of a wide range of domesticated species, from laboratory animals and from autopsy samples of various zoological garden species such as serval *(Felix serval)* and palm squirrel *(Funambulus pennanti)* skins [120].

Relative to the large number of human collagen types that have been described, considerably fewer other mammalian collagen types have been described, even with genome data being notionally available. Those that have been purified are generally only those in good abundance, typically types I, II, III and V collagens. In many cases these non-human collagens have been purified for the development and characterisation of antibodies, particularly monoclonal antibodies that have found application on examining normal tissues and in tracking the host changes to biomedical materials [128, 129]. If type V collagen is specifically required, then placenta provides a good source tissue with pepsin digestion providing soluble collagen [130, 131]. Unlike skin, placenta contains two distinct type V collagen isoforms, $[\alpha1(V)]_2\alpha2(V)$ and $\alpha1[V]\alpha2[V]\alpha3[V]$, that can be separated by chromatography on phosphocellulose [131].

Some collagen types that have been described for mammals can also be identified in other animal species, again typically the most abundant type I collagen. It has been described, for example, in marsupials [125], birds [132], reptiles [133], amphibians [134] as well as from insects [135, 136]. Additionally, a type I-like collagen has been described from various primitive species, including cephalopods [137, 138], echinoderms [139], platyhelminthes [140] and cnidarians (jellyfish) [119]. Of these, there has been recent interest in purification of jellyfish collagen for biomedical applications, as the source material is abundant—in some places a significant nuisance to traditional fisheries—and is seen as being disease free while retaining key biological properties for cell and other interactions [119].

Type I collagen has also been examined from very large number of fish species, where it has been purified from skin, scales, bones and tendon structures. The structure of type I collagen generally appears to differ in bony fish where a third chain type is present in an $\alpha1[I]\alpha2[I]\alpha3[I]$ structure [92]. There has been a recent increase in interest in fish collagens as ideal materials for production of scaffold for tissue engineering and regenerative medicine [141].

Similarly, type V collagen-like collagens have been described and purified from fish and more primitive species [142–144]. On the other hand, of the interstitial collagens, type III collagen has not been identified clearly in primitive species. It is present in chicken, but, for example, has not been purified from fish. A detailed study of the fibrillar collagens from lamprey [142] did not clearly identify a type III-like collagen among the 5 types that were separated.

An alternative form of type I collagen preparation is that of anionic collagen [145]. This collagen variation arises from deamidation of asparagine and glutamine side chains of native collagen, lowering the isoelectric point of the collagen from around a pI of 9 to a pI of around 5. This can be achieved through treatment of the selected tissue by alkali. The material can be used in fibrous form [145] else could be further treated by proteolysis to give a soluble form. Another source of material could be limed hide splits from the leather industry. Alternatively, deamidation of already solubilised collagen can be used [146].

Type IV collagen

An initial and still preferred choice for the production of type IV collagen has been placental tissue [147, 148]. It is readily available and without major ethical issues. It has been used in commercial large-scale production of this collagen [149]. The collagen was obtained from tissue using pepsin digestion. However, the product obtained can reflect the extent of the pepsin digestion. The structure of type IV collagen (chapter 2) has several points that are susceptible to proteolysis. As with other collagens, the C-propeptide domain, the NC1 domain, is readily released. However, the crosslinking, 7S domain, can also be released as a separate entity [150]; it is more resistant to bacterial collagenase digestion than other collagenous components [151]. In addition, there are a limited number of pepsin sensitive sites along the major triple-helical domain [152].

An alternative source for type IV collagen is from Engelbreth–Holm–Swarm (EHS) mouse sarcoma. This tumour expresses large amounts of type IV collagen along with other basement membrane proteins [153]. Together these components can be used as a coating for cell culture, but the ill-defined nature of the mouse protein mix precludes its direct use in biomedical applications. The individual components can, however, be purified and are readily available from chemical suppliers.

Type IV collagen is present throughout the animal kingdom, including in the most primitive species [154, 155]. However, its low abundance has meant that it has only rarely been extracted from primitive species.

Type VI collagen

In the initial time after its identification, type VI collagen was thought to be widely distributed, but often in low amounts [156]. This would be consistent with its involvement as a 'pathfinding' molecule involved in the deposition of new tissues [129]. The transition to a potentially useful material occurred when a method for isolation from skin and other tissues was described [157]. After initial guanidine-HCl extraction, tissues were solubilised by mild pepsin digestion and purified by NaCl fractionation and chromatography [157]. Yields were, for example >500 mg Kg^{-1} wet placenta and > 250 mg Kg^{-1} wet skin [157]. These yields could be sufficient for biomedical device coatings, for not for making devices. Type VI collagen was also identified in the vitreous of the eye [158]. This tissue has proved to be a convenient source of type VI collagen for biomaterial application [159].

Type X collagen

Although type X collagen is seen as a component of hypertrophic cartilage, low levels prevent ready, large-scale purification, especially from mammals [160, 161]. However, a novel source for type X collagen is chicken egg shell membrane, and potentially large amounts are available as a by-product of baking and other industries. Systems have been described to allow isolation of egg shell membrane in large quantities [162, 163], and the product identified for its value as a biomedical material [164]. However, the eggshell membrane is heavily crosslinked and so is hard to solubilise. An aggressive approach, using performic acid oxidation followed by pepsin digestion step gives a soluble product, albeit somewhat modified [165]. Alternatively, solubilisation with 3-mercaptopropionic acid with acetic acid releases a soluble product [166]. The extract typically contains type I and type V collagens as well as type X collagen [167].

8.4 Production of collagen in cell culture

In this section we are considering cells that are derived from tissues rather than manipulated recombinant cells (chapter 9). The yields obtained are generally low and so this approach is preferable for specific applications, for example, production of autologous human collagen. For laboratory studies, collagen yields could be increases by modification of the medium and inclusion of growth factors (e.g., [168]). Cell culture can also be used for production of minor collagens, such as type IV collagen if appropriate cell types are used, including kidney-derived cells [168] or testis-derived cells (figure 8.3). However, better yields could be obtained using culture on beads, and large-scale commercial production using modified cells should also be possible (see [169]).

One aspect in production of collagens in cell culture is that full processing of the procollagen precursors (see figure 3.1) may not occur. Cell culture is quite distinct from natural cell growth in that the media and cell surroundings are very dilute, leading to potential differences from natural behaviours. An approach to change this is by molecular crowding, whereby water-soluble polymeric compounds, such as polyethylene glycols, are added to the media to change the effective concentrations of components, including processing enzymes, in the media and pericellular environment [170]. Various concepts in the field of macromolecular crowding had been developed (see [171], using solution rather than cellular approaches. Initial studies using cells had assumed a simple model of volume exclusion, but later theoretical [172] and experimental studies [173] suggested that additional mechanisms could also be contributing, especially when charged polymers are used. Various studies have significantly extended the use of macromolecular crowding, and have shown that under appropriate conditions and with selected polymer systems enhanced extracellular matrix production by both differentiated and stem

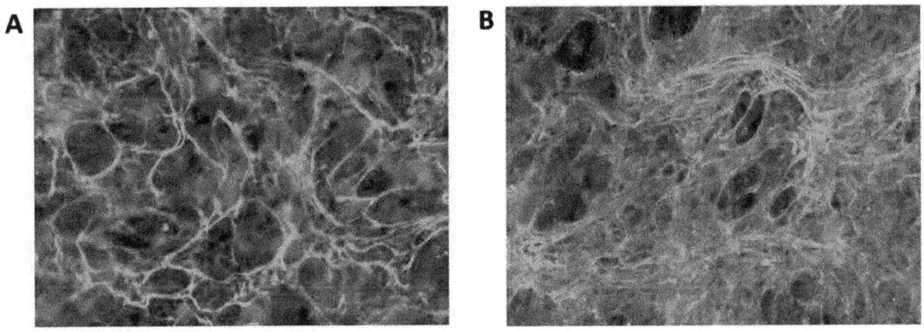

Figure 8.3. Immuno-detection of type IV collagen by testis-derived cells in culture. (A) α2-chain from Sertoli cells, and (B) α1-chain from Myoid cells.

cells [174–177]. From a regulatory perspective it has been suggested that polyvinylpyrrolidone may be an appropriate system to use [177].

8.5 Fractionation of soluble collagens

In most cases the initial soluble collagen extracts, as discussed in the previous section, or recombinant collagens (chapter 9), will need further purification. The standard method that has emerged for purification has been fractionation by NaCl solutions. Early workers (see [178]) explored a range of strategies including salt fractionation. These studies clarified that below around pH 5, collagen is soluble at low ionic strengths, but with increasing ionic strengths collagen precipitation would occur. On the other hand, around neutrality an ionic strength greater than 0.15 M NaCl is needed for solubility [178]. Of interest, when neutral collagen solutions are warmed to 37 °C, collagen fibrillogenesis occurs and a solid collagen gel is formed [88]. While this transformation has been used as a purification strategy, it has found greater application in providing 3-dimensional scaffolds for cell culture studies and particularly in biomedical applications such as soft tissue augmentation [179].

Normally, NaCl fractionation is performed first in acid pH conditions, and then followed by further fractionation at neutral pH (table 8.3). The NaCl concentration required for different collagens varies with type (table 8.3) allowing separation of the different types.

A detailed evaluation of the precipitation of types I, II and III collagens at neutral pH has been presented [78], with evaluation of the differences between rapid (30 min after precipitant addition) and slow (24 h) precipitation. For NaCl precipitation, slow precipitation gave better resolution of the three collagens. The use of $(NH_4)_2SO_4$ was also examined [78]. In this case better resolution was obtained by rapid precipitation, particularly for resolving type III collagen from collagen types I and II (figure 8.4).

Table 8.3. Examples of NaCl precipitation (Reprinted from [180]) with permission from Elsevier.)

Collagen type	Acid conditions (M NaCl)	Neutral conditions (M NaCl)
I	0.7	2.4
and type I trimer		3.5
II	0.7	>4.0
III	0.7	1.7
IV	1.2	
and 7S fragment	1.2	2.0
V	1.2	>4.0
VI	2.0	
IX	2.0	
X	2.0	
XI	1.2	

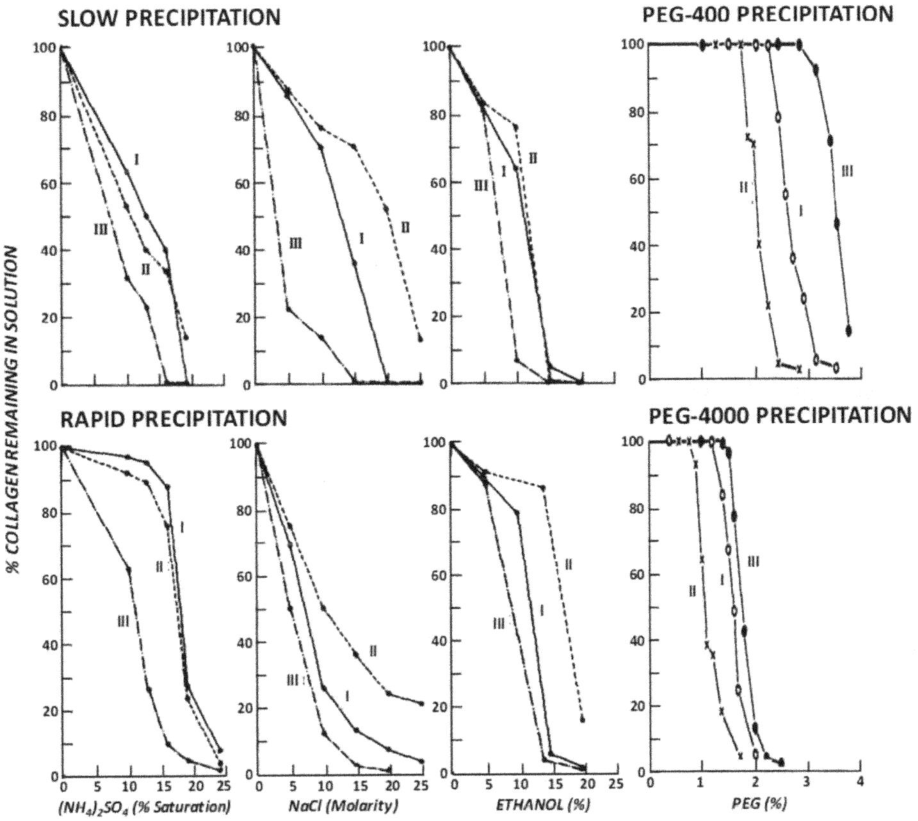

Figure 8.4. The precipitation of 0.1 mg ml^{-1} collagen types I, II and III, from pH 7.6 ionic strength 0.4) solutions, using slow precipitation over 24 h with $(NH_4)_2SO_4$, NaCl, EtOH, PEG-400 and PEG 4000, and using rapid precipitation over 30 min with $(NH_4)_2SO_4$, NaCl and EtOH. (Reprinted from [78] and [181], with permission from Elsevier.)

The use of fractionation systems other than salts has also been examined. Ethanol enables precipitation and fractionation, but the resolution was no better than with salts [78]. An alternative system that has been proposed, but not used extensively, has been the use of neutral polymers such as polyethylene glycols [181]. This system is possibly suitable for isolation of type III from type I collagen.

Specific precipitation and fractionation conditions have been used for certain collagen types. For example, type V collagen can be further purified by precipitation from 50 mM sodium acetate, pH 4.8 [182].

8.6 Further purification of soluble collagens

In many cases, salt fractionation of soluble collagen has proved an effective purification approach, especially for larger scale production. However, in certain circumstances additional steps are included, for example to ensure pepsin removal or

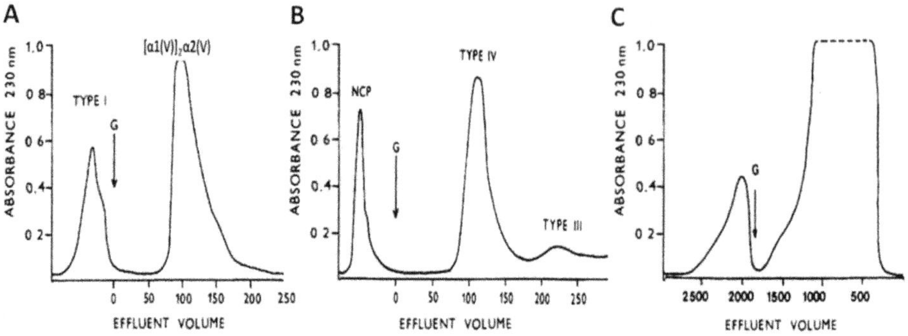

Figure 8.5. Chromatographic purification under non-denaturing conditions of various pepsin-soluble collagens. A. CM-cellulose chromatography of type V collagen using a NaCl gradient at pH 8. B. CM-cellulose chromatography of type IV collagen using a NaCl gradient at pH 4.8. C. DEAE-cellulose chromatography to remove proteoglycans, which are retained on the column. In all cases, G indicates the start of an elution gradient. (Reprinted from [180] with permission from Elsevier.)

glycosaminoglycan removal. Also, for very small preparations, approaches other than salt fractionation are frequently preferred. In these cases, HPLC systems have been adopted as they are suitable for small scale studies and they can provide excellent resolution. Column media selection can include both traditional ion exchange media, while reverse phase systems using hydrophobic media can also be used [183]. These systems are particularly suitable for separation of fragments of collagen, such as tryptic peptides.

8.6.1 Ion exchange chromatography

Ion exchange methods have proved useful for chromatography of native collagens, but have also been used extensively for separation of their individual constituent chains and for purification of proteolytic and CNBr fragments of these chains [100].

Both anion and cation exchange media have been used in the purification of native collagens. For example, CM-cellulose has been used under both acidic and basic buffer conditions [100]. Under basic conditions, the CM-cellulose can be equilibrated in 20 mM tris buffer containing urea and kept at 4 °C to prevent collagen denaturation. Elution can be by NaCl in a linear gradient from 0 to 0.25 M. This system allows, for example, separation of collagens from cornea, a complex tissue with various collagen types. It allowed the resolution of type I and type V collagens from salt fractionated extracts, with type I collagen not binding to the column and a NaCl concentration of more than 0.1 M in the elution gradient being required for the elution of type V collagen [180, 184, 185] (figure 8.5). Under acid conditions, CM-cellulose can be equilibrated to 40 mM sodium acetate, pH 4.8 containing 2 M urea. It was kept at 4 °C to prevent collagen denaturation. Elution was by a linear gradient for 0 to 0.4 M NaCl. This approach allows, for example, the purification of pepsin-soluble type IV collagen, including its separation from type III collagen [100, 147, 180, 185] (figure 8.5).

DEAE-cellulose has been used in the preparation of type V collagen, although it is also applicable to other collagen types. In particular, it has been used for the removal of potential glycosaminoglycan contamination of collagen preparations, especially from cartilaginous tissues [185, 186] (figure 8.5). Collagen samples were dialysed against 50 mM tris buffer, pH 7.5 containing 0.2 M NaCl and loaded onto a DEAE-cellulose column equilibrated in the same buffer. Collagen remains poorly bound or unbound and washes through the column, while the glycosaminoglycan contamination is strongly bound and needs 1.0 M NaCl for elution [185, 186] (figure 8.5).

Other chromatographic media have also been used for specific separations, including hydroxyapatite and phosphocellulose. For example, type V collagen from placenta comprises two different molecules with distinct chain compositions, the $\alpha 1[V]_2\alpha 2[V]$ and $\alpha 1[V]\alpha 2[V]\alpha 3[V]$ [187]. These types can be separated from a salt fractionated starting material by chromatography on phosphocellulose using 40 mM tris buffer, pH 8.2, containing 0.17 M NaCl and 2.0 M urea with elution by a NaCl gradient up to 0.47 M [187].

8.6.2 Gel permeation chromatography

Gel permeation chromatography has found limited application in purification of native collagens. It has, however, found much greater use in the separation of individual collagen chains, CNBr fragments, proteolytic fragments [185] and for removal of minor contaminants [188].

The use of gel permeation chromatography with native collagens has been used to separate monomeric molecules from dimers and higher oligomers [189, 190]. Clearly, as separation is based on molecular size, its utility in separating individual native interstitial collagens or their variants with alternative chain compositions is limited, although in some cases separation can be adventitiously obtained because of interactions between the collagen and the chromatography matrix [191].

8.6.3 Affinity chromatography

For many of the more recently discovered minor collagens, which may be present in very low quantities in native tissue, immuno-affinity approaches are the most appropriate for purification; for example, for collagen types XVII, XXIII and XXV [192–194]. For several collagens there has not been sufficient collagen protein available to raise an antibody and so peptides or fragments of the molecule derived from recombinant expression have been used as immunogens for antibody production [193]. These antibodies can then be used for protein purification as well as for tissue immunohistology.

The other class of affinity absorbents are those based on the specific interactions of collagen with other components of the extracellular matrix [195]. An example, is the interaction of collagens with heparin, such as type V collagen [196]. Another example is native type I collagen which binds to a heparin-sepharose column, while denatured type I collagen chains do not [197].

8.6.4 Other approaches

In certain instances, specific methods have emerged to solve individual problems. If residual pepsin is a likely problem then heat treatment of the collagen in 2 M guanidine-HCl, pH 7.5, results in the precipitation of the contaminating pepsin and this precipitate can then be removed by centrifugation [198]. Another specific problem is the purification of pure type III collagen, free of contamination by type I (and other) collagens. A significant structural property of type III collagen is the presence of two adjacent Cys residues at the C-terminal of the triple-helix [199]. The sequential placing of these two Cys residues enables them to interact with the Cys residues on the adjacent chains, leading to the formation of a disulfide linked barrel or knot that covalently links all of the three type III collagen triple-helix chains [200]. This has been exploited for purification. Types I and III collagen in 2 M guanidine-HCl, pH 7.5 solution can be denatured and the constituent chains separate in solution. Dialysis against deionized water, to remove the guanidine, allows the denatured collagens to refold into their triple-helical structures, and to form into fibrils. For type III collagen, the disulfide knot allows for much more rapid renaturation and fibril formation than for type I collagen whose unlinked chains do not refold and remain in solution [198].

References

[1] Rosenberg N, Gaughran E R L, Henderson J, Lord G H and Douglas J F 1956 *Surg. Forum* **6** 242–46

[2] Lister J 1881 *Brit. Med. J.* **1881** 219–21

[3] Poulose B K, Scholz S, Moore D E, Schmidt C R, Grogan E L, Lao O B, Nanney L, Davidson J and Holzman M D 2005 *J. Surg. Res.* **123** 262–67

[4] Kim M S, Hong K D, Shin H W, Kim S H, Kim S H, Lee M S, Jang W Y, Khang G and Lee H B 2005 *Int. J. Biol. Macromol.* **36** 54–60

[5] Hurst R E, Kyker K D, Bonner R B, Bowditch R D and Hemstreet G P 2003 *Anticancer Res.* **23** 3119–28

[6] Janis A D, Johnson C C, Ernst D M and Brightman A O 2012 *J. Biomater. Appl.* **26** 1013–33

[7] Luo J C, Chen W, Chen X H, Qin T W, Huang Y C, Xie H Q, Li X Q, Qian Z Y and Yang Z M 2011 *Biomaterials* **32** 706–13

[8] Zheng M H, Chen J, Kirilak Y, Willers C, Xu J and Wood D 2005 *J. Biomed. Mater. Res. B Appl. Biomater.* **73** 61–7

[9] Grimes M, Pembroke J T and McGloughlin T 2005 *Biomed. Mater. Eng.* **15** 65–71

[10] Hodde J and Hiles M 2001 *Wounds* **13** 195–201

[11] Bernius J 1975 *J. Chem. Education* **52** 155

[12] Hait M R 1970 *Am. J. Surg.* **120** 330

[13] Battista O A, Erdi N Z, Ferraro C F and Karasinski N Z 1967 *J. Appl. Polym. Sci.* **11** 481–98

[14] Hait M R, Battista O A, Stark R B and McCord C W 1969 *Surg. Forum* **20** 51–3

[15] Erdi N Z, Ferraro C F and Battista O A 1968 *US Patent* 3393080

[16] DeLustro F, Condell R A, Nguyen M A and McPherson J M 1986 *J. Biomed. Mater. Res.* **20** 109–20

[17] Li S-T 1990 *US Patent* 4963146
[18] Davis J S 1910 *Johns Hopkins Hosp. Rep.* **15** 307–96
[19] Sabella N 1913 *Med Rec. NY* **83** 478–80
[20] Stern M 1913 *J. Am. Med. Assoc.* **60** 973–74
[21] Brindeau A 1934 *Gynecol. Obstet.* **29** 385
[22] Trelford J D and Trelford-Sauder M 1979 *Am. J. Obstet. Gynecol.* **134** 833–45
[23] Chao Y C, Humphreys S and Penfield W 1940 *Br. Med. J.* **1** 517
[24] Bose B 1979 *Ann. R. Coll. Surg. Engl.* **61** 444–47
[25] Shortt A J, Secker G A, Lomas R J, Wilshaw S P, Kearney J N, Tuft S J and Daniels J T 2008 *Biomaterials* **30** 1056–65
[26] Wilshaw S P, Kearney J N, Fisher J and Ingham E 2006 *Tissue Eng.* **12** 2117–29
[27] Sanluis-Verdes A, Yebra-Pimentel Vilar M T, García-Barreiro J J, García-Camba M, Ibáñez J S, Doménech N and Rendal-Vázquez M E 2015 *Cell Tissue Bank.* **16** 411–23
[28] Shakouri-Motlagh A, O'Connor A J, Kalionis B and Heath D E 2019 *J. Biomed. Mater. Res.* A **107** 232–42
[29] Gruskin E, Doll B A, Futrell F W, Schmitz J P and Hollinger J O 2012 *Adv. Drug Deliv. Rev.* **64** 1063–77
[30] Harakas N K 1984 *Clin. Orthop. Relat. Res.* **188** 239–51
[31] Urist M R, Iwata H, Ceccotti P L, Dorfman R L, Boyd S D, McDowell R M and Chien C 1973 *Proc. Natl Acad. Sci. USA* **70** 3511–15
[32] Senn N 1989 *Am. J. Med. Sci.* **98** 219–43
[33] Herring G M, Ashton B A and Chipperfield A R 1974 *Prep. Biochem.* **4** 179–200
[34] Levander G 1934 *Acta Chir. Scand.* **74** 425–26
[35] Lacroix P 1945 *Nature* **156** 576–86
[36] Urist M R 1965 *Science* **150** 893–99
[37] Urist M R, DeLange R J and Finerman G A 1983 *Science* **220** 680–86
[38] Wildemann B, Kadow-Romacker A, Pruss A, Haas N P and Schmidmaier G 2007 *Cell Tissue Bank.* **8** 107–14
[39] Badylak S F 2007 *Biomaterials* **28** 3587–93
[40] Gilbert T W, Sellaro T L and Badylak S F 2006 *Biomaterials* **27** 3675–83
[41] Srokowski E M and Woodhouse K A 2017 *Comprehensive Biomaterials II* vol 2 ed P Duychene, K Healy, D W Hutmacher, D W Grainger and C J Kirkpatrick (Amsterdam: Elsevier) pp 452–70
[42] Wong M L and Griffiths L G 2014 *Acta Biomater.* **10** 1806–16
[43] Wong M L, Wong J L, Vapniarsky N and Griffiths L G 2016 *Biomaterials* **92** 1–12
[44] Xu H, Wan H, Zuo W, Sun W, Owens R T, Harper J R, Ayares D L and McQuillan D J 2009 *Tissue Eng. Part* A **15** 1807–19
[45] Badylak S F, Freytes D O and Gilbert T W 2009 *Acta Biomater.* **5** 1–13
[46] Gratzer P F, Harrison R D and Woods T 2006 *Tissue Eng.* **12** 2975–83
[47] DeQuach J A, Mezzano V, Miglani A, Lange S, Keller G M, Sheikh F and Christman K L 2010 *PLoS One* **5** e13039
[48] Singelyn J M, DeQuach J A and Christman K L 2009 *Conf. Proc. IEEE Eng. Med. Biol. Soc.* **2009** 2406–08
[49] Young D A, Ibrahim D O, Hu D and Christman K L 2011 *Acta Biomater.* **7** 1040–49
[50] DeQuach J A, Yuan S H, Goldstein L S and Christman K L 2011 *Tissue Eng. Part* A **17** 2583–92

[51] DeQuach J A, Lin J E, Cam C, Hu D, Salvatore M A, Sheikh F and Christman K L 2012 *Eur. Cell Mater.* **23** 400–12

[52] Gaetani R, Aude S, Demaddalena L L, Straessle H, Dzieciatkowska M, Wortham M, Bender H R, Nguyen-Ngoc K V and Schmid-Schoenbein G W *et al* 2018 *Tissue Eng. Part C Methods* **24** 697–708

[53] Johnson T D, Braden R L and Christman K L 2014 *Methods Mol. Biol.* **1181** 109–20

[54] Hernandez M J and Christman K L 2017 *JACC Basic Transl. Sci.* **2** 212–26

[55] Song M, Liu Y and Hui L 2018 *Mol. Med. Rep.* **17** 138–46

[56] Wilson G J, Yeger H, Klement P, Lee J M and Courtman D W 1990 *ASAIO Trans.* **36** M340–3

[57] Brown A L, Farhat W, Merguerian P A, Wilson G J, Khoury A E and Woodhouse K A 2002 *Biomaterials* **23** 2179–190

[58] Sarig U, Au-Yeung G C, Wang Y, Bronshtein T, Dahan N, Boey F Y, Venkatraman S S and Machluf M 2012 *Tissue Eng. Part* A **18** 2125–37

[59] Grauss R W, Hazekamp M G, Oppenhuizen F, van Munsteren C J, Gittenberger-de Groot A C and DeRuiter M C 2005 *Eur. J. Cardiothorac. Surg.* **27** 566–71

[60] Kheir E, Stapleton T, Shaw D, Jin Z, Fisher J and Ingham E 2011 *J. Biomed. Mater. Res.* A **99** 283–94

[61] Kajbafzadeh A M, Masoumi A, Hosseini M, Borjian M A, Akbarzadeh A and Mohseni M J 2014 *J. Biosci. Bioeng.* **117** 236–41

[62] Livesey S A, del Campo A A, Nag A, Nichols K B and Coleman C 1994 *US Patent* 5336616

[63] Wainwright D J 1995 *Burns* **21** 243–48

[64] Weymann A, Loganathan S, Takahashi H, Schies C, Claus B, Hirschberg K, Soós P, Korkmaz S and Schmack B *et al* 2011 *Circ. J.* **75** 852–60

[65] Maghsoudlou P, Totonelli G, Loukogeorgakis S P, Eaton S and De Coppi P 2013 *J. Vis. Exp.* **80**

[66] Lin P, Chan W C W, Badylak S F and Bhatia S N 2004 *Tissue Eng.* **10** 1046–53

[67] Bonvillain R W, Scarritt M E, Pashos N C, Mayeux J P, Meshberger C L, Betancourt A M, Sullivan D E and Bunnell B A 2013 *J. Vis. Exp.* **82** e50825

[68] Courtman D W, Pereira C A, Kashef V, McComb D, Lee J M and Wilson G J 1994 *J. Biomed. Mater. Res.* **28** 655–66

[69] Baert Y, Stukenborg J B, Landreh M, De Kock J, Jörnvall H, Söder O and Goossens E 2015 *Hum. Reprod.* **30** 256–67

[70] Caplan A 2011 *Am. J. Bioeth.* **11** 1–5

[71] Lin S S, Rich L, Pal J D and Sade R M 2012 *Ann. Thorac. Surg.* **93** 1773–79

[72] Tonti-Filippini N and Zeps N 2011 *Med. J. Aust.* **194** 263–65

[73] Pirnay J P, Vanderkelen A, Zizi M, De Vos D, Rose T, Laire G, Ectors N and Verbeken G 2010 *Bull. World Health Organ.* **88** 870–72

[74] Gillette E P 1872 *Du tissu conjonctif ou lamineux. Concours d'agrégation (anatomie et physiologie)* (Paris: Parent)

[75] Zachariades P A 1900 *Compt. Rend. Seances Soc. Biol. Fil.* **52** 182–84

[76] Nageotte J 1927 *Compt. Rend. de l'Académie des sciences* **184** 115–17

[77] Nageotte J 1927 *Compt. Rend. Seances Soc. Biol. Fil.* **96** 172–74

[78] Trelstad R L, Catanese V M and Rubin D F 1976 *Anal. Biochem.* **71** 114–18

[79] Ju H, Dan D, Dan W and Liu X 2014 *Chinese Patent* 104189944-A

[80] Ramshaw J A M 2016 *J. Biomed. Mater. Res. B-Appl. Biomater.* **104** 665–75
[81] Werkmeister J A and Ramshaw J A M 2012 *Biomed. Mater.* **7** 012002
[82] Tustanovskii A A 1947 *Biokhimiya* **12** 285–90
[83] Orekhovich V N, Tustanovskii A A, Orekhovich K D and Plotnikova N E 1948 *Biokhimiya* **13** 55–60
[84] Gross J, Highberger J H and Schmitt F O 1955 *Proc. Natl Acad. Sci. USA* **41** 1–7
[85] Jackson D S and Fessler J H 1955 *Nature* **176** 69–70
[86] Bornstein P 1974 *Annu. Rev. Biochem.* **43** 567–603
[87] Rajan N, Habermehl J, Cote M-F, Doillon C J and Mantovani D 2006 *Nat. Protoc.* **1** 2753–58
[88] Gross J and Kirk D 1958 *J. Biol. Chem.* **233** 355–60
[89] Veis A and Payne K 1988 *Collagen* vol 1 ed M E Nimni (Boca Raton, FL: CRC Press) pp 113–37
[90] Peterova E, Mrkvicova A, Podmolikova L, Rezacova M and Kanta J 2016 *Physiol. Res.* **65** 661–72
[91] Bell E, Ivarsson B and Merrill C 1979 *Proc. Natl Acad. Sci. USA* **76** 1274–78
[92] Ramshaw J A M, Werkmeister J A and Bremner H A 1988 *Arch. Biochem. Biophys.* **267** 497–502
[93] Rigby B J, Mitchell T W and Robinson M S 1977 *Biochem. Biophys. Res. Commun.* **79** 400–05
[94] Fauzi M B, Lokanathan Y, Aminuddin B S, Ruszymah B H I and Chowdhury S R 2016 *Mater. Sci. Eng. C-Mater. Biol. App.* **68** 163–71
[95] Jackson D S, Leach A A and Jacobs S 1958 *Biochim. Biophys. Acta* **27** 418–20
[96] Gross J 1958 *J. Exp. Med.* **107** 247–63
[97] Wood G C and Keech M K 1960 *Biochem. J.* **75** 588–98
[98] Lowther D A, Green N M and Chapman J A 1961 *J. Biophys. Biochem. Cytol.* **10** 373–88
[99] Gallop P M, Seifter S and Meilman E 1957 *J. Biophys. Biochem. Cytol.* **3** 545–57
[100] Miller E J and Rhodes R K 1982 *Meth. Enzymol.* **82** 33–64
[101] Geiger B J, Steenbock H and Parsons H T 1933 *J. Nutrit.* **6** 427–42
[102] Bachhuber T E, Lalich J J, Angevine D M, Schilling E D and Strong F M 1955 *Proc. Soc. Exp. Biol. Med.* **89** 294–97
[103] Bachhuber T E and Lalich J J 1954 *Science* **120** 712–13
[104] Lalich J J, Barnett B D and Bird H R 1957 *Arch. Pathol.* **64** 643–48
[105] Levene C I and Gross J 1959 *J. Exp. Med.* **110** 771–90
[106] Narayanan A S, Siegel R C and Martin G R 1972 *Biochem. Biophys. Res. Commun.* **46** 745–51
[107] Sage H, Trueb B and Bornstein P 1983 *J. Biol. Chem.* **258** 3391–401
[108] Nimni M E 1977 *Proc. Roy. Soc. Med.* **70** 65–72
[109] Walshe J M 1956 *Am. J. Med.* **21** 487–95
[110] Ewald A and Kühne W 1887 *Verb. d. Naturhist. Med. Ver. Heidelberg, N. S.* **1** 451
[111] Thomas A W and Seymour-Jones F L 1923 *J. Am. Chem. Soc.* **45** 1515–22
[112] Hodge A J, Highberger J H, Deffner G G J and Schmitt F O 1960 *Proc. Natl Acad. Sci. USA* **46** 197–206
[113] Gottschall G Y and Kies M W 1942 *Food Res.* **7** 373–81
[114] Miyada D S and Tappel A L 1956 *Food Res.* **21** 217–25

[115] Nishihara T and Miyata T 1962 *Collagen Symposium* vol 3 (Japan: Collagen Research Society) pp 66–93

[116] Kühn K, Fietzek P and Kühn J 1963 *Naturwissenschaften* **50** 444–5

[117] Morimoto K, Kunii S, Hamano K and Tonomura B 2004 *Biosci. Biotech. Biochem.* **68** 861–7

[118] Nishihara T 1962 *US Patent* 3034852

[119] Addad S, Exposito J-Y, Faye C, Ricard-Blum S and Lethias C 2011 *Marine Drugs* **9** 967–83

[120] Werkmeister J A, Ramshaw J A M and Ellender G 1990 *Eur. J. Biochem.* **187** 439–43

[121] Miller E J, Finch J E, Chung E, Butler W T and Robertson P B 1976 *Arch. Biochem. Biophys.* **173** 631–37

[122] Wang H M, Chan J, Pettigrew D W and Sodek J 1978 *Biochim. Biophys. Acta* **533** 270–77

[123] Welgus H G, Burgeson R E, Wootton J A M, Minor R, Fliszar C and Jeffrey J J 1985 *J. Biol. Chem.* **260** 1052–59

[124] Peng Y, Glattauer V, Werkmeister J A and Ramshaw J A M 2004 *J. Cosmet. Sci.* **55** 327–41

[125] Stephens L J, Werkmeister J A, Caine J M and Ramshaw J A M 1988 *Das Leder* **39** 88–93

[126] Stephens L J, Werkmeister J A, Tebb T A and Ramshaw J A M 1991 *Das Leder* **42** 41–4

[127] Vogeli G, Horn E, Carter J and Kaytes P S 1986 *FEBS Lett.* **206** 29–32

[128] Werkmeister J A, Peters D E and Ramshaw J A M 1989 *J. Biomed. Mater. Res.* **23** 273–83

[129] Werkmeister J A, Tebb T A, White J F and Ramshaw J A M 1993 *J. Histochem. Cytochem.* **41** 1701–6

[130] Rhodes R K and Miller E J 1978 *Biochemistry* **17** 3442–48

[131] Sage H and Bornstein P 1979 *Biochemistry* **18** 3815–22

[132] Peng Y Y, Glattauer V, Ramshaw J A M and Werkmeister J A 2010 *J. Biomed. Mater. Res.* **93A** 1235–44

[133] Wood A, Ogawa M, Portier R J, Schexnayder M, Shirley M and Losso J N 2008 *Comp. Biochem. Physiol. B-Biochem. Mol. Biol.* **151** 246–49

[134] Sai K P and Babu M 2001 *Comp. Biochem. Physiol. B-Biochem. Mol. Biol.* **128** 81–90

[135] Ashhurst D E and Bailey A J 1980 *Eur. J. Biochem.* **103** 75–83

[136] Francois J, Herbage D and Junqua S 1980 *Eur. J. Biochem.* **112** 389–96

[137] Kimura S, Takema Y and Kubota M 1981 *J. Biol. Chem.* **256** 13230–234

[138] Cozza N, Bonani W, Motta A and Migliaresi C 2016 *Int. J. Biol. Macromol.* **87** 504–13

[139] Tomita M, Kinoshita T, Izumi S, Tomino S and Yoshizato K 1994 *Biochim. Biophys. Acta* **1217** 131–40

[140] Torre-Blanco A and Toledo I 1981 *J. Biol. Chem.* **256** 5926–30

[141] Yamada S, Yamamoto K, Ikeda T, Yanagiguchi K and Hayashi Y 2014 *BioMed Res. Int.* **2014** 302932

[142] Kelly J, Tanaka S, Hardt T, Eikenberry E F and Brodsky B 1988 *J. Biol. Chem.* **263** 980–7

[143] Tillet E, Franc J M, Franc S and Garrone R 1996 *Comp. Biochem. Physiol. B-Biochem. Mol. Biol.* **113** 239–46

[144] Wang L, Liang Q, Wang Z, Xu J, Liu Y and Ma H 2014 *Food Chem.* **148** 410–14

[145] Goissis G, Piccirili L, Goes J C, de Guzzi Plepis A M and Das-Gupta D K 1998 *Artif. Organs* **22** 203–9

[146] Rauterberg J and Kühn K 1968 *Hoppe Seylers Z. Physiol. Chem.* **349** 611–22

[147] Glanville R W, Rauter A and Fietzek P P 1979 *Eur. J. Biochem.* **95** 383–89

[148] Sage H, Woodbury R G and Bornstein P 1979 *J. Biol. Chem.* **254** 9893–900

[149] Tayot J L and Tardy M 1995 *US Patent* 5436135
[150] Risteli J, Bächinger H P, Engel J, Furthmayr H and Timpl R 1980 *Eur. J. Biochem.* **108** 239–50
[151] Dixit S N, Stuart J M, Seyer J M, Risteli J, Timpl R and Kang A H 1981 *Coll. Relat. Res.* **1** 549–56
[152] Mayne R, Sanderson R D, Wiedemann H, Fitch J M and Linsenmayer T F 1983 *J. Biol. Chem.* **258** 5794–97
[153] Kleinman H K, McGarvey M L, Liotta L A, Robey P G, Tryggvason K and Martin G R 1982 *Biochemistry* **21** 6188–93
[154] Exposito J Y and Garrone R 1990 *Proc. Natl Acad. Sci. USA* **87** 6669–73
[155] Fowler S J, Jose S, Zhang X, Deutzmann R, Sarras M P and Boot-Handford R P 2000 *J. Biol. Chem.* **275** 39589–599
[156] Cescon M, Gattazzo F, Chen P and Bonaldo P 2015 *J. Cell Sci.* **128** 3525–31
[157] Trüeb B, Schreier T, Bruckner P and Winterhalter K H 1987 *Eur. J. Biochem.* **166** 699–703
[158] Bishop P, Ayad S, Reardon A, McLeod D, Sheehan J and Kielty C 1996 *Graefes Arch. Clin. Exp. Ophthalmol.* **234** 710–13
[159] Lekskul M, Burrows R, Kublin C L and Cintron C 2000 *Cornea* **19** 194–203
[160] Quarto N, Cancedda R and Descalzi-Cancedda F 1985 *Eur. J. Biochem.* **147** 397–400
[161] Ayad S, Kwan A P and Grant M E 1987 *FEBS Lett.* **220** 181–6
[162] Thoroski J H 2003 *US Patent* 6649203
[163] Vlad V 2009 *US Patent* 7584909
[164] Sah M K and Rath S N 2016 *Mater. Sci. Eng. C Mater. Biol. Appl.* **67** 807–21
[165] Takahashi K, Shirai K, Kitamura M and Hattori M 1996 *Biosci. Biotech. Biochem* **60** 1299–302
[166] Yi F, Guo Z X, Zhang L X, Yu J and Li Q 2004 *Biomaterials* **25** 4591–99
[167] Arias J L, Carrino D A, Fernández M S, Rodríguez J P, Dennis J E and Caplan A I 1992 *Arch. Biochem. Biophys.* **298** 293–302
[168] Ziyadeh F N, Sharma K, Ericksen M and Wolf G 1994 *J. Clin. Invest.* **93** 536–42
[169] Bomgardner M 2016 *C&EN* **94** 13
[170] Bateman J F, Cole W G, Pillow J J and Ramshaw J A M 1986 *J. Biol. Chem.* **261** 4198–203
[171] Laurent T C 1995 *Biophys. Chem.* **57** 7–14
[172] Minton A P 2013 *Biopolymers* **99** 239–44
[173] Breydo L, Reddy K D, Piai A, Felli I C, Pierattelli R and Uversky V N 2014 *Biochim. Biophys. Acta* **1844** 346–57
[174] Lareu R R, Subramhanya K H, Peng Y, Benny P, Chen C, Wang Z, Rajagopalan R and Raghunath M 2007 *FEBS Lett.* **581** 2709–14
[175] Chen C, Loe F, Blocki A, Peng Y and Raghunath M 2011 *Adv. Drug Deliv. Rev.* **63** 277–90
[176] Benny P, Badowski C, Lane E B and Raghunath M 2015 *Tissue Eng. Part* A **21** 183–92
[177] Rashid R, Lim N S, Chee S M, Png S N, Wohland T and Raghunath M 2014 *Tissue Eng, Part C Methods* **20** 994–1002
[178] Piez K A 1967 *Treatise on collagen* vol 1 ed G N Ramachandran (London: Academic) pp 207–52
[179] Knapp T R, Kaplan E N and Daniels J R 1977 *Plast. Reconst. Surg.* **60** 398–405
[180] Deyl Z and Adam M 1989 *J. Chromatog. Biomed. App.* **488** 161–97
[181] Ramshaw J A M, Bateman J F and Cole W G 1984 *Anal. Biochem.* **141** 361–65
[182] Hong B S, Davison P F and Cannon D J 1979 *Biochemistry* **18** 4278–82

[183] Skinner S J M, Grego B, Hearn M T W and Liggins G C 1984 *J. Chromatog.* **308** 111–19
[184] Welsh C, Gay S, Rhodes R K, Pfister R and Miller E J 1980 *Biochim. Biophys. Acta* **625** 78–88
[185] Deyl Z, Macek K, Adam M and Horáková M 1982 *J. Chromatog.* **230** 409–14
[186] Miller E J 1971 *Biochemistry* **10** 1652–59
[187] Rhodes R K and Miller E J 1981 *Collagen Rel. Res.* **1** 337–43
[188] Peng Y Y, Yoshizumi A, Danon S J, Glattauer V, Prokopenko O, Mirochnitchenko O, Yu Z, Inouye M and Werkmeister J A *et al* 2010 *Biomaterials* **31** 2755–61
[189] Noelken M E and Bettin B D 1983 *Anal. Biochem.* **134** 374–81
[190] Condell R A, Hanko V P, Larenas E A, Wallace G and McCullough K A 1993 *Anal. Biochem.* **212** 436–45
[191] ChandraRajan J and Klein L 1980 *J. Chromatog.* **182** 94–9
[192] Hirako Y, Usukura J, Nishizawa Y and Owaribe K 1996 *J. Biol. Chem.* **271** 13739–45
[193] Banyard J, Bao L and Zetter B R 2003 *J. Biol. Chem.* **278** 20989–994
[194] Hashimoto T *et al* 2002 *EMBO J.* **21** 1524–34
[195] Sweeney S M, Orgel J P R O, Fertala A, McAuliffe J D, Turner K R, Di Lullo G A, Chen S, Antipova , O and Perumal , S *et al* 2008 *J. Biol. Chem.* **283** 21187–97
[196] Ricard-Blum S, Beraud M, Raynal N, Farndale R W and Ruggiero F 2006 *J. Biol. Chem.* **281** 25195–204
[197] Mizuno K and Hayashi T 1994 *J. Biochem.* **116** 1257–63
[198] ChandraRajan J 1978 *Biochem. Biophys. Res. Commun.* **83** 180–86
[199] Glanville R W, Allmann H and Fietzek P P 1976 *Hoppe-Seyler's Z. Physiol. Chem.* **357** 1663–65
[200] Barth D, Kyrieleis O, Frank S, Renner C and Moroder L 2003 *Chemistry* **9** 3703–14

IOP Publishing

John A M Ramshaw and Veronica Glattauer

Chapter 9

Production of recombinant collagens

Recombinant technology enables the production of all types of human collagens, even those that occur in very low amounts naturally. The products are of high purity and are disease free and can be produced in sufficient yields for use in biomaterial applications. The commercial availability of synthetic genes makes it possible to produce new structures, based on natural collagens or innovative designs, with specific functionality. These novel materials may be particularly useful in areas such as tissue engineering.

9.1 Recombinant animal collagen production

The ongoing advances in molecular biology methods that have made expression of proteins routine have now been applied to a wide range of different collagens. Recombinant forms of collagen were slower to appear than recombinant globular proteins, due in part to various technology challenges that were typically not encountered in the production of most other proteins. These included, for example, the highly repetitive collagen sequence with frequent Gly and Pro residues, where the complimentary GGN and CCN codon sequences give the potential for unwanted gene modifications. Another challenge is the need for effective secondary modification to provide hydroxylation, so as to ensure protein triple-helical stability. Collagen lacking hydroxylation has a melting temperature, T_m, more than 10°C lower than native hydroxylated collagen [1]. The more efficient microbial expression systems (*E. coli* and yeasts) do not have the collagen post-translational enzyme systems that are found in animal cells and so development of functional, intra-cellular Pro hydroxylation was a significant additional step.

Almost all studies on recombinant collagens have been on various human collagen types [2]. In many cases this has allowed samples for identification, characterisation and biological studies, but in other cases the aims have been to provide materials in sufficient quantity for use in biomedical materials. As a result, a

doi:10.1088/978-0-7503-2096-2ch9

wide selection of recombinant animal collagen genes has been examined using a wide range of different expression systems (table 9.1). These studies have been the subject of several reviews (including [2]).

Recombinant collagens offer significant advantages compared to animal-based collagens for use in biomedical applications [2]. These include uniformity in purity and quality, freedom from infectious agents and ease of isolation and purification by standardised methods. Development of human collagen products minimises the potential for adverse immunological responses to collagen biomaterials. Also, collagen types that may be in low abundance in nature are equally as available as any other more abundant collagen type. Beyond the use for production of natural structures, recombinant technology also readily enables production of novel variations on natural protein structures, novel structures based on repeating

Table 9.1. Examples of expression systems for recombinant collagens. (Adapted from [2], with permission from IOP Publishing.)

(a) Expression in systems with endogenous prolyl-4-hydroxylase (P4H) activity				
System	Host	Collagen type	Hydroxylation	
Cell culture	HT1080 cells	Type 1 α1	Endogenous	[3]
	CHO cells	Type 4	Endogenous	[4]
	HEK293 cells	Type 10	Endogenous	[5]
	Sf9 with baculovirus	Type 1	Endogenous	[6]
		Type 2	Augmented	[7]
		Type 3	Endogenous	[8]
		Type 9	Augmented	[9]
		Type 13	Endogenous	[10]
Transgenic animals	Mouse	Type 1 α1	Endogenous	[11]
	Silkworm	Type 3 fragment	Endogenous	[12]
(b) Expression in systems with no P4H activity				
Transgenic plants	Tobacco leaves	Type 1 α1	No	[1, 13]
	Maize seed	Type 1 α1	No	[14, 15]
	Barley		No	[16]
Micro organisms	*E. coli*	Type 2 CB8, CB10	No	[17]
		Type I	No	[18]
(c) Expression in systems with added P4H activity				
Transgenic plants	Tobacco leaves	Type 1 α1	Yes	[19]
		Type 1		[13]
	Maize seed	Type 1 α1	Yes	[14]
Micro organisms	*S. cerevisiae*	Type 1	Yes	[20]
	P. pastoris	Types 1 & 3	Yes	[21]
	H. polymorpha		Yes	[22]
	E. coli	Type 3	Yes	[23]

functional sequences and development of novel structures based on chimeric constructs with other collagens or other molecule types, such as growth factors. This directed modification of animal collagen sequences allows for the design of targeted collagen products for specific applications.

However, there are difficulties too, particularly the inclusion of active secondary modification enzymes. While the general expression process leads to uniformity in the primary product, the potential for a variable extent of secondary modification re-introduces potential structural variations within the product. The other major consideration for biomaterial applications is the production of recombinant collagens in commercially relevant yields and within competitive cost constraints.

9.1.1 Bacterial expression

Various systems have been examined for collagen expression (table 9.1), with initial studies focused on use of *E. coli*, where the extensive experience and tools that had been developed from studies of globular proteins (for example, [24]) could be exploited. Although this system allowed the production of gelatin-like chains based on designed collagen-like sequences [25, 26] the difficulties at that point in introducing active prolyl-4-hydroxylase (P4H) activity limited the progress with this system.

Despite the difficulties in hydroxylation, there have been several studies on production of human-like collagen, lacking hydroxylation, in fed-batch cultures [27]. Various studies have examined approaches to increase the yields obtained, including analysis of metabolic flux [28], use of CO_2 pulses and gas control [29], control of acetic acid assimilation [30] and genetic modification of the expression host [31]. Overall, these improvements led to yields exceeding 12 g l^{-1} [32].

Direct inclusion of Hyp during expression in *E. coli* has also been shown using co-translational incorporation during fermentation through modification of the media immediately prior to expression [33]. This approach, however, does not select between the Xaa and Yaa positions for incorporation, and Hyp would be incorporated into any non-collagenous segments that were present. While inclusion in the Yaa position would enhance stability, incorporation in the Xaa position may be slightly destabilising [34].

Subsequently, approaches that allow specific modification to produce Hyp in the Yaa position have been studied. Initially, an active human P4H was incorporated in *E. coli*, using Origami B(DE3) cells which have a relatively oxidising cytosol [35]. Subsequently, the host genes were adapted to introduce a biosynthetic shunt that was able to produce ascorbate-like molecules as well as expressing heterologous human P4H [36]. This new strain produced high levels of Hyp in collagenous sequences, including a series of short $(Pro\text{-}Pro\text{-}Gly)_n$ constructs ($n = 5, 7, 10$) [36], a short collagen-like segment in adiponectin [37] and a polymeric collagen-like fragment, $(GAPGAPGSQGAPGLQ)_{16}$, that showed 10% hydroxylation [17].

9.1.2 Animal cell expression

As a result of the initial limitations with *E. coli*, a range of other host systems were examined. These included various mammalian cells [3–5], including CHO, COS,

HEK293 and HT1080 (see [2]) and insect cells, principally using baculoviral transfection [8], where an endogenous P4H activity was present. This P4H activity is greater in the mammalian cells, so in some insect cell studies this activity was also augmented with introduced enzyme [7]. These systems proved useful in producing a range of different collagen types (table 9.1), including those that tested new concepts on designed structures [38]. However, although good for small scale trial experiments, these cell systems were not suited to commercial biomaterial applications as the yields of protein were too low.

9.1.3 Yeast expression

Various studies indicated that yeast systems with introduced P4H could provide a commercially viable biomaterial. *Saccharomyces cerevisiae* has been used successfully to produce hydroxylated heterotrimeric collagen type I in good quantity [20], and subsequently also for type III collagen [39]. The most detailed work has used expression in *Pichia pastoris,* with a particular emphasis on obtaining hydroxylation at native levels [40]. This recombinant, purified collagen behaves as native collagen, for example forming banded fibrils, and shows potential in clinical applications. While most reports have used expression that accumulates the collagen product within the cell, *Pichia* can also be used to produce a secreted product [41], but the yields were not sufficient to establish a purification strategy.

9.1.4 Transgenic expression

Animal systems

A further approach to recombinant collagen production has been the establishment of transgenic animal where the collagen is expressed into the milk, using the endogenous hydroxylation system [11]. In mice, a very good yield of up to 8 mg ml^{-1} was obtained of type I $\alpha1$trimer, mostly, if not all, as triple-helical protein, with a T_m of around 31 °C, indicating good hydroxylation of about 45% [11]. Again, the overall amount of protein was not sufficient to establish a purification strategy, and despite this initial success, the technology was not established in larger animals that would have given enough protein for purification.

Transgenic silkworms have also been investigated as a convenient bioreactor system. This system has the potential advantage that cultivation and production of silkworm cocoons and the subsequently processing are all well understood in the large-scale silk industry. Initial studies used a type III collagen fragment as part of a fibroin light chain and green fluorescent protein construct. Transfection with this construct using a *piggyBac* vector system led to cocoons containing the fluorescently labelled collagen fragment, which could then be readily purified [42]. Subsequently, full-length type I collagen $\alpha1$-chain was produced in cocoons. Although initial yields were low, various improvements allowed production of 4.2 mg per cocoon (about 8%). This product was readily purified. It contained no Hyp and CD spectra showed that it was a gelatin-like material with no triple-helical content. Nevertheless, this material proved useful as a cell culture substrate [12].

Plant systems
Various transgenic plant systems have been developed, typically based on chromosomal integration of the collagen system genes, although the use of transient expression [43] is also an option. Initial studies focused on tobacco leaf as the production tissue expressing a type I collagen α1-chain to give a homotrimer product, lacking targeted hydroxylation [1, 44]. Subsequent studies moved to include mammalian P4H activity, using transient expression, giving a good level of hydroxylation and a T_m of 37 °C [19]. Further development of this approach has led to a complete type I collagen heterotrimeric collagen, that included full prolyl hydroxylation (by human P4H) and also lysyl hydroxylation (by human LH3) [13]. This material possibly represents the present pinnacle of recombinant collagen production and has led to commercial, regulatory approved biomedical products.

Expression of recombinant collagen into various grain seeds has also been examined. Grains were seen as an opportunity for large-scale production as growth and harvesting are clearly already established and process by-products can be used as biomass for further products. The host species have been rice [45], barley [16] and maize (corn) [14, 15]. These studies also looked at producing a type II collagen α1-chain fragment [45] and type I α1-chains (and fragments) to give a homotrimer product. Initial studies did not include added, targeted hydroxylation, but low levels of endogenous hydroxylation seemed to occur [15, 16], although the plant hydroxylation specificity may be broader than the highly specific P4H [46]. A later study used co-expression with human P4H [14] which gave very good hydroxylation levels, possibly augmented by some less-specific endogenous activity.

9.1.5 Purification and quality

Prior to purification, a principle consideration has been the extraction of the collagen. The wide variety of different expression hosts (table 9.1) has in turn resulted in a wide range of yields, thus necessitating different approaches. Processing of cells and tissues to release soluble collagen for purification has generally followed standard methods, as outlined in various manuals (e.g., [24]). These include, for example, use of ultrasonic and extrusion (French press) methods. It is possible that in some cases, where nominal production yields are good, that further variations in extraction approaches may still give enhanced yields of purified collagens in future. For example, there were significant studies through the 1960s on the large scale, high yield extraction of plant protein from leaves, to develop as a food additive or substitute [47]. It is possible that adaptation of these methods may yet be of value to the efficient large-scale extraction of recombinant collagen from leaves. In other cases, more indirect extraction approaches have been developed. For example, for recombinant collagen expressed in barley seeds, the seeds were germinated and extracted based on the approach to producing malt for the brewing industry [48]. For these and other seeds either dry or wet milling can be used.

In some expression system, for example using *Pichia* or *Saccharomyces*, it may be possible to produce a secreted product. This would generally be expected to be a moderately dilute solution with a different set of impurities, this time derived from the media components as opposed to cell derived components. Hence, after extraction or collection, a suitable initial concentration and fractionation step is probably needed in many if not all approaches. Subsequently, purification of the collagens has generally followed the established precipitation and chromatographic techniques (chapter 8).

A further consideration during purification is the potential for a range of distinct components within the preparation, albeit that the intention is uniformity. This variation arises almost entirely from the extent of secondary modification that has occurred. In some cases, no attempts to obtain prolyl hydroxylation have been made [49], while on the other hand there are examples where a high degree of hydroxylation, essentially equivalent to natural collagens, have been obtained [50]. In other cases, hydroxylation of lysine has also been introduced [13]. Whenever any hydroxylation is not complete, then a potential range of partially modified proteins could be present.

9.2 Recombinant bacterial collagen production

Bacterial collagens are ideally suited for recombinant expression. Their stability despite the lack of Hyp means that these materials are readily expressed in *E. coli* as the expression host. This expression system is well understood and has been a dominant system in pharmaceutical production [51]. These collagens (table 9.2), but principally the Scl2 gene from *S. pyogenes,* have been produced and purified in small scale experiments for use in laboratory studies of the collagens and of variations in collagen structure.

Some groups have used modified structures to assist in efficient production (figure 9.1). The changes that have been introduced include the addition of a His_6-tag sequence, normally, but not exclusively, at the N-terminal of the constructs, which was then used in conjunction with immobilised metal affinity chromatography (IMAC) with a Ni^{2+} charged nitriloacetic acid-based column in small scale purification [57]. This tag can also be added via a vector used during production. Also, a protease domain can be added between the V and CL domains to assist in removal of the V-domain, taking advantage of the typical stability of the triple-helix to proteolysis. Finally, the non-helical tail is deleted as it is not usually required.

The most frequently used expression vector system has been the high yield pCold (Takara Bio) cold-shock expression system [59], but other vectors including pET are

Table 9.2. Expression of bacterial collagens, where no P4H activity is required for stability.

Host	Collagen type		Host	Collagen type	
E. coli	*S. pyogenes* Scl1	[52]	*E. coli*	*C. perfringens*	[55]
	S. pyogenes Scl2	[52]		*S. usitatus*	[55]
	B. anthracis BclA	[53]		*R. palustris*	[55]
	L. pneumophilia Lcl	[54]		*Methylobacterium sp.*	[55]
	Bacteriophage ECs2717 (from *E. coli* O157:H7)	[56]			

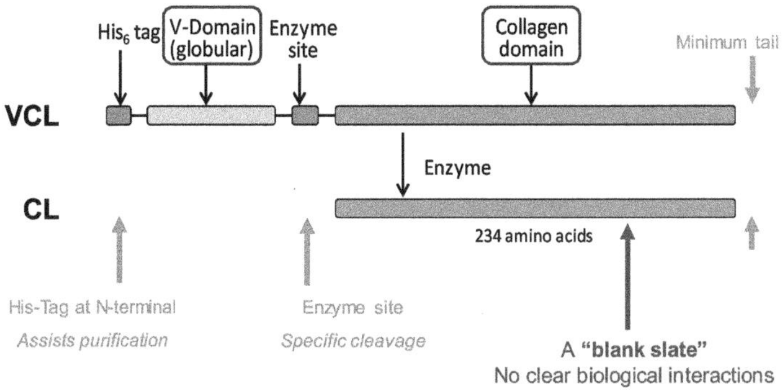

Figure 9.1. A schematic representation of an *S. pyogenes* Scl2 construct that has been used for collagen production (Reprinted from [58] with permission from IOP Publishing.)

Table 9.3. Fermentation conditions and protein yields for V-CL in a pCold vector using defined medium. (Reprinted from [60] with permission from Springer Nature.)

Process	Growth phase	Induction phase	Final OD (600 nm)	Wet cell paste ($g\ L^{-1}$)	Volumetric productivity ($g\ L^{-1}$)
Flask	24 h at 37 °C	25 °C for 10 h 15 °C for 14 h	5.6	8.3	0.2
Batch	5 h at 37 °C	25 °C for 7 h	5.5	5.3	0.7
Fed-batch	24 h at 37 °C	25 °C for 10 h	88	113	9.5
Fed-batch	24 h at 37 °C	25 °C for 10 h 20 °C for 5 h 15 °C for 9 h	>96	Up to 148	Up to 19.3 (Av(6) = 13.4)

also entirely suitable and could possibly give better commercial yields. Small scale purifications of *S. pyogenes* Scl2 protein have used shake flasks for production, but this approach only gives low yields of recombinant product, < 1 g L^{-1} [60] (table 9.3).

Bacterial collagens have significant potential for development of new biomedical materials. Initial studies have shown that they are non-cytotoxic and non-immunogenic [61]. However, for commercial applications, production and purification at a larger scale that is compatible with possible regulatory requirements is required along with ease of scale up and cost-effective processing.

Large-scale production in *E. coli* has been achieved through systematic studies of fermenter technology, typically in 2 L vessels [60]. More relevant yields of up to 10 g L^{-1} were obtained with a high cell density fed-batch process and defined media. The use of defined media specifically avoids any animal-derived components. Further variations including an extended 24 h production time (table 9.3) led to production yields of up to 19 g L^{-1} for V-CL constructs [60]. These yields compare favourably with the average, non-proprietary yields of around 14 g L^{-1} that have been reported for other bacterial expression studies [62]. A good fermentation yield, nevertheless, needs to be translated into a good purified protein yield, although if the bacterial collagen is to be used as a coating for a medical device or as a composite, the production needs could be less than that if it is used as a single component.

The use of the His_6-tag protocol, generally followed by gel permeation chromatography using a Superdex™ 200 column, is well suited to small scale laboratory purifications [57, 61]. Enzyme digestion, for example, using trypsin, proved efficient for removal of the V-domain, if this is required. However, this approach is not suitable for large-scale commercial preparations.

For large-scale production an effective strategy has been described that included removal of the V-domain and gave overall yields of at least 10 g L^{-1} [58]. The use of IMAC and many other chromatographic approaches are not suitable for large-scale production. In a large-scale process, host *E. coli* proteins were substantially removed through use of isoelectric precipitation. This was followed by a proteolytic digestion step that degraded residual host proteins and also removed the V-domain from the V-CL product. The bacterial collagens, like animal collagens, share the high proteolytic resistance of the triple-helical domain, so the proteolytic step normally leaves the bacterial CL domain intact (figure 9.2) [58]. Final steps can include use of a cross-flow filtration membrane [58] and a polishing step if required. As part of this, another consideration that is important is that the protocol should control the extent

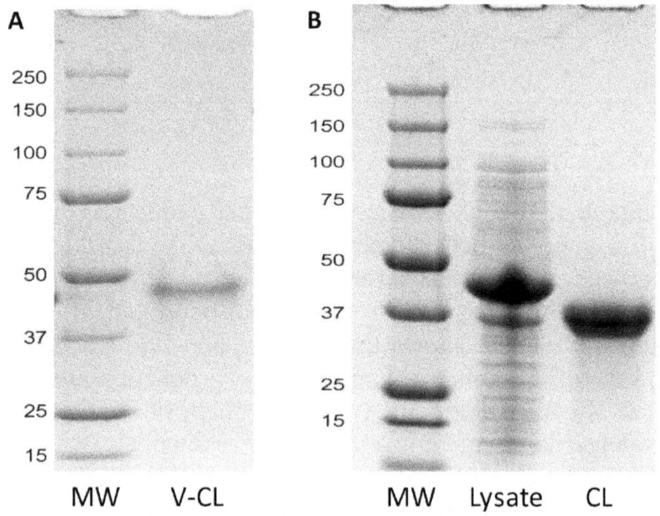

Figure 9.2. SDS-PAGE of samples from purifications of recombinant bacterial collagen from *S. pyogenes.* (A) Small, laboratory scale extraction and purification using a His_6-tag affinity method. (B) Large-scale extraction and purification using precipitation and proteolysis. MW are molecular weight standards (kDa), Lysate is the *E. coli lysate* and V-CL and CL are the purified *S. pyogenes* collagen components. (Gel photographs courtesy of Yong Peng, CSIRO.)

of endotoxin contamination [63], which is more readily achieved in an industrial rather than a laboratory environment.

9.3 Recombinant 'bioengineered' adaptations to collagen structures

9.3.1 Adaptations to animal collagens

The number of variations that have been made to animal collagens is fairly limited as these collagens are not easily expressed as recombinant proteins, and, from a biomaterials perspective, it is unlikely that many of these products will have commercial value.

Recombinant collagens with repeating triple-helix segments

Stably transfected tumour cell line HT1080 constructs were used for initial production of recombinant collagens [64]. This approach allowed production of type II collagen that formed a stable triple-helix and D-periodic fibrils [65]. Subsequently, a 'cassette model' was developed to allow a further understanding of the functionality of this collagen. Each cassette was of a D-period, 234 amino acids as well as the shorter overlap sequence comprising 0.4 of a full D-period. Collagen structures. each lacking one of the D-periods were expressed (figure 9.3). Of these five constructs, only those that contained D1, D4 and 0.4 D sequences formed fibrils with a periodic banding pattern by TEM [66]. These studies also showed that the D4 period is the sequence element that allows adhesion, spreading and migration of chondrocytes [67], while also containing the DDR cell receptor [65, 67–71]. Together, these data were used to design a novel molecule based around 3 tandem D4 repeats (figure 9.3) that showed better performance as a coating than unmodified type II collagen [38].

A construct based on repeating sequences from type III collagen was also reported. In this study, 1, 2 and 3 repeats of an integrin binding region ware made and expressed

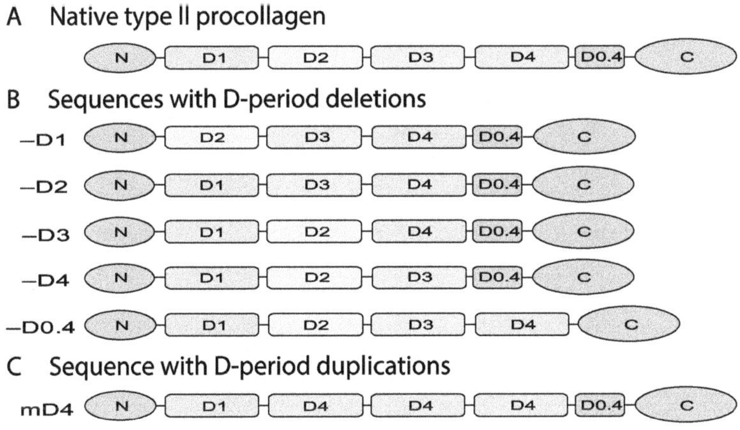

Figure 9.3. Diagram of native (A) type II procollagen showing its D-periodic segments and propeptides and (B) recombinant constructs with one segment (-D1, -D2, -D3, -D4 or -D0.4) deleted. (C) a novel construct comprising repeats of the D4 segment. (Reprinted from [72] by permission from Springer Nature.)

using *Saccharomyces,* along with co-expressed prolyl hydroxylase. In this study, however, the extent of Hyp formation was very limited [73].

A modular system has been developed for the construction of intact collagen molecules and for the facile insertion of sequence modifications but has not been exploited to make structures based on repeating structural element [39].

Examining the principle matrix metalloproteinase (MMP) site

A limited number of local mutation sites have been introduced into recombinant animal (human) collagens around the principle MMP cleavage site. Initially these studies involved modification of a full-length collagens [74, 75]. These studies used mutations to a cloned type I collagen α1-chain transfected into the mouse *Mov13* fibroblast system [74] or mutations to type III collagen expressed in *Pichia* [75] (figure 9.4). More recently the relevant region of the human type III sequence has been incorporated with various mutations between two *S. pyogenes* collagen domains and expressed in *E. coli* [76] (figure 9.4). These studies all showed that mutations can lead to loss of activity, especially introduction of Pro residues and that activity is dependent on an extended unmodified sequence being present. Inclusion of a mutation identified using the *Mov13* system into a mouse by gene targeting led to mice that showed normal development until juvenile, but with

A: Wu et al., 1990: α1 chain, Type I

Construct	P4	P3	P2	P1	↓	P1'	P2'	P3'	P4'	P5'	P6'	P7'	P8'	P9'	P10'	P11'	Collagenase Cleavage
	-772-	773-	774-	775-	↓	776-	777-	778-	779-	780-	781-	782-	783-	784-	785-	786-	
	-Gly-	Pro-	Gln-	Gly-	↓	Ile-	Ala-	Gly-	Gln-	Arg-	Gly-	Val-	Val-	Gly-	Leu-	Pro-	Yes
I						-Pro-											0
II			-Pro-				-Pro-					-Ala-	Ala-				0
III												-Ala-	Ala-				Slow
IV			-Pro-			-Met-	Pro-					-Ala-	Ala-				0
V						-Met-						-Ala-	Ala-				Slow

B: Williams & Olsen, 2009: α1 chain, Type III

Construct		P4	P3	P2	P1	↓	P1'	P2'	P3'	P4'	P5'	P6'	P7'	P8'	P9'	P10'	P11'	Collagenase Cleavage
		781-	782-	783-	784-	↓	785-	786-	787-	788-	789-	790-	791-	792-	793-	794-	795-	
FG-5016	(Wild type)	Gly-	Pro-	Leu-	Gly-	↓	Ile-	Ala-	Gly-	Ile-	Thr-	Gly-	Ala-	Arg-	Gly-	Leu-	Ala-	+++
FG-5015						↓	Pro-											+
FG-5029				Pro-		↓		Pro-			Pro-			Pro-			Pro-	0
FG-5021				Pro-		↓	Pro-	Pro-			Pro-			Pro-			Pro-	0

C: Yu et al., 2012: α1 chain, Type III

Construct	P7	P6	P5	P4	P3	P2	P1	↓	P1'	P2'	P3'	P4'	P5'	P6'	P7'	P8'	P9'	P10'	P11'	Collagenase Cleavage
	778-	779-	780-	781-	782-	783-	784-	↓	785-	786-	787-	788-	789-	790-	791-	792-	793-	794-	795-	
P4-P8'	*Gly-*	*Lys-*	*Pro-*	Gly-	Pro-	Leu-	Gly-	↓	Ile-	Ala-	Gly-	Ile-	Thr-	Gly-	Ala-	Arg-	*Gly-*	*Pro-*	*Arg-*	0
P7-P11'	*Gly-*	*Lys-*	*Pro-*	Gly-	Pro-	Leu-	Gly-	↓	Ile-	Ala-	Gly-	Ile-	Thr-	Gly-	Ala-	Arg-	Gly-	Leu-	Ala-	+
P4-P11'	Gly-	Ala-	Pro-	Gly-	Pro-	Leu-	Gly-	↓	Ile-	Ala-	Gly-	Ile-	Thr-	Gly-	Ala-	Arg-	*Gly-*	*Pro-*	*Arg-*	+
P7-P8'	Gly-	Ala-	Pro-	Gly-	Pro-	Leu-	Gly-	↓	Ile-	Ala-	Gly-	Ile-	Thr-	Gly-	Ala-	Arg-	Gly-	Leu-	Ala-	0
P1-P11'	*Gly*	*Gln-*	*Pro-*	*Gly-*	*Lys-*	*Pro-*	Gly-	↓	Ile-	Ala	Gly-	Ile-	Thr-	Gly-	Ala-	Arg-	Gly-	Leu-	Ala-	0

Figure 9.4. Examples of modifications to the principle MMP cleavage site in mammalian collagens, examined (A, B) by changes to the animal sequences [74, 75] and (C) by changes to a mammalian sequence insert in a bacterial collagen framework [76] leading to changes in the extent of cleavage.

increasing age various changes were noted including marked dermal fibrosis [77]. Lessening collagen turnover has potential for improved collagen filler products.

Changes of individual residues
A modular system was developed for expression of type III collagen in *S. cerevisiae*. In this system the triple-helical domain was established by combining 12 separate segments, each optimised for codon utilisation [39]. The expression also included introduced P4H activity. The opportunity for 'mix and match' was used to fabricate different combinations, including ones where individual Cys residues have been inserted at specific locations within the triple-helical domain [39]. This strategy allows, among many opportunities, further chemical modifications of the site-specific Cys residues to introduce other functions, or inclusion of other amino acids such as Tyr and Trp that are not normally found in collagens.

Manipulations of integrin binding
Further development of the modular assembly approach for constructing and expressing type III collagen [39], has been used to examine and modify integrin sites [78]. This study removed 5 known integrin binding sites within the type III sequence (GROGER, GAOGER, GLOGEN, GLKGEN and GMOGER), with the result that cell binding was eliminated. Re-introduction of a GFOGER sequence [79] restored cell binding, with introduction of additional GFOGER sites further enhancing the cell binding [78].

9.3.2 Adaptations to bacterial collagens

Exchange of V-domains
The various bacterial collagens that have been characterised (table 9.2) all contain a V-domain in addition to the triple-helical domain. This V-domain module is seen as assisting in the registration and initiation of triple-helix formation and may be present at either the N- terminal or the C-terminal of the triple-helical domain [55, 80]. In some cases, these domains can be exchanged in constructs with different triple-helical domains, still allowing production of a triple-helical products in good yield, whereas in other cases little if any product is formed [55, 58, 80].

Repeating segments
To date, other adaptations to bacterial collagens have almost exclusively been made to the *S. pyogenes* collagen-like Scl2 protein [81], with a handful of other similar examples, but with a different species, being described in the patent literature.

The various bacterial collagens examined to date have all been much shorter than an animal interstitial collagen. The *S. pyogenes* collagen-like Scl2 protein consists of a helical domain of 234 residues which is just under a quarter of the length of animal interstitial collagens, which are just over 1000 residues long. Therefore, longer constructs with 2 to 4 copies of the bacterial collagen have been made [57, 82]. The stability of the longer forms, produced in 2 L fermentation studies, did not change substantially with the increase in length [82]. However, from a biomaterial perspective, the longer constructs are of less value since the production yields drop significantly with the increase in length [82] (figure 9.5).

The converse, that is constructs of shorter segments of the *S. pyogenes* Scl2 protein, have also been considered, but principally for understanding the protein properties, particularly contributions to triple-helix stability and folding, rather than as potential biomaterials [82, 83]. The full sequence was dissected into three essentially equal sized segments, of 25–27 triplets each, that were studied as individual products and also when each was recombined into dimers and trimers; the small, one-third sequences were less stable than the full sequence [83]. So, for biomaterial applications, yield and stability considerations suggest that monomers are probably the best product to develop.

A particular application for the dimer constructs has been for the study of the functions of biological sequences. These sequences can be inserted between the two dimer segments. The most notable study has looked at the principle MMP collagen site from various interstitial collagens [76]. These studies indicated that for efficient cleavage a minimum of one triplet preceding and three triplets succeeding the cleavage triplet was required, while the absence of modified Hyp residues did not inhibit this activity [76]. Subsequent studies indicated that the α2-chain in heterotrimeric type I collagen destabilises the cleavage site enhancing activity [84].

Introduction of specific location changes involving multiple residues

Changes to sequences have involved substitutions to a native sequence, retaining the same size [85, 86], introductions of new sequences within a native sequence, leading to longer constructs [87] or additions of further residues at the N- or C-terminal of a native sequence [57, 88]. In some cases, this last group of changes arise as a consequence of the addition of a proteolytic site to allow ready removal of the V-domain [57], else are there to specifically add new functionality [88] (figure 9.6).

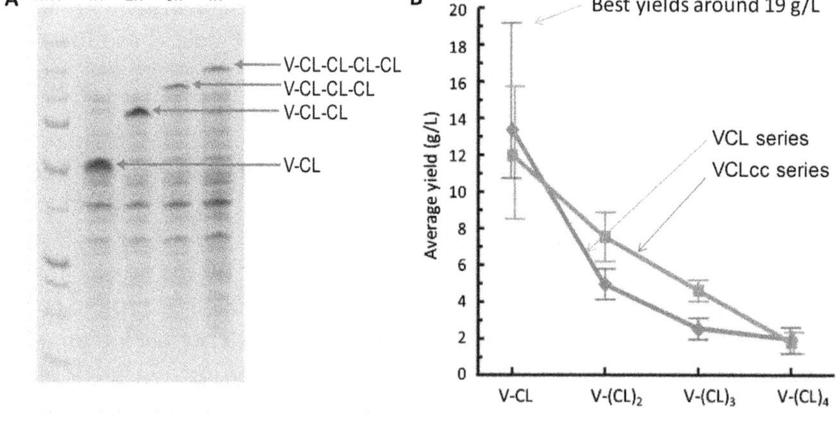

Figure 9.5. Production of monomer to tetramer constructs of *S. pyogenes* Scl2 protein, showing (A) different sizes in lysates by SDS-PAGE and (B) the lower yields with increase in size. (From [82] with permission.)

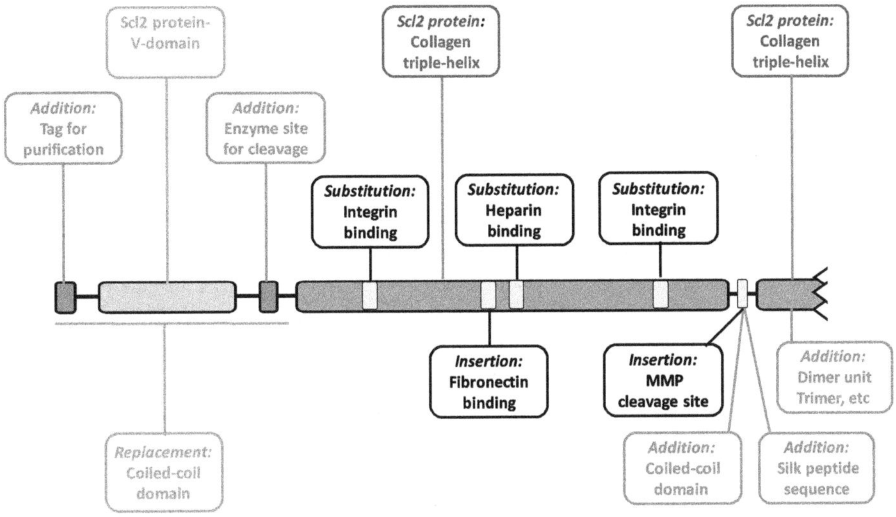

Figure 9.6. A summary of examples of various sequence modifications to the *S. pyogenes* Scl2 collagen-like protein.

The bacterial collagen Scl2.28 from *S. pyogenes* is well suited for study of functional domains and for design of new collagens for biomedical applications as it does not contain any known ligand binding sites and so behaves as a 'blank slate'. Animal collagens and collagen-like sequences contain a wide variety of functional interaction domains (see chapter 7). Type I collagen alone contains more than 50 biologically functional domains [89]. The various domains interact with four members of the integrin family ($\alpha_2\beta_1$, $\alpha_1\beta_1$, $\alpha_{10}\beta_1$, $\alpha_{11}\beta_1$), with three members of the immunoglobulin superfamily of receptors (GPVI, LAIR1, OSCAR) as well as two discoidin domain receptors (DDRs) [71]. A triple-helical conformation is required for these binding functions. Thus, it has been found that incorporation of functional sequences for various interactions into the triple-helix domain of a bacterial collagen, leads to the expected biological activity in the expressed protein, although the extent of the interaction may be limited in some cases by the lack of hydroxyproline (figure 9.6).

The most studied interaction has been that with integrins. Various studies, especially using peptides, have identified a range of triple-helical integrin binding sequences in mammalian collagens (see chapter 7). The first was GFOGER from the α1[I] chain, which was key for platelet interactions [79]. This was among the initial sequences, as GFPGER, that were inserted into the *S. pyogenes* Scl2 bacterial collagen, singly or in combination with a different binding sequence [85, 86] (figure 9.6). The addition of this sequence transforms the 'blank slate' of the native protein into a construct that binds cells, albeit with a reduced strength that was attributed to the presence of Pro rather than Hyp [86], in accord with the previous peptide chemistry data [79] and confirming that Hyp was not essential. Changing single Gly residues within and adjacent to this sequence led to a triple-helix, but with some small conformation perturbation. Changes to Gly residues C-terminal to the binding sequence did not affect integrin binding. However, changes to the two Gly residues in the binding sequence led to a complete loss of cell binding, while changes N-terminal to the site, up to four triplets prior, led to a decrease in cell binding [90].

In addition, other integrin binding sequences (GLPGER, GFPGEN, GERGFP, GERGVE) have been introduced into the *S. pyogenes* collagen [85, 91] (figure 9.6). GLPGER was shown, like GFPGER, to bind to both $\alpha_1\beta_1$ and $\alpha_2\beta_1$ integrins [91]. On the other hand, introduction of a GFPGEN sequence showed selective adhesion to endothelial cells compared with smooth muscle cells that bound to GFPGER, in accord with predicted specificity differences to the α1 and α2 I domains [91]. This specificity difference is exciting for development of new biomedical products as, for example, coatings could be beneficial for developing blood contact materials [85].

Another binding function that could be valuable for development of biomedical materials is introduction of fibronectin (Fn) binding, which could assist in new matrix accumulation. This was examined using a region derived from human collagen type II, both as an insert within a CL domain or between the two CL domain repeats [87, 92] (figure 9.6). Introduction of a GLAGQRGIVGLPGQRGER sequence led to collagen binding fibronectin through its gelatin binding domain. Denaturation of the modified collagen increased the fibronectin binding [87]. In this case, the presence of Hyp was not essential for Fn binding. Instead of being a 'blank slate', the modified construct showed good cell responses, supporting human mesenchymal stem cell (hMSC) spreading and proliferation as well as megakaryocyte attachment and differentiation [87]. In this case, disruption of the site through changes to Gly residues led to more noticeable structural disruptions and decreased binding [93].

In contrast to the integrin and fibronectin binding, full interactions to DDR may be dependent on the presence of Hyp [94]. Previous studies had shown strongest binding for DDR by GVMGFO [95]. Including this sequence in a 6-triplet insert into the *S. pyogenes* collagen (figure 9.6) led to a construct with specific receptor binding, but without the ability to stimulate DDR autophosphorylation, probably due to the absence of key Hyp content [94].

Finally, other functional domains that have been introduced into a bacterial collagen sequence include the 4 triplet GRPGKPGKQGQK sequence, corresponding to a heparin binding site [96] (figure 9.6) that has been incorporated by sequence substitutions of collagen, with activity shown by binding to fluorescently labelled heparin [86].

Introduction of specific single site changes

The available data on how amino acid sequence can influence triple-helical stability [97] allows ready changes to the stability of bacterial collagens. Typically, bacterial collagens have a stability close to body temperature [81]. This could decrease with side chain modifications but can be increased with crosslinking in a final biomedical product. Nevertheless, it could be advantageous to make substitutions, either to increase stability or to minimise steric effects on binding domains if side chain reactions are intended. A significant example of this approach was the introduction of around 22 changes, mainly Lys to Arg, which increased stability by around 5 °C and minimised interference from Lys modifications [98]. Introduction of a Cys-Cys sequence at the C-terminal, equivalent to that found in type III collagen, leads to only a small increase in stability, about 2 °C, if the V-domain was still present [82].

A further use for point mutations is to introduce opportunities for specific introduction of new functionalities. The native sequence contains various residues that can be modified, for example Lys residues, but as these are spread along the entire sequence so single site-specific modification is difficult. However, modifying the DNA to include unusual residues such as Tyr and particularly Cys allows introduction of sites that are not normally present for site-specific modification [88]. Introduction of these sites allows different strategies for crosslinking reactions [88, 99].

Introduction of unusual amino acids

The use of *E. coli* as the expression system provides opportunities for introducing unusual amino acids. An example where this has been achieved is with the introduction of Hyp residues by means of a simple method that utilises co-translational incorporation during fermentation [33]. Inclusion of Hyp in the expression media leads to incorporation of Hyp into sites usually occupied by Pro [100]. The disadvantage is that the Hyp may be incorporated in either the Xaa or Yaa positions, with slight stability issues arising when the Hyp is in the Xaa position [34]. The introduction of Hyp into the *S. pyogenes* collagen led to a reduction in solubility [100]. A specific approach for specific introduction of Hyp into the Yaa position could be to make sequence replacements to avoid Pro residues in the Xaa position. Alternatively, Hyp and a wide range of other unusual amino acids could be incorporated using the emerging systems that the use expression where the normal protein synthesis and/or codon usage has been manipulated. This could be, for example, a system such as non-canonical amino acid mutagenesis, where the cellular protein synthesis is 'evolved' so that the tRNA and tRNA synthase systems are adapted to allow incorporation of Hyp or other unusual amino acids [101, 102].

9.4 Recombinant chimeric fusion proteins

The option of making animal collagen-based hybrid proteins that would not be found in nature that included a bioactive component was initially described in various patents [103]. The collagen could be intact, or a smaller fragment that did not favour lateral association, and the active segment could, for example, be bone morphogenic protein or transforming growth factor β1.

Subsequently, constructs of collagen with epidermal growth factor (EGF) and interleukin-2 were described, with expression in insect cells [104, 105] (figure 9.7). Additional sequence changes at the junction allowed control of proteolysis in this region. In both cases, the collagen retained its triple-helical structure and demonstrated the expected biological activities, and so could be considered for potential clinical applications [104, 105].

The opportunity also exists for making chimeric constructs involving bacterial collagens, although these have been limited to date. One example is a structure where a short consensus silk sequence $(GAGAGS)_n$ was incorporated at the C-terminal of the bacterial collagen [92]. This added sequence enables bacterial

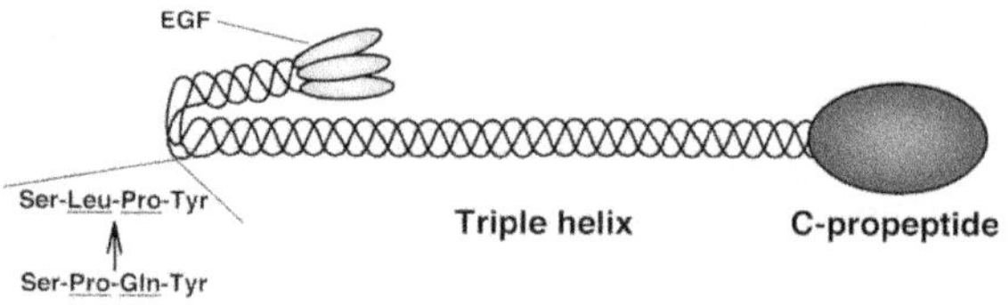

Figure 9.7. An example of a recombinant fusion protein that combines collagen and EGF in a single sequence, with telopeptide substitutions to control proteolytic cleavage. (Reprinted from [104] with permission from Elsevier.)

collagen to be annealed to fabricated silk scaffolds by non-covalent hydrogen bonding. This chimeric protein allows stable structures to be formed without introduced chemical crosslinking and provides opportunities for products that combine the excellent mechanical properties of silk along with various functional domains incorporated in one or more engineered bacterial collagen [92].

For bacterial and designed collagens, the other option for chimeric constructs is to enable more efficient recombinant production. For example, the trimer-forming domains from different bacterial collagen sources species can be swapped to enable better yields of soluble protein [55]. Other options could include using folding domains from animal collagens, the foldon domain, a trimerization domain from bacteriophage T4 [106–108] or natural or designed coiled-coil sequences [109, 110].

A further concept for recombinant chimeric constructs was the development of hydrogels based on small blocks with thermo-reversible trimerising collagen-based domains with an intervening non-trimerising segment [111]. Formation of trimers by the terminal collagen-like domains leads to formation of a gel, where the melting temperature can be controlled through the sequence of the collagen domain. This concept has allowed the development of useful materials, generally based on a silk-like repetitive sequence as the intervening non-trimerising segment [112].

9.5 *De novo* designed structures

A range of novel molecules have been constructed based on sequential removal of D-period modules from full-length sequences and construction of a new molecule consisting principally of the D1 modules followed by repeats of the D4 module [65, 68]. These studies were for type II collagen, but a novel molecule based on the α1-chain of type I collagen, with the central 2 D-periods deleted, has also been described [113]. This truncated construct formed a stable triple-helix, but with a lower T_m by about 3 °C.

Although there is now a wide range of information available on how amino acid sequence contributes to triple-helical stability [97, 114], there has been limited progress utilising these data to design *de novo* novel functional molecules. An example is a construct that after expression and purification from *E. coli* gave axially periodic fibrils [108]. This protein construct was based on 3 repeats of a 36-triplet module with a strong triple-helix forming propensity, with each repeat separated by a $(Gly\text{-}Pro\text{-}Pro)_4$ sequence. A C-terminal foldon sequence was used to support trimerization, and this module was removed subsequently by proteolysis during the purification. Although Hyp was absent, and hence not providing any triple-helical stability or assisting in lateral aggregation [44, 97] this construct, nevertheless, gave a stable triple-helix that formed into a lateral aggregate, fibril-like structure with an axial period of about 35 nm [108]. Specific functional domains, such as cell binding motifs could also be included in such structures.

References

[1] Ruggiero F, Exposito J Y, Bournat P, Gruber V, Perret S, Comte J, Olagnier B, Garrone R and Theisen M 2000 *FEBS Lett.* **469** 132–36

[2] Werkmeister J A and Ramshaw J A M 2012 *Biomed. Mater.* **7** 012002

[3] Geddis A E and Prockop D J 1993 *Matrix* **13** 399–405

[4] Fukuda K, Hori H, Utani A, Burbelo P D and Yamada Y 1997 *Biochem. Biophys. Res. Commun.* **231** 178–82

[5] Frischholz S, Beier F, Girkontaite I, Wagner K, Pöschl E, Turnay J, Mayer U and von der Mark K 1998 *J. Biol. Chem.* **273** 4547–55

[6] Tomita M, Kitajima T and Yoshizato K 1997 *J. Biochem.* **121** 1061–69

[7] Nokelainen M, Helaakoski T, Myllyharju J, Notbohm H, Pihlajaniemi T, Fietzek P P and Kivirikko K I 1998 *Matrix Biol.* **16** 329–38

[8] Tomita M, Ohkura N, Ito M, Kato T, Royce P M and Kitajima T 1995 *Biochem. J.* **312** 847–53

[9] Pihlajamaa T, Perälä M, Vuoristo M M, Nokelainen M, Bodo M, Schulthess T, Vuorio E, Timpl R, Engel J and Ala-Kokko L 1999 *J. Biol. Chem.* **274** 22464–68

[10] Snellman A, Keränen M R, Hägg P O, Lamberg A, Hiltunen J K, Kivirikko K I and Pihlajaniemi T 2000 *J. Biol. Chem.* **275** 8936–44

[11] Toman P D, Pieper F, Sakai N, Karatzas C, Platenburg E, de Wit I, Samuel C, Dekker A and Daniels G A *et al* 1999 *Transgenic Res.* **8** 415–27

[12] Adachi T, Wang X, Murata T, Obara M, Akutsu H, Machida M, Umezawa A and Tomita M 2010 *Biotechnol. Bioeng.* **106** 860–70

[13] Stein H, Wilensky M, Tsafrir Y, Rosenthal M, Amir R, Avraham T, Ofir K, Dgany O, Yayon A and Shoseyov O 2009 *Biomacromolecules* **10** 2640–45

[14] Xu X, Gan Q, Clough R C, Pappu K M, Howard J A, Baez J A and Wang K 2011 *BMC Biotechnol.* **11** 69

[15] Zhang C, Baez J, Pappu K M and Glatz C E 2009 *Biotechnol. Prog.* **25** 1660–68

[16] Eskelin K, Ritala A, Suntio T, Blumer S, Holkeri H, Wahlström E H, Baez J, Mäkinen K and Maria N A 2009 *Plant Biotechnol. J.* **7** 657–72

[17] Tang Y, Yang X, Hang B, Li J, Huang L, Huang F and Xu Z 2016 *Appl. Biochem. Biotechnol.* **178** 1458–70

[18] Guo J, Luo Y, Fan D, Yang B, Gao P, Ma X and Zhu C 2010 *Biotechnol. Appl. Biochem.* **57** 55–62

[19] Merle C, Perret S, Lacour T, Jonval V, Hudaverdian S, Garrone R, Ruggiero F and Theisen M 2002 *FEBS Lett.* **515** 114–18

[20] Toman P D, Chisholm G, McMullin H, Giere L M, Olsen D R, Kovach R J, Leigh S D, Fong B E and Chang R *et al* 2000 *J. Biol. Chem.* **275** 23303–309

[21] Myllyharju J, Nokelainen M, Vuorela A and Kivirikko K I 2000 *Biochem. Soc. Trans.* **28** 353–57

[22] de Bruin E C, Werten M W, Laane C and de Wolf F A 2002 *FEMS Yeast Res.* **2002** 291–98
[23] Rutschmann C, Baumann S, Cabalzar J, Luther K B and Hennet T 2014 *Appl. Microbiol. Biotechnol.* **98** 4445–55
[24] Sambrook J and Russell D W 2001 *Molecular cloning: a laboratory manual* 3rd edn (Cold Spring Harbor, NY: Cold Spring Harbor Laboratory)
[25] Williams J I, Salerno A J, Goldberg I and McAllister W T 1992 *US Patent* 5089406
[26] Capello J and Ferrari F A 1996 *US Patent* 5496712
[27] Fan D D, Luo Y, Mi Y, Ma X X and Shang L 2005 *Biotechnol. Lett.* **27** 865–70
[28] Luo Y, Zhang T, Fan D, Mu T, Xue W, Hui J and Ma X 2014 *Biotechnol. Appl. Biochem.* **61** 237–47
[29] Xue W J, Fan D D, Shang L, Zhu C H, Ma X X and Yu Y Y 2009 *Biotechnol. Lett.* **31** 221–26
[30] Xue W, Fan D, Shang L, Zhu C, Ma X, Zhu X and Yu Y 2010 *J. Biosci. Bioeng.* **109** 257–61
[31] Luo Y E, Fan D D, Shang L A, Shi H J, Ma X X, Mi Y and Zhao G F 2008 *Biotechnol. Lett.* **30** 637–43
[32] Chi L, Fan D D, Ma X X, Mi Y, Luo Y E, Zhu C H, Zhu X L and Xue W J 2011 *Biotechnol. Appl. Biochem.* **58** 175–84
[33] Buechter D D, Paolella D N, Leslie B S, Brown M S, Mehos K A and Gruskin E A 2003 *J. Biol. Chem.* **278** 645–50
[34] Persikov A V, Ramshaw J A M, Kirkpatrick A and Brodsky B 2003 *J. Am. Chem. Soc.* **125** 11500–501
[35] Kersteen E A, Higgin J J and Raines R T 2004 *Protein. Expr. Purif.* **38** 279–91
[36] Pinkas D M, Ding S, Raines R T and Barron A E 2011 *ACS Chem. Biol.* **6** 320–24
[37] Ding S, Pinkas D M and Barron A E 2012 *Biomacromolecules* **13** 1035–42
[38] Ito H, Steplewski A, Alabyeva T and Fertala A 2006 *J. Biomed. Mater. Res. A* **76** 551–60
[39] Chan S W, Hung S P, Raman S K, Hatfield G W, Lathrop R H, Da Silva N A and Wang S W 2010 *Biomacromolecules* **11** 1460–69
[40] Olsen D, Yang C, Bodo M, Chang R, Leigh S, Baez J, Carmichael D, Perälä M and Hämäläinen E R *et al* 2003 *Adv. Drug. Deliv. Rev.* **55** 1547–67
[41] He J, Ma X, Zhang F, Li L, Deng J, Xue W, Zhu C and Fan D 2015 *Biotechnol. Appl. Biochem.* **62** 293–99
[42] Tomita M, Munetsuna H, Sato T, Adachi T, Hino R, Hayashi M, Shimizu K, Nakamura N, Tamura T and Yoshizato K 2003 *Nat. Biotechnol.* **21** 52–6
[43] Sainsbury F and Lomonossoff G P 2014 *Curr. Opin. Plant Biol.* **19** 1–7
[44] Perret S, Merle C, Bernocco S, Berland P, Garrone R, Hulmes D J S, Theisen M and Ruggiero F 2001 *J. Biol. Chem.* **276** 43693–698
[45] Hashizume F, Hino S, Kakehashi M, Okajima T, Nadano D, Aoki N and Matsuda T 2008 *Transgenic Res.* **17** 1117–29
[46] Robinson D G, Andreae M and Blankenstein P 1990 *Organisation and Assembly of Plant and Animal Extracellular Matrix* ed W S Adair and R P Mecham (San Diego, CA: Academic Press-Harcourt Brace Jovanovich) pp 283–99
[47] Morrison J E and Pirie N W 1961 *J. Sci. Food Agric.* **12** 1–5
[48] Cook A H 2013 *Barley and Malt* (Amsterdam: Elsevier Science)
[49] Wang Y, Cui F Z, Hu K, Zhu X D and Fan D D 2008 *J. Biomed. Mater. Res. B Appl. Biomater.* **86** 29–35

[50] Báez J, Olsen D and Polarek J W 2005 *Appl. Microbiol. Biotechnol.* **69** 245–52
[51] Sanchez-Garcia L, Martín L, Mangues R, Ferrer-Miralles N, Vázquez E and Villaverde A 2016 *Microb. Cell Fact.* **15** 33
[52] Xu Y, Keene D R, Bujnicki J M, Höök M and Lukomski S 2002 *J. Biol. Chem.* **277** 27312–18
[53] Boydston J A, Chen P, Steichen C T and Turnbough C L 2005 *J. Bacteriol.* **187** 5310–17
[54] Vandersmissen L, De Buck E, Saels V, Coil D A and Anné J 2010 *FEMS Microbiol. Lett.* **306** 168–76
[55] Xu C, Yu Z, Inouye M, Brodsky B and Mirochnitchenko O 2010 *Biomacromolecules* **11** 348–56
[56] Ghosh N, McKillop T J, Jowitt T A, Howard M, Davies H, Holmes D F, Roberts I S and Bella J 2012 *PLoS One* **7** e37872
[57] Yoshizumi A, Yu Z, Silva T, Thiagarajan G, Ramshaw J A M, Inouye M and Brodsky B 2009 *Protein Sci.* **18** 1241–51
[58] Peng Y Y, Stoichevska V, Madsen S, Howell L, Dumsday G J, Werkmeister J A and Ramshaw J A M 2014 *Appl. Microbiol. Biotechnol.* **98** 1807–15
[59] Qing G, Ma L C, Khorchid A, Swapna G V, Mal T K, Takayama M M, Xia B, Phadtare S and Ke H *et al* 2004 *Nat. Biotechnol.* **22** 877–82
[60] Peng Y Y, Howell L, Stoichevska V, Werkmeister J A, Dumsday G J and Ramshaw J A M 2012 *Microb. Cell. Fact.* **11** 146
[61] Peng Y Y, Yoshizumi A, Danon S J, Glattauer V, Prokopenko O, Mirochnitchenko O, Yu Z, Inouye M and Werkmeister J A *et al* 2010 *Biomaterials* **31** 2755–61
[62] Adrio J L and Demain A L 2010 *Bioeng. Bugs* **1** 116–31
[63] Zhang H, Fan D, Deng J, Zhu C, Hui J and Ma X 2014 *Mater. Sci. Eng. C Mater. Biol. Appl.* **42** 124–29
[64] Fertala A, Sieron A L, Ganguly A, Li S W, Ala-Kokko L, Anumula K R and Prockop D J 1994 *Biochem. J.* **298** 31–7
[65] Fertala A, Han W B and Ko F K 2001 *J. Biomed. Mater. Res.* **57** 48–58
[66] Steplewski A, Hintze V and Fertala A 2007 *J. Struct. Biol.* **157** 297–307
[67] Arnold W V, Sieron A L, Fertala A, Bächinger H P, Mechling D and Prockop D J 1997 *Matrix Biol.* **16** 105–16
[68] Arnold W V, Fertala A, Sieron A L, Hattori H, Mechling D, Bächinger H P and Prockop D J 1998 *J. Biol. Chem.* **273** 31822–28
[69] Majsterek I, McAdams E, Adachi E, Dhume S T and Fertala A 2003 *Protein Sci.* **2003** 2063–72
[70] Leitinger B, Steplewski A and Fertala A 2004 *J. Mol. Biol.* **344** 993–1003
[71] Leitinger B and Hohenester E 2007 *Matrix Biol.* **26** 146–55
[72] Brodsky B and Ramshaw J A M 2017 *Subcell. Biochem.* **82** 601–29
[73] Peng Y Y, Werkmeister J A, Vaughan P R and Ramshaw J A M 2009 *Biomed. Mater.* **4** 015006
[74] Wu H, Byrne M H, Stacey A, Goldring M B, Birkhead J R, Jaenisch R and Krane S M 1990 *Proc. Natl Acad. Sci. USA* **87** 5888–92
[75] Williams K E and Olsen D R 2009 *Matrix Biol.* **28** 373–79
[76] Yu Z, Visse R, Inouye M, Nagase H and Brodsky B 2012 *J. Biol. Chem.* **287** 22988–997
[77] Liu X, Wu H, Byrne M, Jeffrey J, Krane S and Jaenisch R 1995 *J. Cell Biol.* **130** 227–37

[78] Que R A, Chan S W, Jabaiah A M, Lathrop R H, Da Silva N A and Wang S W 2015 *Biomaterials* **53** 309–17
[79] Knight C G, Morton L F, Peachey A R, Tuckwell D S, Farndale R W and Barnes M J 2000 *J. Biol. Chem.* **275** 35–40
[80] Yu Z, Mirochnitchenko O, Xu C, Yoshizumi A, Brodsky B and Inouye M 2010 *Protein Sci.* **19** 775–85
[81] Yu Z, An B, Ramshaw J A M and Brodsky B 2014 *J. Struct. Biol.* **186** 451–61
[82] Peng Y Y, Stoichevska V, Howell L, Madsen S, Werkmeister J A, Dumsday G J and Ramshaw J A M 2014 *Bioengineered* **5** 378–85
[83] Yu Z, Brodsky B and Inouye M 2011 *J. Biol. Chem.* **286** 18960–968
[84] Mekkat A, Poppleton E, An B, Visse R, Nagase H, Kaplan D L, Brodsky B and Lin Y S 2018 *J. Struct. Biol.* **203** 247–54
[85] Cosgriff-Hernandez E, Hahn M S, Russell B, Wilems T, Munoz-Pinto D, Browning M B, Rivera J and Höök M 2010 *Acta Biomater.* **6** 3969–77
[86] Peng Y Y, Stoichevska V, Schacht K, Werkmeister J A and Ramshaw J A M 2014 *J. Biomed. Mater. Res.* **102** 2189–96
[87] An B, Abbonante V, Yigit S, Balduini A, Kaplan D L and Brodsky B 2014 *J. Biol. Chem.* **289** 4941–51
[88] Stoichevska V, Peng Y Y, Vashi A V, Werkmeister J A, Dumsday G J and Ramshaw J A M 2017 *J. Biomed. Mater. Res. A* **105** 806–13
[89] Sweeney S M, Orgel J P R O, Fertala A, McAuliffe J D, Turner K R, Di Lullo G A, Chen S, Antipova O and Perumal S *et al* 2008 *J. Biol. Chem.* **283** 21187–97
[90] Yigit S, Yu H, An B, Hamaia S, Farndale R W, Kaplan D L, Lin Y S and Brodsky B 2016 *J. Biol. Chem.* **291** 19196–207
[91] Seo N, Russell B H, Rivera J J, Liang X, Xu X, Afshar-Kharghan V and Höök M 2010 *J. Biol. Chem.* **285** 31046–54
[92] An B, DesRochers T M, Qin G, Xia X, Thiagarajan G, Brodsky B and Kaplan D L 2013 *Biomaterials* **34** 402–12
[93] Chhum P, Yu H, An B, Doyon B R, Lin Y S and Brodsky B 2016 *J. Biol. Chem.* **291** 27073–86
[94] An B, Abbonante V, Xu H, Gavriilidou D, Yoshizumi A, Bihan D, Farndale R W, Kaplan D L and Balduini A *et al* 2016 *J. Biol. Chem.* **291** 4343–55
[95] Konitsiotis A D, Raynal N, Bihan D, Hohenester E, Farndale R W and Leitinger B 2008 *J. Biol. Chem.* **283** 6861–68
[96] Doss-Pepe E, Deprez P, Silva T, Inestrosa N C, Kirkpatrick A, Ramshaw J A M and Brodsky B 2004 *Biochim. Biophys. Acta* **1698** 187–95
[97] Persikov A V, Ramshaw J A M and Brodsky B 2005 *J. Biol. Chem.* **280** 19343–49
[98] Cereceres S, Touchet T, Browning M B, Smith C, Rivera J, Höök M, Whitfield-Cargile C, Russell B and Cosgriff-Hernandez E 2015 *Adv. Wound Care (New Rochelle)* **4** 444–56
[99] Stoichevska V, An B, Peng Y Y, Yigit S, Vashi A V, Kaplan D L, Werkmeister J A, Dumsday G J and Ramshaw J A M 2016 *J. Biomed. Mater. Res. A* **104** 2369–76
[100] Peng Y Y, Nebl T, Glattauer V and Ramshaw J A M 2018 *Acta Biomater.* **80** 169–75
[101] Liu C C and Schultz P G 2010 *Annu. Rev. Biochem.* **79** 413–44
[102] Lieblich S A, Fang K Y, Cahn J K B, Rawson J, LeBon J, Ku H T and Tirrell D A 2017 *J. Am. Chem. Soc.* **139** 8384–87
[103] Gruskin E A, Buechter D D, Zhang G and Connolly K 1998 *US Patent* 5821089-A

[104] Hayashi M, Tomita M and Yoshizato K 2001 *Biochim. Biophys. Acta* **1528** 187–95
[105] Hayashi M, Tomita M and Yoshizato K 2002 *Protein Eng.* **15** 429–36
[106] Stetefeld J, Frank S, Jenny M, Schulthess T, Kammerer R A, Boudko S, Landwehr R, Okuyama K and Engel J 2003 *Structure* **11** 339–46
[107] Du C, Wang M, Liu J, Pan M, Cai Y and Yao J 2008 *Appl. Microbiol. Biotechnol.* **79** 195–202
[108] Kaur P J, Strawn R, Bai H, Xu K, Ordas G, Matsui H and Xu Y 2015 *J. Biol. Chem.* **290** 9251–61
[109] Hoppe H J, Barlow P N and Reid K B M 1994 *FEBS Lett.* **344** 191–95
[110] Yoshizumi A, Fletcher J M, Yu Z, Persikov A V, Bartlett G J, Boyle A L, Vincent T L, Woolfson D N and Brodsky B 2011 *J. Biol. Chem.* **286** 17512–520
[111] Werten M W, Teles H, Moers A P, Wolbert E J, Sprakel J, Eggink G and de Wolf F A 2009 *Biomacromolecules* **10** 1106–13
[112] Rombouts W H, de Kort D W, Pham T T, van Mierlo C P, Werten M W, de Wolf F A and van der Gucht J 2015 *Biomacromolecules* **16** 2506–13
[113] Zafarullah K, Sieron A L, Fertala A, Tromp G, Kuivaniemi H and Prockop D J 1997 *Matrix Biol.* **16** 245–53
[114] Beck K, Chan V C, Shenoy N, Kirkpatrick A, Ramshaw J A M and Brodsky B 2000 *Proc. Natl Acad. Sci. USA* **97** 4273–78

IOP Publishing

John A M Ramshaw and Veronica Glattauer

Chapter 10

Evaluation of the quality of collagen preparations

Any collagen-based biomedical device that moves to clinical application will need to achieve regulatory approval. While laboratory-based design and development studies will typically be done with a range of readily available tests for initial quality evaluation of the collagen component, further application in fabricated devices will also require a range of quality, safety and efficacy tests. This section includes a range of chemical and biophysical assays.

Collagen preparations and prepared tissue materials that are used as the raw materials for biomedical device fabrication need evaluation to ensure that the material is what was intended and that in multiple preparations the process is reproducible. Further testing, looking at different characteristics including the effects of stabilisation chemistry and of sterilisation will be needed for final products.

There are a wide variety of tests that are available, some used more regularly than others, with the majority being more suitable for soluble collagens than for tissue based materials. Many tests are regularly used in laboratories, particularly electrophoresis, while others may come into play to meet the requirements of regulators. A range of specialist accredited laboratories are available to meet these more demanding studies.

This chapter covers a selection of chemical and biophysical methods. The biological testing of products, including acute and chronic toxicity, teratogenic effects, immunological properties, and endotoxin, microbial and viral contamination are not considered here. Immunological testing can examine the presence of impurities in a collagen preparation [1] and the intrinsic immunogenicity of the collagen itself. This is particularly important for collagens from various different species, for example avian, piscine and more primitive species, including bacterial collagens [2, 3].

doi:10.1088/978-0-7503-2096-2ch10

10.1 Collagen quantitation

A range of methods are available for collagen quantitation, with many applicable to either soluble or tissue-based products, while some are format dependent. In all cases, these methods can give erroneous values for collagen content if any non-collagenous proteins are present. In all methods of quantitation, consideration must be given to the involvement of water in the structure of collagen (see chapter 2). Certain water molecules form an integral part of the structure and cannot be readily removed from native collagens by drying, so that a method such as amino acid analysis will give a protein only weight while weight determinations give a protein plus water value. Maintaining collagen samples under standard humidity conditions allows comparison of weight to weight samples. However, the use of dry weight is not always suitable to determine collagen content as some preparations contain a significant salt content, as can be seen from the ash content or conductivity of samples. In most cases, the easiest weight quantitation is of freeze dried samples that have been extensively dialysed against a fully volatile solvent, such as dilute acetic acid.

Total nitrogen content, determined by Kjeldahl analysis, can be used to give a collagen concentration. The conversion factor should be calculated from the known structures [4] and take into account the high Gly content of collagens compared to globular proteins as well as the variation in Pro content between collagen types. Most collagens do not contain Gly in the Xaa or Yaa positions, but for some collagens, e.g., type III collagen, additional Gly residues are present and must also be taken into account.

A more specific indirect test for collagen quantitation involves analysis for Hyp which is an imino acid that is characteristic of collagenous proteins. Hyp can be determined by amino acid analysis, by a specific colorimetric assay [5, 6] or by mass spectrometry [7]. In this case the specific collagen type(s) present must be considered as different types have different Hyp contents. For example, type III collagen has a much greater content than type I collagen, while collagens that have more recently been suggested for biomaterial applications, such as from marine sources, have a lower Hyp content.

For more direct quantitation, the concentration of fully soluble collagen with concentration above 2 mg ml^{-1} can be rapidly determined using a Biuret assay, once called Piotrowski's test [8]. This is a colorimetric test based on detection of peptide bonds using copper(II) ions in alkaline solution [9]. In many applications this test has been used for its simplicity, rapidity and reliability. For this test, appropriate collagen standards must be used as globular proteins give much greater colour yields and would, therefore, always interfere in impure samples. For lower concentrations, various other colorimetric assays have been suggested. These include variations to the copper binding approach, such as reaction with the Folin and Ciocalteau's reagent (containing phospho-molybdic-tungstic acid) [10, 11] and subsequent further modifications. These enhanced assays generally include reactions with Tyr and Trp residues, which are typically in low abundance or absent in collagens. Other colorimetric assays use Coomassie Blue or Sirius Red dye binding. The 'Bradford' assay is based on the binding of Coomassie Brilliant Blue G-250 to proteins, where the colour change on protein binding provides a rapid reaction with excellent sensitivity and colour stability [12]. The method works well with collagens in combination with appropriate calibration and collagen standards, but the high sensitivity may actually be a disadvantage for larger scale preparative studies, unless large dilutions are used. An intermediate sensitivity assay can be achieved using Sirius Red, which has a reasonable selectivity for collagens. This reagent can be adapted to spectroscopic solution [13, 14] and dot-blot [15] assays. Numerous variations and other approaches, have been described in the literature, along with the limitations, such as interference, to particular methods. The reagents for popular methods are now available as kits from commercial suppliers, and the instructions and safety sheets include further information.

Immunological methods, especially using monoclonal antibodies, allow the identification and quantitation of different collagen types in samples. This is best achieved via an indirect ELISA assay [16]. Monoclonal antibodies provide excellent reagents that are highly type specific and often species specific for this assay. However, estimates of collagen quantities from immuno-histological slides is not appropriate as the visible intensity depends on a variety of factors such as antibody affinities, tissue accessibility and crosslinking [17]. Conversely, immunological methods can be used to detect specific impurities, such as serum albumin for the tissue source or pepsin used during the sample extraction.

Quantitation by full amino acid analysis data, not just for Hyp analysis, is also useful as it can confirm the collagenous nature of a sample through the high Gly content and allow an estimate of purity by comparison of the known structural data to the observed amino acid compositions values. The amount of Tyr present can also show the extent of telopeptides cleavage, as this amino acid is usually only found in these regions, as is the case for type I collagen [4]. However, type III collagen does contain a low level of Tyr within the helical domain [4]. The hydrolysis step for amino acid analysis, typically overnight prior to the chromatographic analysis may be considered too slow by some. In such cases, more rapid methods are available, including the use of a microwave system [18].

Various other methods that are principally used for quality and purity evaluation, for example viscosity measurements, can also be adapted to quantitation.

10.2 Solution properties

Solution properties would generally apply to soluble collagen preparations, but in some circumstances aqueous extracts of tissue collagens may be appropriate samples for analysis.

In laboratory studies, it would be normal to measure the pH value for preparation. If any salts are present, they may affect the pH as they may be buffering the collagen solution. The observed pH may affect the solubility of the collagen preparation as samples that contain oligomers are more soluble at acidic pH, while at neutral pH the collagen oligomers become less soluble and may precipitate [19]. The presence of salts can be determined by ash content. The presence of salts will also be shown by conductivity analyses.

One issue that many methods do not readily distinguish is whether some of the collagen in a sample has become denatured and is present as gelatin. Amino acid analysis, Hyp analysis and dry weight, for example, would not show this distinction. A solution method that is available uses the property that native and denatured collagen separate out from solution in alcohols, such as methanol or ethanol, as different phases allowing each form to be quantified [20].

There are several other analyses that would not necessarily be used regularly in laboratory work but would generally be needed for regulated products. These include determinations that show that heavy and toxic metals are absent. Well documented tests are available [21] and for laboratories that do not normally perform these tests certified laboratories are readily available.

10.3 Electrophoretic methods

Different electrophoretic techniques are suitable for analysis of collagen purity and detection of any breakdown. The methods are typically cheap and readily and reproducibly performed, particularly if commercially available pre-cast gels are used. The stain systems that are used to detect the collagenous components in each of the electrophoretic methods are typically not collagen specific, but rather they stain all proteins and thus allow detection of any protein contaminants. Very low amounts of impurities can be seen if the sensitive silver staining method is used [22]. If the sample is pre-treated with a bacterial collagenase to digest all the collagen present then the impurity bands become more obvious. It is very important that control proteins and the highest-grade enzyme are used as any non-collagenase proteolytic activity would lead to spurious results [23].

The most commonly used electrophoresis approach is polyacrylamide gel electrophresis with sodium dodecyl sulfate (SDS-PAGE) [24]. This method gives the molecular weight distribution of the individual components after denaturation, so triple helical collagen will run as individual chains, except where native crosslinks are present, giving dimer (β-), trimer (γ-) or higher polymer species (figure 10.1). In the case of type III collagen, disulfide bonds are present holding the chains together as a single trimeric unit. In this specific case, reduction of these disulfide bonds, for example using 2-mercaptoethanol (BME), dithiothreitol (DTT) or tris(2-carboxyethyl)phosphine (TCEP) leads to monomer rather than trimer chains (figure 10.1).

The single chain components, the α-chains, are the fastest moving components from intact collagens. Any bands moving faster than the α-chains, if present, could indicate impurities or degradation products, although they may also be hydrolysis products that can occur if the sample is heated in sample buffer at 100 °C for >2 min prior to electrophoresis. Cross-linked chain components, for example the dimer β- and trimer γ-chains, move more slowly on the gel (figure 10.1). The cross-linked components could arise either from either intramolecular or intermolecular bonds. Generally, however, they indicate the presence of residual telopeptides, which could add to the immunogenicity of a product.

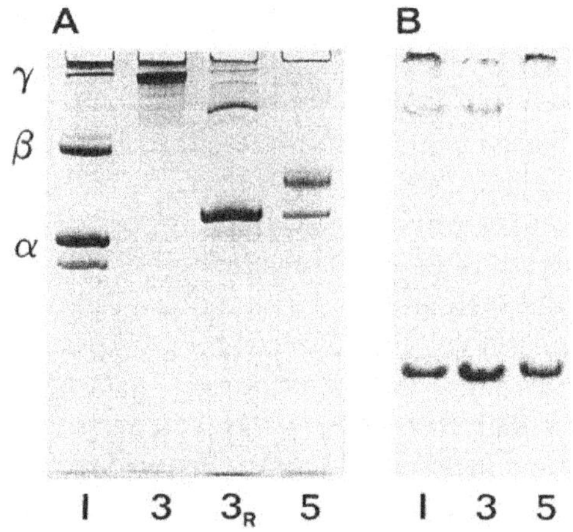

Figure 10.1. Separation of different human collagen types by (A) SDS-PAGE, showing monomer (α) dimer (β) and trimer (γ) chains, and (B) non-denaturing-PAGE, showing intact molecules. The different collagen types are shown below each lane. (Reprinted from [25] with permission from Elsevier.)

In some cases chains from different collagen types run to similar positions on a gel, for example α1(I) and α1(III) chains. Separation of these chain in collagen mixtures, such as may be present in unfractionated skin collagens, may be achieved by interrupted electrophoresis, in which the reduction of the type III collagen trimer is delayed until a short while after the electrophesis has commenced [26]. Alternatively, for some animal species (e.g., bovine) but not others, different separations can be acheived by addition of urea to the separation buffers [27]. SDS-PAGE can also used for characterisation of collagens through analysis of CNBr fragments. This reagent cleaves the collagen at Met residues, where the specific locations in different collagen chains lead to characteristic patterns.

A disadvantage of SDS-PAGE is that the collagen molecules are denatured. So immuno-blotting (Western Blotting) using conformation dependent antibodies, such as most monoclonal antibodies, cannot be achieved, especially for heterotrimers where the chains move to different locations in the gel during the analysis. An alternative system of electrophoresis under non-denaturing conditions, using highly porous gels, can be used for immuno-blotting [28]. Collagen dimers enter the gel, but insoluble or highly polymeric materials do not. However, this separation does not resolve intact chains of similar sizes, such as the interstitial collagens (figure 10.1).

Non-denaturing electrophoresis can also be used to give an indication of the isoelectric point of collagen preparations. Native collagens have a high pI, typically around pH 9, which is consistent with the calculated values that can be determined from amino acid sequence data. If composition data from amino acid analysis are used, then the deamidation that occurs for asparagine and glutamine residues during hydrolysis must be considered if correct pI values are to be determined. The direction of migration of collagen in native gels at different pH values can give an indication of the pI of the collagen. On the other hand, collagens that have been prepared from alkali treated material, for example limed collagen from leather processing, will have lost the amides from asparagine and glutamine residues and will thus have a low pI values [29].

A further option is to run 2-dimensional separations, generally with CNBr fragments as larger fragments will not be readily separated. This separation normally uses isoelectric focussing as the first dimension, followed by SDS-PAGE in the second dimension [30]. As well as separating the different fragments by size, these gels also show variations in charge within fragment populations, possibly arising from deamidation or other modifications.

A range of other electrophertic systems for analysis of collagen and collagen fragments have also been used for characterisation, including capillary electrophoresis systems [31, 32]. Capillarly electrophoresis separation is now frequently coupled with mass spectrometry analysis [33].

10.4 Optical methods

10.4.1 UV and visible spectroscopy

A good quality preparation of soluble collagen should give a clear and colourless dilute (< 5 mg ml^{-1}) solution. If any collagen is not soluble, or has formed aggregates then a white turbidity will be present. Any insoluble collagen that is present can be removed by centrifugation, but the resulting solution is then no longer representative of the total sample.

Use of UV/visible spectroscopy also allows determination of the extent of any turbidity that may be present. It also enables detection and quantitation of any UV absorbing preservatives that may be present in an externally sourced material.

10.4.2 IR spectroscopy

The IR spectrum of collagen conforms with the expectation of a protein spectrum. It also will show the presence of any other organic materials including organic buffers or any organic preservatives. Raman spectroscopy is also now being used to examine collagens, although more usually for tissues and composite materials than for purified collagen samples.

At high resolution, however, IR spectra provide some useful structural information. Relative differences in the intensities of the bands at 1660 cm^{-1} and 1633 cm^{-1} will arise if any denaturation of the collagen to gelatin has occurred. The 1660 cm^{-1} peak declines and 1633 cm^{-1} peak increases as the extent of denaturation and gelatin content increases [34]. However, other approaches may be more readily available to assess gelatin content.

10.4.3 CD and ORD spectroscopy

Initial studies of collagens used ORD, but use of this approach has declined with CD spectrometry becoming readily available. Nevertheless, ORD can still provide a quick and easy method for collagen quantitation using the characterised rotation values that are available [35].

CD spectroscopy can also be used to determine native collagen concentration. Collagen has a distinctive CD spectrum, with the triple-helix giving a maximum near 222 nm (mean residue ellipticity ~ 5000 deg cm^2 $dmol^{-1}$), and a large minimum near 198 nm (mean residue ellipticity ~ 60 000 deg cm^2 $dmol^{-1}$) (see figure 6.1).

In addition, CD spectroscopy allows determination of the T_m of collagen. Increasing temperature leads to loss of the signal at 222 nm, with the T_m value taken at the temperature when 50% of the signal is lost (see figure 6.1) [36, 37]. Care is needed to avoid UV damage to the protein sample, which could affect accuracy [37].

10.4.4 Microscopy

Microscopy methods have some, but limited, use in examining the quality of collagen preparations. SEM can indicate the nature and extent of any fibrous or particulate components in a preparation. Depending of the specific method used for sample preparation it is possible that lipid contamination may appears as spots in an image. The main use of SEM, and environmental SEM has been as a method to evaluate the porosity of collagen sponges [38] and other fabricated materials such as electrospun membranes [39]. TEM may give additional information on any aggregates that are present in a preparation, while rotary shadowing may be useful in studying novel materials, such as bacterial collagens. TEM may also be used to establish whether collagen fibrillar structures still retain a D-periodic banding pattern. Atomic force microscopy can identify inidividual fibrils, but is probably most useful for examining the surface characteristics of collagen coated materials [40].

10.4.5 Refractive index

Most of the previous methods in this section are principally for the study of soluble collagens. However, insoluble tissue derived materials, such as insoluble comminuted collagen and fibre dispersions are less readily studied. An approach that has been proposed, which could also be used on reconstituted collagen products such as sponges, is the use of refractive index (RI) [41]. Collagen has a theoretical RI of between 1.544 and 1.548. Its RI can be estimated using a Becke test, where it is compared with mixtures of nitrobenzene (1.550) and methylsalicylate (1.522) until a match is observed. Differences from the expected value increase when impurities are present [41].

10.5 Biophysical methods

10.5.1 Calorimetry

Differential scanning calorimetry has been used to examine the thermal properties of collagen samples, including those derived from soluble collagen, from tissue samples and those after chemical modification [42]. Calorimetry identifies the thermal transition that occur during duration, and others that may be associated with fibril associations. T_m values are given as the temperature at the mid-point of the thermal transition. Thermal stabilities must be compared using defined standard conditions. For example, the pH must be defined as changes in pH leads to variation of T_m. A key factor that must be recorded is the rate of heating. In normal circumstances a fairly high rate of heating is used, for example 5 °C per minute, but fast rates of heating lead to higher values that obtained by other methods. Hence the need for standardised, defined conditions.

10.5.2 Other methods

Viscosity

While a standard approach for determination of a viscosity value for edible gelatin has been well established [43] the same is not the case for collagen preparations. Perhaps because of its complexity, determination of collagen viscosity has not become a frequent standard method for definition of final collagen preparations but has more generally been used to provide insights into collagen structure and solution properties [44]. However, the emerging interest in bioprinting, where more viscous solutions are needed, has established viscosity as an important parameter to allow predictable and reproducible performance.

Many different types of viscometers and rheometers can be used to measure viscosity, including viscosity cups, falling ball, glass capillary (e.g., Ubbelhode, Cannon-Fenske), vibration and rotational viscometers.

Rotational viscometers are seen as the most versatile and most suitable for collagen studies. Collagen solutions/gels and fibre dispersions behave as non-Newtonian, shear-thinning fluids, with viscosities that decrease with increasing the shear rate [45, 46]. Further, the viscosity of collagen solutions is dependent on other factors including pH, temperature and buffer salts [44]. Current rotational viscometers allow close control of temperature and other parameters, allowing examination of parameters including a storage modulus, G′, and a loss modulus, G″, helping to understand the nature and extent of any microstructure. Despite these difficulties, viscosity measurement has some specific applications. For example, the transition change (decrease) in viscosity with increase in temperature allows determination of T_m [47]. Also, for large scale preparations of pepsin soluble collagen, the use of a viscosity cup, such as a Zahn cup (albeit a paint industry approach), can be used with the increase in efflux time indicating the increase in viscosity as the collagen is solubilised. This can enable a rapid determination of the digestion progress by following the viscosity increase to meet established values.

Chromatography

A range of chromatographic techniques for collagen separation and purifications have been described (for example, [48]) (see chapter 8), most of which could be adapted as necessary for quality control applications. For example, while these separations are designed for purification, conversely they can be used for detecting the presence of impurities. This could include samples after bacterial collagenase digestion to specifically digest the collagen, or on urea or guanidine extracts that release any bound impurities.

Mass spectroscopy
Mass spectrometry continues to undergo significant improvements on top of the many instrumental and computer analysis tools that have emerged over the last two decades. Initially it has been used typically for analysis and characterisation of collagen fragments, generally in conjunction with a prior separation technique [49, 50] but the continuing increase in high mass capabilities will enhance its use in analysis of larger fragments and intact collagen molecules. Mass spectrometry can already be used to examine monomers and dimers of bacterial collagens [51]. It has also been used for characterisation of the site of proteolysis for V-domain removal from intact *S. pyogenes* Scl2 collagen molecules, and for the extent of introduced Hyp to be monitored [52].

Water binding
For some applications of collagens their ability to hold water is a useful characteristic. For dry sponges, the extent of this property can be assessed by measuring water uptake in a defined humidity chamber. However, this approach needs caution as any salts present in preparations can lead to false interpretations. Materials based on soluble collagens generally bind more water than fibre based or fibre containing samples [29].

Shrinkage temperature
The thermal transition of tissues immersed in water or buffer can be assessed, the shrinkage temperature, but is normally applied to assessing the changes that occur on chemical modification of collagen samples (see chapter 12).

X-ray photoelectron spectroscopy (XPS, ESCA)
XPS allows for an elemental analysis of the top ~10 nm surface of a material. For collagen materials this may enable certain impurities to be shown and quantified in samples, such as higher S content from glycosaminoglycans, P from DNA, or other atoms that are not expected to be present. Also, the method can be used to examine the extent of coverage of a collagen coating on a material [40] or modifications where a hetero atom has been incorporated.

Contact angle
Contact angle measurements [53] are principally of use for studying collagen coatings on polymer or other surfaces. Variations from a normal range [54] may then be due to incomplete coverage or to hydrophobic compounds in the collagen preparation.

References

[1] DeLustro F, Condell R A, Nguyen M A and McPherson J M 1986 *J. Biomed. Mater. Res.* **20** 109–20
[2] Peng Y Y, Glattauer V, Ramshaw J A M and Werkmeister J A 2010 *J. Biomed. Mater. Res. A* **93** 1235–44

[3] Peng Y Y, Yoshizumi A, Danon S J, Glattauer V, Prokopenko O, Mirochnitchenko O, Yu Z, Inouye M and Werkmeister J A *et al* 2010 *Biomaterials* **31** 2755–61
[4] Kadler K 1995 *Protein Profile* **2** 491–619
[5] Huszar G, Maiocco J and Naftolin F 1980 *Anal. Biochem.* **105** 424–29
[6] Hofman K, Hall B, Cleaver H and Marshall S 2011 *Anal. Biochem.* **417** 289–91
[7] Chan S W, Greaves J, Da Silva N A and Wang S W 2012 *BMC Biotechnol.* **12** 51
[8] Piotrowski G 1857 *Proc. Imp. Acad. Phil. Vienna (Math. Nat. Sci.)* **24** 335–37
[9] Robinson H W and Hogden C G 1940 *J. Biol. Chem.* **135** 707–25
[10] Wu H 1922 *J. Biol. Chem.* **51** 33–9
[11] Lowry O H, Rosebrough N J, Farr A L and Randall R J 1951 *J. Biol. Chem.* **193** 267–75
[12] Bradford M M 1976 *Anal. Biochem.* **72** 248–54
[13] Marotta M and Martino G 1985 *Anal. Biochem.* **150** 86–90
[14] Lareu R R, Zeugolis D I, Abu-Rub M, Pandit A and Raghunath M 2010 *Acta Biomater.* **6** 3146–51
[15] Rodríguez-Rodríguez P, Arribas S M, de Pablo A L, González M C, Abderrahim F and Condezo-Hoyos L 2013 *Anal. Bioanal. Chem.* **405** 6863–71
[16] Rennard S I, Berg R, Martin G R, Foidart J M and Robey P G 1980 *Anal. Biochem.* **104** 205–14
[17] Ramshaw J A M, Casagranda F, White J F and Werkmeister J A 1994 *Clin. Mat.* **16** 9–13
[18] Chen S T, Chiou S H, Chu Y H and Wang K T 1987 *Int. J. Pept. Protein Res.* **30** 572–76
[19] Miller E J and Rhodes R K 1982 *Methods Enzymol.* **82** 33–64
[20] Gopinath A, Shanmugam G, Madhan B and Rao J R 2017 *Int. J. Biol. Macromol.* **102** 1156–65
[21] Rice E W, Baird R B and Eaton A D 2017 *Standard Methods for the Examination of Water and Wastewater* 23rd edn (Washington, DC: American Public Health Association)
[22] Merril C R, Dunau M L and Goldman D 1981 *Anal. Biochem.* **110** 201–7
[23] Peterkofsky B 1982 *Meth. Enzymol.* **82** 453–71
[24] Laemmli U K 1970 *Nature* **227** 680–85
[25] Ramshaw J A M, Werkmeister J A and Peters D E 1990 *Current Perspectives on Implantable Devices* vol 2 ed D F Williams (London: JAI Press) pp 151–220
[26] Sykes B, Puddle B, Francis M and Smith R 1976 *Biochem. Biophys. Res. Commun.* **72** 1472–80
[27] Hayashi T and Nagai Y 1979 *J. Biochem.* **86** 453–59
[28] Ramshaw J A M and Werkmeister J A 1988 *Anal. Biochem.* **168** 82–7
[29] Peng Y, Glattauer V, Werkmeister J A and Ramshaw J A M 2004 *J. Cosmet. Sci.* **55** 327–41
[30] Benya P D 1981 *Coll. Relat. Res.* **1** 17–26
[31] Chalmers M, Huckle J and Cotton N J 1999 *J. Chromatogr. Sci.* **37** 443–47
[32] Miksík I, Sedláková P, Mikulíková K and Eckhardt A 2006 *J. Chromatogr. B Analyt. Technol. Biomed. Life Sci.* **841** 3–13
[33] Mikulíková K, Eckhardt A, Pataridis S and Miksík I 2007 *J. Chromatogr.* A **1155** 125–33
[34] Payne K J and Veis A 1988 *Biopolymers* **27** 1749–60
[35] Nagelschmidt M and Viell B 1987 *J. Biomed. Mater. Res.* **21** 201–9
[36] Piez K A and Sherman M R 1970 *Biochemistry* **9** 4134–40
[37] Hayashi T, Curranpatel S and Prockop D J 1979 *Biochemistry* **18** 4182–87
[38] Yang Y, Li C, Song W, Wang W and Qian G 2016 *Int. J. Biol. Macromol.* **89** 344–52

[39] Buttafoco L, Kolkman N G, Engbers-Buijtenhuijs P, Poot A A, Dijkstra P J, Vermes I and Feijen J 2006 *Biomaterials* **27** 724–34
[40] Sharan J, Koul V, Dinda A K, Kharbanda O P, Lale S V, Duggal R, Mishra M, Gupta G and Singh M P 2018 *Colloids Surf. B Biointerfaces* **161** 1–9
[41] Jayabalan M 1985 *Leather Science* **32** 140–42
[42] Wallace D G, Condell R A, Donovan J W, Paivinen A, Rhee W M and Wade S B 1986 *Biopolymers* **25** 1875–95
[43] Briefer M and Cohen J H 1927 *Indust. Eng. Chem.* **19** 252–7
[44] Bianchi E, Conio G, Ciferri A, Puett D and Rajagh L 1967 *J. Biol. Chem.* **242** 1361–69
[45] Ananthanarayanan S and Veis A 1972 *Biopolymers* **11** 1365–77
[46] McPherson J M, Ledger P W, Sawamura S, Conti A, Wade S, Reihanian H and Wallace D G 1986 *J. Biomed. Mater. Res.* **20** 79–92
[47] Safandowska M and Pietrucha K 2013 *Int. J. Biol. Macromol.* **53** 32–7
[48] Meredith S C and Kézdy F J 1981 *Biochim. Biophys. Acta* **668** 357–69
[49] Nimptsch A, Schibur S, Ihling C, Sinz A, Riemer T, Huster D and Schiller J 2011 *Cell Tissue Res.* **343** 605–17
[50] DiChiara A S, Taylor R J, Wong M Y, Doan N D, Rosario A M and Shoulders M D 2016 *ACS Chem. Biol.* **11** 1408–21
[51] Stoichevska V, An B, Peng Y Y, Yigit S, Vashi A V, Kaplan D L, Werkmeister J A, Dumsday G J and Ramshaw J A M 2016 *J. Biomed. Mater. Res. A* **104** 2369–76
[52] Peng Y Y, Nebl T, Glattauer V and Ramshaw J A M 2018 *Acta Biomater.* **80** 169–75
[53] Kwok D Y, Gietzelt T, Grundke K, Jacobasch H-J and Neumann A W 1997 *Langmuir* **13** 2880–94
[54] Tamada Y and Ikada Y 1994 *J. Biomed. Mater. Res.* **28** 783–89

John A M Ramshaw and Veronica Glattauer

Chapter 11

Fabrication of biomedical products

While purified collagen provides the basic biomedical material, fabrication to form it into a useful medical device is a key process. A variety of approaches to fabrication are available so that the biomaterial can be presented in the most appropriate form for clinical application. In most cases, these approaches are suitable for microfibres as well as fully soluble collagen.

Reviews and texts from the last three to four (and more) decades indicate that many of the presently used fabrication options were already well established [1–4]. These included, for example, collagen sponges for wound healing and as haemostats, injectable gels for tissue augmentation and membranes as barriers. These products have typically undergone improvements and new processes and products have also been introduced. In many cases the broad principles of fabrication are described in the scientific literature. However, in many cases the details are not disclosed and are part of confidential commercial know-how. The patent literature can provide some information as the monopoly provided by the patent is notionally repaid by the description in the patent being sufficient to teach others how to repeat the process or product when the monopoly expires. Often, however, know-how has overtaken these descriptions.

A section on sterilisation has been included at the end of this chapter. This does not mean that consideration of sterilisation should be left to the end. In fact, it is quite the opposite. Sterilisation will almost certainly change the properties of the product somewhat, or in some cases considerably. In some cases, such as certain chemical approaches it may add to the durability of the product, whereas in others, such as irradiation, some degradation may occur. Thus, it is essential that sterilisation be considered from the beginning and throughout the development of collagen-based materials.

doi:10.1088/978-0-7503-2096-2ch11

11.1 Gels and hydrogels

In considering gels and hydrogels a question could be 'what is a gel (or an aqueous gel) compared to a hydrogel' as on occasions these terms seem to be used interchangeably? So, a collagen gel is often seen as a quite viscous, yet still mobile solution of collagen, such as would be found for a 3% or higher concentration solution for injection as a filler [5]. Yet a polyacrylamide gel, as used for SDS-PAGE, is undoubtedly a solid and not mobile gel. The term 'hydrogel' undoubtedly became more popular in the 1960s and onwards, when used to describe water absorbing synthetic polymer systems [6, 7] and emerging biomedical materials [8]. However, hydrogel as a term was introduced in 1864 [9] to distinguish the 'gelatinised' form of aqueous silicic acid. Later, gelatin was referred to as a hydrogel, more than a century ago [10], while the liquid form was a hydrosol. Gel, as an isolated term as opposed to a suffix, seems to have appeared later [11] for a solid or semi-solid material.

In present times, a definition that hydrogels are hydrophilic, three-dimension networks, which are able to imbibe large amounts of water or biological fluids, and thus resemble, to a large extent, a biological tissue has been proposed [12, 13]. This type of gel is insoluble because of crosslinking between chains. The crosslinks can be covalent, or through interactions [13] giving junction domains, such as the reforming of triple-helical segments in set gelatin. So, set gelatin is a hydrogel, which on heating comes apart giving a gelatin solution, a hydrosol.

In the present case, for collagen preparations, we have chosen to consider viscous, solutions, such as could be injected, as gels, while preparations that have sufficient introduced crosslinking, enough to provide dimensional stability, as hydrogels.

For making gels in the laboratory, moderately viscous gels can be made by re-dissolving freeze-dried collagen to give a high concentration [14, 15], which is suitable for solutions up to around 1.5% to 2% [5, 16]. Another approach is to re-dissolve a low ionic strength collagen precipitate that has been concentrated by centrifugation [5]. A further alternative is by placing an acidic collagen solution in a dialysis bag and the excess water allowed to evaporate through the membrane or other approaches for evaporative ultrafiltration [17] or drawn out, for instance covering the membrane with high molecular weight polyethylene glycol, as a solid or concentrated solution [18, 19], else with dry gel-filtration resin [18]. At small scale and very high concentrations, > 175 mg ml^{-1}, molecular order is initiated [20, 21]. On a larger, commercial scale processes present additional challenges and cost of goods becomes important [22]; adaptations and additional approaches are described in patents.

Collagen hydrogels can be made by a thermal treatment, heating to 37 °C at neutral pH, where an entangled fibril network forms from the collagen solution [23] with reaction conditions and collagen source determining the formation and properties of the material [24, 25]. However, these hydrogels lack the durability and stability that are required for many applications. So, for making collagen hydrogels, where collagen is the main or sole component a crosslinking strategy is typically used. There are other hydrogel systems that include collagen, but these are generally hydrogel polymers with added collagen to improve biological properties [26, 27] or as interpenetrating gel network systems where there are mechanical benefits [28, 29].

There are a wide range of crosslinking strategies that could be used (see chapter 12), with the final properties of the hydrogel depending on collagen concentration and crosslink density. Simple bifunctional reactive crosslinkers, such as GA, can be added to collagen solutions leading to hydrogel formation. For a neutral pH collagen solution, hydrogel formation by GA can take place in 10 minutes. Other systems, for example, include the formation of non-enzymatic glycosylation cross-links using ribose where the gel properties can be controlled by sugar concentration [30]. Alternatively, crosslinking can be provided using various oxidised polysaccharides, with the resulting aldehydes crosslinking the collagen. One approach to controlling the extent of crosslinking, and hence the properties of the product, is by addition of competing amino acids to the reaction system [31].

The reactive groups for crosslinking can also be placed at the ends of spacer molecules, such as PEG, where the properties and size of the spacer can contribute to the hydrogel properties. Various bifunctional and multifunctional PEGs are readily available from commercial suppliers. Methacrylated collagen is also available, allowing direct formation of gels, and formation of gels using thiol-ene approaches. The use of peptide-based crosslinking, where the peptide is sensitive to proteolysis such as by an *in situ* MMP provides a mechanism for hydrogel formation. Such gels can be used for cell delivery and subsequently degrade at the application or injection site [32]. Further modification of gels and hydrogels by plastic compression enables denser matrices to be produced, including ones that already contain cells [33], with the use of a biaxial approach providing highly aligned and permanently densified collagens structures [34].

It is also possible to consider making self-crosslinking hydrogels that could be used for cell encapsulation [35] or other *in situ* applications, with type II collagen being suitable in some circumstances as an alternative to type I collagen [36].

In a different concept, rather than using highly purified collagen preparations, collagen-based hydrogels can be prepared from decellularized matrix in which other ECM components are still present [37].

11.2 Foams and sponges

In foams, the gas bubbles are isolated and lack linkages, whereas in a sponge the pores are linked in a contiguous manner, allowing easy transfer of liquid through the material and the potential for cellular infiltration to occur. The use of these terms seems variable. So, using the descriptions above, for a biomaterial product it is likely that a sponge is the more useful and commonly used term. However, in foams the wall thickness between individual bubbles may be so thin as to be unstable in use or be rapidly biodegraded so that a continuous porous material is there in practise.

For formation of foams a collagen solution would be required. However, for sponges either collagen solutions or dispersions of fine collagen fibre segments can be used. Indeed, many commercial products have used collagen fibre preparations [38, 39].

11.2.1 Foams

Foams can be made through external aeration of a collagen solution or gel or by including a chemical or biochemical system within this collagen to generate gas bubbles. Technology was initially established for gelatin solutions [40, 41], albeit that medical applications were not originally envisaged. It has, nevertheless, not been widely applied to collagen. Generally, the resulting foams are freeze-dried, which could, along with subsequent handling, break the bubble walls. External aeration could use sterile air, and be supplemented by vigorous agitation (blending) of the mixture [42]. Internal gas generation has been used for other proteins, such as keratose mixtures, through addition of catalase and H_2O_2 which lead to immediate gas production [43]. The alternative method that has been envisaged is the use of low molecular weight poragens, which, when included in the mixture, subsequently dissolve to introduce cavities in the structure [44]. The extent of interconnectivity, if any, will depend on the size, amount and distribution of the added poragen particles.

11.2.2 Sponges

The use of freeze drying to produce porous protein scaffolds is well established [45] and was applied to collagen solutions or fibre dispersions many decades ago [46]; it is a major approach for manufacture of collagen-based biomedical products.

Pore size has emerged as a key issue in the fabrication of sponges. Freeze drying produces larger more uniform pores than other methods, including air drying, from either water or solvent, or critical point drying [47]. Among things that are important include the effects on transport of nutrient and oxygen to cells, while allowing removal of catabolites and waste products [48]. It was shown that the pores of a collagen-based sponge for wound dressing needed to be between 20 and 125 μm if effective healing was to be obtained [49], although can vary with the cell type [50]. A feature that is also determined by the pore size is the surface area available for cellular interactions, and the two are well correlated [51]. Matrix stiffness is also important for cell response [52, 53], and may also be adapted through pore properties, including any crosslinking present [54]. Also, the fibrous microarchitecture of the collagenous surfaces is advantageous as fibrous scaffolds perform better than non-fibrous polymer surfaces [55].

Control of pore size can be determined by several approaches [56]. The pore size varies with the concentration of the collagen (figure 11.1). Frequently, the collagen solution or dispersion is acidic, which, for suspensions, increases the swelling present, with the pH influencing pore size [50]. Freezing conditions are important, and should be closely controlled for uniform and reproducible results. It has been suggested that altering the pan size and its stiffness can be used to control pore size, while preparation using a controlled, constant cooling rate provides more uniformity than a rapid quenching approach or a variable cooling rate [57]. A lower final

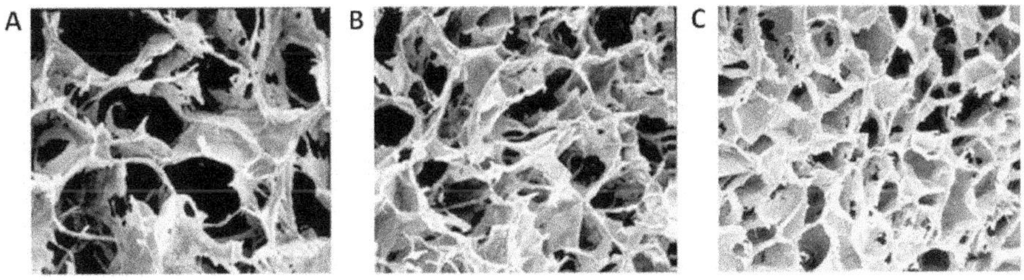

Figure 11.1. Freeze-dried collagen sponges made from different strength collagen solutions showing variation in pore size. (A) 20 mg ml^{-1}, (B) 40 mg ml^{-1}, and (C) 80 mg ml^{-1}.

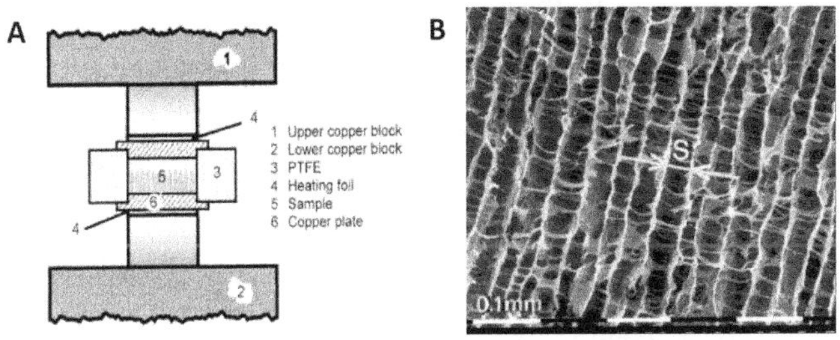

Figure 11.2. (A) Schematic of an apparatus for gradient freezing a collagen solution to provide a freeze-dried sponge with aligned pores. (B) Aligned pores in a freeze-dried collagen sponge, with uniform widths (S). (Republished from [63] with permission of John Wiley and Sons Inc., through the Copyright Clearance Centre Inc.)

freezing temperature leads to a smaller pore size [51, 58]. However, a subsequent annealing step can raise the pore size [59].

Macroporous gels can be made by controlled freezing and gelation at sub-zero temperatures [60], and this approach has been used successfully with various biological polymers, including gelatin [61]. These materials made by cryogelation can be used for chromatography of solutions and of cells, as either particles or as monoliths, or as substrates for cell culture. Opportunities for collagen have not yet been fully explored.

Another aspect of freeze-dried sponges that has emerged is orientation of the pores. In the examples above (figure 11.1), the pores are not oriented in any particular direction, yet for certain tissues, such as cartilage, the orientation of the pores in a scaffold may be advantageous. A system was devised using controlled, directional freezing that allowed highly orientated pores to be formed in collagen sponges [62, 63] (figure 11.2). A device that allows unidirectional solidification during freezing is used in which stable ice crystals form from the cooled lower surface of the device and progress upwards without side branches, while excluding the collagen which accumulates between the ice columns (figure 11.2) [63]. The characteristics of the ice columns can be adjusted by the solvent, with acetic acid and ethanol additions preferred [64], with the type and concentration of ions present also having an impact [65]. Final freeze-drying leads to a collagen sponge with long, unidirectional pores (figure 11.2). A wide range of conditions can be developed for this technique which allow the production of a porous sponge, with characteristics such as pore size and mechanical properties to match the desired product applications [66]. Clearly, the permeability of these materials is radically different along the axis of the pores compared to across this axis. The approach has been used for various cell types, including growth muscle cells where the long, oriented pores are particularly useful [67]. It is also suitable for composites with other proteins, such as elastin [68].

11.2.3 Measurement of pore size and porosity

Measurement of pore size is not necessarily a simple task nor a comparable measure between laboratories. An SEM photograph of a cut surface is often used for measuring pore size, but this can illustrate several issues. For example, has the pore being measured been cut at its maximum size or is it just at an edge? Does the pore have fairly uniform dimensions or is it elongated, perhaps significantly? And if the pore has had good walls around it, what is the size of the entry/exit site for materials or cells migrating in or out of the sponge? Is this aperture size the size of the pore or the larger compartment dimensions? And are there difficulties in moving between pores if the path is particularly tortuous? And there is a large number, >20, of other parameters that could be considered [69]. In a practical sense, it is the accessibility of pores that defines their function, which is the permeability of the material. So, when a pore size is quoted it is important to indicate or read what is being measured and described. And beyond that, the other feature of pores that is important can be their internal microstructure, which will vary in composites of the same components but in different ratios [70]. For any specific case, a system of measurement that assures reproducibility is needed.

Porosity and permeability can be determined by various approaches [69, 71] and these have been adapted to less regular structures, such as collagen-based materials, and those that are highly ordered, such as those obtained by bioprinting methods [71]. Frequently used methods include a gravimetric approach where the included fluid allows the pore volume to be estimated, which often uses mercury as the fluid [71]. For permeability, a pump can be used to provide a steady, non-pulsed continuous flow which leads to a pressure drop between the input and output sides of the sample. The pressure drop at defined flow rates allows the permeability to be calculated. This approach is complex, but has been used with synthetic polymer scaffolds [72], being clearly harder than examining an SEM photograph. Other approaches to produce flow include gravity flow with either a constant head or with a falling head of fluid, with the former being more readily interpreted, and have been used for collagen composite materials [73].

Advanced mathematical approaches now allow a wide range of parameters relating to pore size and permeability to be computed from micro computed tomography images [74, 75], as well as for a range of emerging experimental and other imaging techniques. Mathematical models are useful for regular structures such as bioprinted structures.

11.3 Reconstituted fibres

A wide variety of fabrication systems have been developed for fibre production from various categories of materials [76, 77]. For collagen, reconstituted fibres and fibre-based materials have mostly been made by various approaches to wet spinning and electrospinning. Most recently, bioprinting is being used and is gaining considerable attention.

11.3.1 Wet spinning

Wet spinning of polymer solutions to produce fibres has a long history, with numerous patents, derived mainly from the emerging synthetic textile industry. For example, a process had been developed for producing a soluble cellulose, cellulose xanthate, with the viscose process in which cellulose pulp is initially treated with aqueous NaOH followed by reaction with CS_2 [78]. Treatment with a mineral acid, such as sulfuric acid, precipitates the cellulose again, releasing CS_2, so that threads, films and similar can be formed through extrusion of the viscose solution into a coagulation bath, containing acid or ammonium chloride or alternatives (e.g., [79, 80]). Later, considerably safer and less toxic methods for making cellulose solutions emerged. At the same period, gelatin threads became available, through extrusion of hot gelatin onto a cold surface followed by fixation, for example by using chrome alum, tannic acid or 'formic aldehyde' (formaldehyde), with plasticising by small amounts of glycerin [81].

Over time, a wide range of improvements to these processes have been described, including, for example, systems for making tubes [82]. However, it seems a while before extrusion of collagen threads and other products emerged, not least because of the date at which ready supplies of soluble collagen or fine fibrous dispersions became available. Nevertheless, by the mid-1930s various patents on the fabrication of reconstituted collagen materials appeared, normally using extrusion of a swollen collagen fibre mass, including for manufacture of threads and tubes (mainly for the sausage casing industry) (e.g., [83–85]). It has been reported that subsequently, in the USA, from 1942, collagen extrusion was extensively studied to make sutures and sheet and tube materials for burn and wound management for use by the military (see [86]). In the late 1940s and for the next two decades various further developments for threads made from extruded collagen fibre paste included introduction of consistent stretching after the coagulation bath and collagen fixation using chromic salts and formaldehyde [87]. Stretching of up to 700% was reported [88] giving a better and stronger product. Commercial production rates of around 30 m min^{-1} of 110 μ fibre [87], 1300 feet in 6 h for 5/0 thread [89] and 1200 m h^{-1} for 60 mm tubing (casing) [90] were reported. The resulting monofilament threads could be merged into a larger diameter strand or tape during production [89] or finished thread further manufactured into multifilament threads by twisting or braiding [91], or woven or knitted into fabric [92].

The introduction of collagen solution for extruded thread occurred later still [93], with water soluble polymers, polyethylene glycol (PEG) and Dextran, being introduced as agents for the coagulation bath. Collagen solutions at about 5 mg ml^{-1}

and 20%–35% PEG(8000) were preferred, along with aldehyde or carbodiimide stabilisation [93]. Fibres could be modified by passing a newly coagulated fibre through a particulate suspension, leading to a coating by the particulates [94], while strength can be improved by incorporating a plasticiser and then stretching the fibre and allowing the load to dissipate through stress-relaxation; in addition, bundles of over 500 fibres were made [95].

These various reports over time have frequently included diagrams of the particular experiment arrangement of the extrusion, coagulation and other steps (e.g., [82, 84, 87, 93]), some being quite extensive [89]. All are variations on a similar theme (e.g., figure 11.3) with a series of baths and other equipment, including a pumped extrusion nozzle, a draw (stretching) unit and a module for drying and spooling thread. The order, number and content of these baths and units varies between studies, other than the coagulant bath is next to the extrusion nozzle (figure 11.3).

Various laboratory studies have extended the approach of early commercial systems as the interest on using collagen fibres in many medical devices emerged. In many cases, short lengths of material were made, for example only 500 mm because of fragility [96], but frequently so that a wide variety of differences in conditions, including collagen concentration and form (solution or comminuted), coagulant (PEG, salts, alcohols), the extent of draw, the presence or absence of crosslinking or a plasticiser, could be evaluated [97–101]. Choice of conditions to match the intended application is thus possible. In some instances, large scale production was reported with multiple spools of material produced, suitable if needed for further processing such as knitting or weaving [102]. In most cases a simple extrusion nozzle was used, but a counter-rotating system, analogous to that used for casing manufacture, has also been described which allows control of fibre orientation in tubes [103], that could be adapted for fibres.

Most recently, microfluidics has been developed for collagen microfibre production [104]. The device enables a collagen solution to be mixed with a PEG solution,

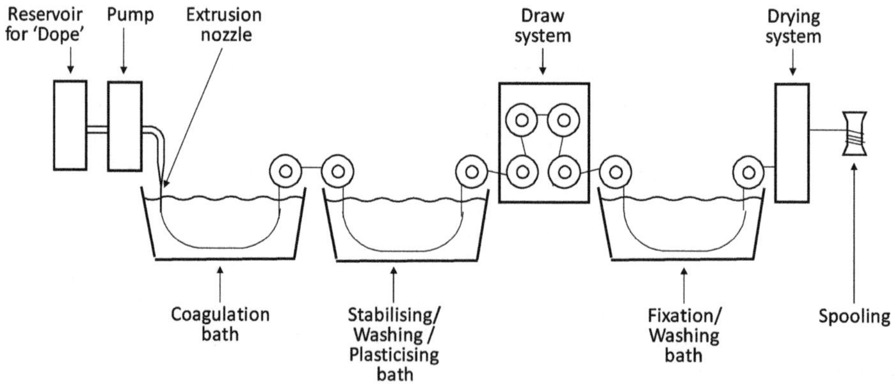

Figure 11.3. A schematic of a typical system for wet spinning of collagen fibres. The configuration, uses and order of the various baths and additional equipment may vary between different experimental and production systems.

leading to defined fibres. The flow rates can be used to control the fibre diameters, where examples of uniform, dry fibres between 3 μm up to 150 μm were obtained. When these fibres where formed under selected conditions it was possible to collect them, in quantity, on a spool. The mechanical properties of fibres were generally seen as better than those reported for laboratory fibres produced by wet spinning, consistent with the good collagen orientation in the fibre that was observed [104].

11.3.2 Electrospinning

The origins of the observations and technologies that contribute to the process of electrospinning go back over several centuries [105], while the introduction of technology that led to practical electrospinning dates from 1900 [105]. Various developments continued [106], notably by Formhals with designs that include nozzle development (e.g., [107]), Norton with descriptions of melt-electrospinning [108], and by several Russian scientists who developed various commercial filter materials [105]. The principle theoretical developments occurred in the 1960s with studies by Taylor, particularly on the shape of the cone formed by a fluid droplet travelling in an electric field [109], now termed the 'Taylor Cone'. Electrospinning has now become a standard method [110]. In its simplest form, a polymer solution is pumped through an orifice (needle) towards a collection target, which can be stationary or rotating. A high voltage power supply provides an electric field between the orifice and the collection target to accelerate a jet of polymer towards the target. If the polymer is in a sufficiently volatile solvent this leads to the deposition of stable dry fibres (figure 11.4). Various improvements to this system are now used, including systems with more complex orifice configurations, including a coaxial format [111] and development of systems and configurations that allow deposition of numerous fibres simultaneously (e.g., [112]).

Electrospinning of collagen emerged significantly in the early 2000s. Initial studies using calf skin type I collagen dissolved in 1,1,1,3,3,3-hexafluoro-2-propanol (HFIP) produced 100 nm fibres in fibrous mats (figure 11.4) that showed a 67 nm banding pattern by TEM [113]. Various parameters were optimised, including collagen concentration, the voltage potential difference and the airgap distance. It was also

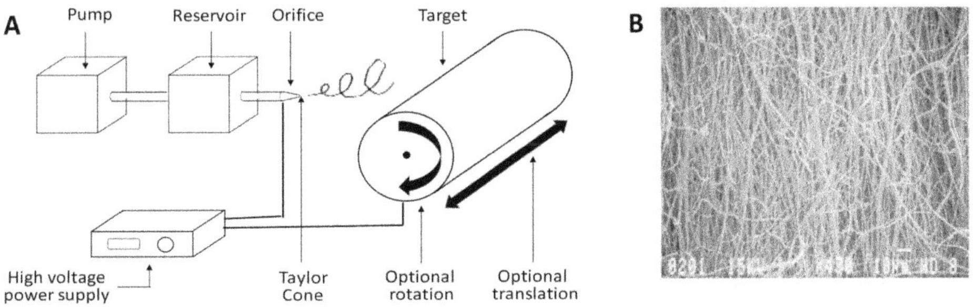

Figure 11.4. (A) A schematic of a basic electrospinning arrangement. In many examples a syringe pump is used for the pump and reservoir. (B) An aligned electrospun collagen fibrous mat ([113], reprinted from the American Chemical Society, Copyright 2002).

found that the type I collagen source, skin or placenta, led to different outcomes, as well as if there was any type III collagen present. Purified type III collagen also produced a fibrous product [114]. Rotation of the collector mandrel led to a high level of fibre alignment (figure 11.4) [113]. The quality and solubility of the collagen used was shown to be important in determining the quality of the electrospun product [115]. Further studies showed that another interstitial collagen, type II collagen, could also be formed into electrospun fibrous membranes [116]. As well as animal derived collagens, recombinant [117] and fish collagens have also been examined [118].

Subsequently, CD analysis of re-dissolved, electrospun collagen suggested that about 50% of the collagen was denatured [119], while other studies suggested that the collagen could be 100% denatured due to the effects of the HFIP solvent [115]. Another suggested solvent, 2,2,2-trifluoroethanol, has also been shown to be unsuitable as it is denaturing [120]. Despite these reports to exercise caution in selection of solvent, several (many) studies have continued to use them, particularly HFIP. On the other hand, others have realised the need to develop benign solvents. Successful systems include, for example, phosphate buffered saline (PBS) with different salt concentrations mixed 1:1 with EtOH [121] and an acetic acid/DMSO system [118]. In another approach, collagen in acetic acid was used and an NaOH aerosol was placed between the orifice and the target to neutralise and precipitate the collagen as the fibres are formed [122].

In many cases, the electrospun scaffold is stabilised after production. This has included use of glutaraldehyde (GA) [123] as well as N-[3-(dimethylamino)propyl]-*N′*-ethylcarbodiimide (EDC), genipin, UV light and transglutaminase [124] and photo-crosslinking using a Rose Bengal catalyst [125]. However, the effectiveness of the transglutaminase may well reflect the presence of gelatin [126] (chapter 12).

Several variations to collagen electrospinning have been described. These include production of composite, mixed fibres, with biopolymers such as chitosan [127] and synthetic polymers such as poly(ethylene oxide) [117], as well as with a hydroxyapatite sol [128]. Other composite fibres can be made through using a coaxial delivery orifice [129], while composite fibrous mats can be prepared by successive electrospinning of collagen and different materials onto the target [130]. Alternatively, a collagen fibre can be produced by layer-by-layer coating of an electrospun fibre [131] where resorption of the carrier leaves collagen only.

Various studies have reported on cellular interaction with collagen electrospun substrates, with beneficial results being reported. The mechanism concepts, however, may need review in these studies to resolve whether the observed effects are due to triple-helical collagen or to denaturation, giving gelatin.

11.3.3 Printing

Continuous extrusion and stabilisation of a collagen fibre is the basis for the rapidly developing, predominantly in the last decade, technology of bioprinting. An alternative approach could be the assembly of numerous contiguous drops of material to generate a continuous structure [132]. A range of different approaches have emerged [133, 134], with printing of polymers as the main focus, along with composite materials [135]. Generally, the printing is done onto a defined surface associated with a printing machine in the laboratory, but can also be achieved with a hand-held printing device that a surgeon could use to directly apply a collagen-based fibre or gel to a surgical site [136]. Similarly, delivery of a film is possible [137].

Collagen alone is difficult to use in printing, as solutions have inappropriate viscosity with low collagen content, and the scaffold fibres that are formed lack strength to fully support themselves and may contract when cells are present [138]. Nevertheless, strategies for printing collagen alone have been reported. This printing has mostly used purified type I collagen, although alternatives include collagen along with other ECM components from solubilised acellular matrix [139], type II collagen [140] and type IV collagen in Matrigel [141]. The success of these constructs is dependent on immediate fixation during printing, for example freezing [142] prior to later processing, or stabilisation, either physical or chemical, during the printing process. This can include fibril formation from a collagen solution, by temperature and/or pH changes, with bicarbonate being a suitable neutralising agent [143]. Also, it can include UV or visible light induced crosslinking, using sensitising agents such as flavin mononucleotide [144] or riboflavin [145]. Chemically, stabilisation can be achieved by addition of a crosslinking agent, such as genipin [146]. This may require use of a modified delivery system, using a coaxial or multiaxial delivery nozzle [147]. Otherwise, the collagen can be modified, for example as a methacrylated derivative, which can then be used to facilitate rapid crosslinking [148, 149].

Collagen mixtures with other biomolecules have proved effective to enable scaffolds with a significant collagen content to be printed. These other components include hyaluronic acid (HyA) [150], agarose [151], and alginate [152]. Agarose can be removed, if wanted, after collagen stabilisation to allow a collagen only scaffold [153, 154]. Methacrylated gelatin can also be added to collagen to provide a stabilising network for the collagen after crosslinking [155], although gelatin composites can be rapidly degraded [156]. The alternative approach is to make a composite, often a coating, with a synthetic polymer, where the polymer provides the mechanical strength for the construct [138].

Initially, bioprinted devices and tissues will likely be used as tissue models, for example for drug evaluation and assessing drug suitability for a patient, rather than as implantable, tissue replacement materials [157, 158].

11.4 Films and membranes

Membranes can be made using some of the approaches in the proceeding sections, including by compressing a foam or sponge of soluble (e.g., [159]) or fibrous collagen (e.g., [160]), or by isolating and stabilising an electrospun film. The principles of the extrusion process used to make threads can also be extended to make films and membranes [161]. Otherwise, air drying of a collagen solution can lead to a film, although these can be very brittle without plasticiser present.

Some other approaches can also be used. For example, a collagen precipitate, or a dispersion can be formed into a film by centrifugation in a basket centrifuge,

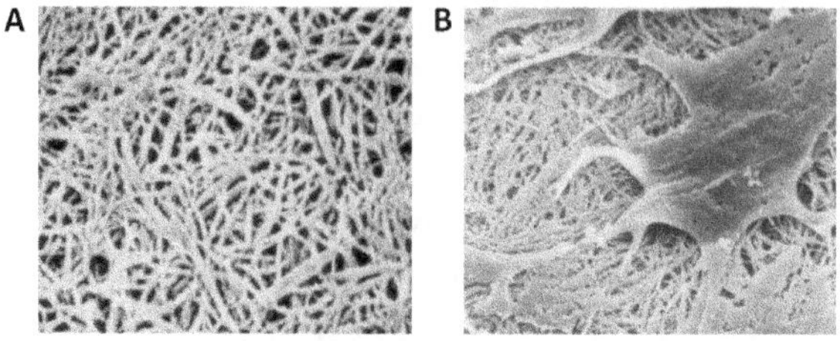

Figure 11.5. An example (A) of a fibrillar collagen membrane that shows (B) a porosity that prevents ready cell migration. (A: Republished from [162] with permission of John Wiley and Sons Inc., through the Copyright Clearance Centre Inc.)

followed by stabilisation by gamma irradiation, which also provides sterilisation [162]. In this case, and in others where the membrane is being used as a barrier, such as for treating periodontal disease [163], the fibre density is important and it should not allow easy migration of cells (figure 11.5).

Other approaches to membrane formation include the formation of a film comprising collagen and an oppositely charged polymer, such as hyaluronic acid [164] or chemically modified collagens [165] and the use of filtration moudling [166].

A variety of commercial products are available based on a range of these approaches, for use in dermal, dural and other repair applications (e.g., [167, 168]. In addition, amnion membrane, which was used as a medical material more than a century ago [169], is also now commercially available as an acellular, off-the-shelf product.

11.5 Beads and particles

Collagen beads and particles have proved useful in biomedical applications as tissue fillers, especially in hard tissue repair [170], as well as for large scale propagation of cells in spinner culture [171, 172]. Large scale manufacture of gelatin beads is clearly feasible as a variety of commercial products are readily available. For example, emulsion methods enable gelatin beads to be made in a variety of sizes and formats with selection of the extents of crosslinking used to control biodegradation rates [173]. Production of beads from soluble collagen has proved harder, due to the highly hydrated state of collagen, the high viscosity of solutions of >1%, and the lower concentration of the solutions that can be made.

11.5.1 Using purified, soluble collagen

There are few reports of beads made using purified collagen, with the majority of reports using dispersions of fibrous collagen (see below). There are various reports of collagen composites, mainly with synthetic polymers for drug delivery [174], or mixed with hydroxyapatite (HAp) for bone repair [175], with the manufacturing often being a proprietary process. These also include beads that have been given a coating of soluble collagen [172].

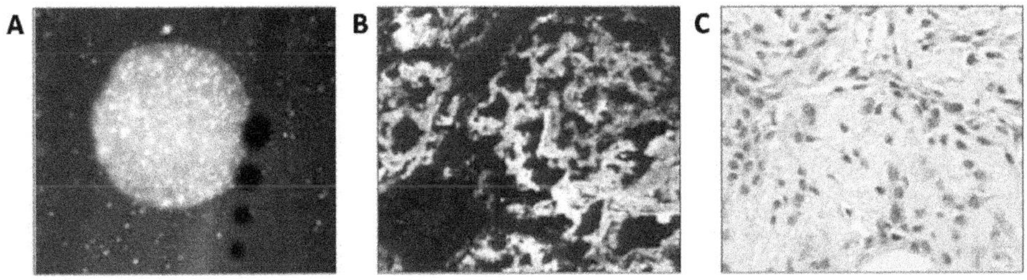

Figure 11.6. Porous beads made using soluble collagen in an alginate carrier. (A). A bead after alginate removal, about 300 μm in diameter, (B) the porous structure shown by immunohistology with an anti-type collagen antibody (white is positive), and (C) Haematoxylin and Eosin staining showing chondrocytes after 12 d in culture in a group of beads. (Reprinted from [153] with permission from Springer Nature.)

One use of composites is as an intermediate for preparation of collagen beads from soluble collagen, which could allow different collagen types to be used that could match a cell or clinical use. One example is using agarose as the intermediate. Agarose and collagen-agarose mixtures can be readily prepared as defined beads by drop-wise addition to a $CaCl_2$ solution. The collagen in the beads can then be stabilised by crosslinking, for example by GA (chapter 12) and the alginate carrier then removed through extensive washing in NaCl solutions [153]. This approach gives stable highly porous collagen beads that are readily infiltrated by cells when in culture (figure 11.6).

A further approach, which is yet to be explored in detail, is the use of superhydrophobic surfaces on which the collagen, and any added cells of solutes, can bead up and then be stabilised as needed [176].

11.5.2 Using collagen fibre dispersions

A range of approaches has been used to make collagen beads from dispersions of uniform, fine collagen fibres, normally obtained by extensive comminution of tendon or dermis. For very fine dispersions, suitable for spraying, comminution of tissue in a selected alcohol, in a ball-mill can be used [177]. For example, in one approach the dispersion is passed through a vibrating small diameter tube forming drops that fall into liquid nitrogen. Subsequent freeze drying gives collagen beads which can be stabilised by crosslinking [178]. For cell culture applications a heavier bead may be needed, so this system was modified to include Ti nanoparticles in the collagen dispersion, prior to particle formation in liquid nitrogen [179]. Alternatively, the collagen dispersion can be freeze-dried to give a microporous structure that is then subsequently milled to generate particles [180]. Various other processes exist, some in patents, but others as proprietary processes that have not been fully disclosed.

11.5.3 Using collagen tissue

An alternative approach that provides particles is the comminution of tissues. Dermis, in the form of freeze-dried acellular dermal matrix, can be converted into particles through cryogenic homogenising of short cut strips until the desired particle size is achieved [181, 182]. Other tissues that can be used include endosteum [183] and seminiferous tubules [184], both of which provide a type IV collagen substrate for cell culture. The other major source, however, is demineralised bone particles (figure 11.7). They can be used for spinner cell culture, where the possible residual growth factors lead to excellent cell proliferation properties, superior to gelatin beads [185] as well as being used frequently as a tissue filler for bone augmentation or repair [170].

11.6 Fibrous capsule materials

The production of a fibrous capsule around an implanted device has been used to develop collagen-based tissue devices. A particular area of interest has been production of tubes, for vascular application, but it is also possible to make devices in different formats, either by opening the tube to obtain a sheet or by modifying the implant shape.

In the 1950s, following on from various earlier studies that had shown that polyethylene could stimulate fibrous tissue formation [186], the principle was extended to grow fibrocollagenous tubes for vascular use [187]. Further animal studies on mandrel grown fibrocollagenous tubes confirmed this area of opportunity, but also noted that in aortic applications an unacceptable incidence of aneurysms occurred [188]. This led to the suggested inclusion of a Dacron mesh around the mandrel [189, 190] and the further suggestion that an autologous tube could be prepared *in situ* adjacent to the intended application site [190].

A further development was the use of a metal rather than a polymer mandrel, as a component in an assembly that could be implanted into the patient. This 'tissue die' (the Sparks Mandrel) also incorporated a polymer reinforcing mesh [191]. The next development was to use a silicone rod as the mandrel, which again included either a polyester (Dacron) or polypropylene reinforcing mesh [192]. This advance was adapted and included in an updated approach to the Sparks Mandrel method for *in situ* autologous vessel production [193]. This approach gained some popularity, until in subsequent years an increasing number of adverse events were noted, mainly

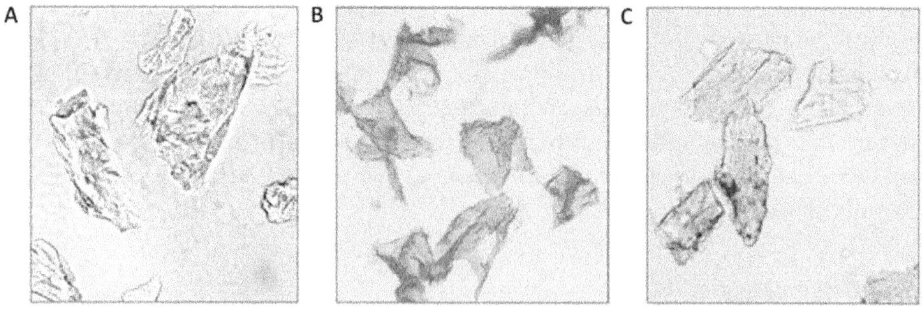

Figure 11.7. Tissue-based collagen particles: (A) decalcified endosteal particles, (B) seminiferous tubule particles, and (C) demineralised bone. (From [185] with permission of InTech Publishing.)

arising from lack of durability leading to degradation and aneurysms [194; reviewed by [195]]. Based on these data, the clinical use of autogenous mandrel grown fibrocollagenous tubes ceased. Despite these problems, the concept with variations still appears on occasion. For example, autologous mandrel grown tubes prepared in the patient's peritoneal cavity have been proposed [196].

Also, in the early 1950s, several surgeons were looking at the development of vascular devices made from synthetic polymers [197; reviewed by [198]]. Over time, Dacron became the preferred fibre [199], in a variety of weights and textures [198]. Early studies showed clearly the importance of good tissue integration for effective performance [200]. In 1975, ePTFE devices were introduced, and despite an initial lack of enthusiasm, by 1992 had become the predominant device, 99%, for lower flow axillofemoral and femoropopliteal locations [198], although Dacron-based devices remain for larger diameter applications. This experience with Dacron also led to work on adding strength and durability, through use of a surrounding (supporting) mesh for human umbilical vein-based devices [201, 202]. While use of an unmodified vessel was not successful, the development of a GA stabilised device showed initial promise, but over longer time scales the extent of adverse events, notably aneurysms and dilatations, led to the low patencies [203]. The possibility of using GA to stabilise natural vessels from xenogeneic species was also proposed [204].

11.6.1 Biosynthetic materials

A specific case of a material made through formation of a fibrous capsule has been the development of biosynthetic materials [205, 206]. This fabrication approach is distinct from biofabrication, which is generally seen as an automated, engineering based *in vitro* approach to form biologically functional products [207].

The various background technologies had established various innovative strategies for vascular prostheses, but each showed specific difficulties that limited their use. These data had established that a silicone mandrel could be used to produce a fibrocollagenous tube, and that this tube could be made more durable by inclusion of a polyester mesh during the tissue deposition phase. This meant that the mesh was fully integrated within the collagenous structure rather than a loosely attached covering. GA fixation of collagen showed that better durability could be obtained, as well as minimising any immunological response allowing xenogeneic material to be used. The combination of all 4 of these key strategies was first shown by Ketharanathan, with sheep as the xenogeneic species [205, 206], and has enabled a clinically successful, off-the-shelf vascular prosthesis product to be developed. This device is known as the Omniflow™ Vascular Prosthesis (OVP) (BioNova International, North Melbourne, later acquired by LeMaitre). The same technology is also suitable for other clinical applications.

While there are other composite collagen–polymer composite materials (see chapter 18), in a biosynthetic material, there is a highly integrated structure that shows characteristics of tissue-based materials, together with the durability of a permanent, fully integrated synthetic component. The chemical stabilisation, by GA, gives durability and a long turnover time, while masking any immunogenicity.

These properties have been shown initially through animal trials, and subsequently through extensive clinical use.

In summary, the material is manufactured by implanting a silicone mandrel covered with polyester mesh, in sheep. This implant is removed after about 12 weeks when sufficient capsule has formed. After trimming and collagen stabilisation, removal of the silicone mandrel leaves a biosynthetic, polymer mesh reinforced collagen tube [206].

Immunohistological studies indicate that various collagen types, particularly types I, III, V and VI, are present in the capsule tissue [208]. Biochemical analyses indicate that the lowest collagen density comes in the zone of contact with the mandrel, but also had the highest amount of type III collagen relative to type I collagen, while type V collagen was present at low levels throughout the thickness of the material [209]. TEM confirmed the low collagen density close to the mandrel, and also showed that there were many small defined fibrils proximal to the Dacron mesh. The body of the capsule consisted of larger well-defined fibrils of uniform size (figure 11.8) [210]. These main fibrils were highly oriented along the axis of the device [210, 211], which would give a better compliance to a tube compared to if they were perpendicular to the axis. The observed fibril orientation may be due to the geometry of the mesh and the micro-motion between the mesh and the silicone mandrel as when there is no polyester mesh, the collagen around the mandrel is randomly aligned [211].

11.7 Other technologies

11.7.1 Ionic liquids

There is an emerging interest in the use of ionic liquids (ILs) for processing collagen. Preferably, this would use neat ionic liquids, but some studies also examine aqueous solutions of ILs. ILs are salts that exist as liquids at or near room temperature with negligible vapour pressures, and which can be used to dissolve natural polymers, especially cellulose but also collagen [212, 213]. ILs that are suitable for dissolving collagen include 1-butyl-3-methylimidazolium chloride ([BMIM]Cl), [213] and 3,3′-[1,2-ethanediylbis (oxy-2,1-ethanediyl)]-bis[1-methyl-imidazolium]-dibromide ($[C_6O_2(mim)_2][Br]_2$) [214]. At high temperature, 100 °C,

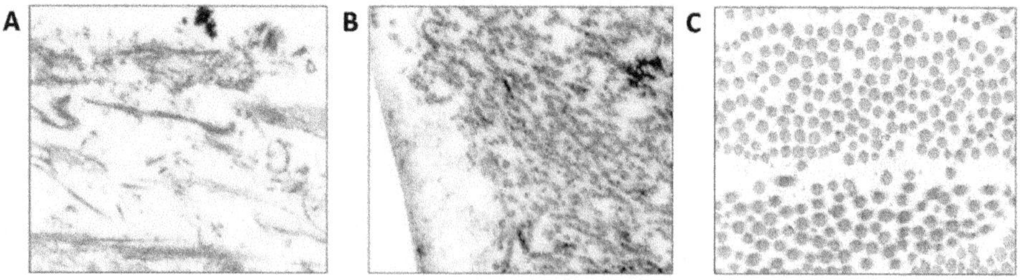

Figure 11.8. TEM of the collagen within the capsule wall of an OVP: (A) the capsule adjacent to the silicone mandrel (top), (B) fine collagen fibrils adjacent to a Dacron fibre (left), and (C) aligned, larger collagen fibrils within the main body of the capsule. (From [210] with permission from Elsevier.)

[BMIM]Cl dissolved collagen fibres over 2h to give clear, viscous solutions. After cooling to 4 °C, collagen could be reconstituted as films or fibres by adding to a precipitation liquid, such as water or EtOH, in which collagen was not soluble, with accompanying washing out of the IL [213]. In the absence of excess water, the collagen remained stable in part during the heating step, suggesting that the reconstituted films could still contain some native-triple-helical collagen [213]. The partial loss of collagen triple-helical structure was also reported for a similar protocol using $[C_6O_2(mim)_2][Br]_2$ [214]. This potential instability of the collagen has also been seen in other studies, in this case using two other ILs, bis-choline sulfate and 1-butyl-3-methylimidazolium dimethyl phosphate [215]. This seems due to the changes in the hydrogen bonding environment and hydration dynamics in the ILs destabilising the collagen, with the extent depending on the ionic composition [215, 216]. For tissue samples, this effect leads to swelling [215, 216].

11.7.2 Alignment technologies

In some cases, such as extrusion, the flow characteristics and/or mechanical forces such drawing can align the collagen. In other instances, additional options to align the collagen can be used. These include the use of magnetic fields [217] and electrochemical methods [218]. Preparations of collagen at high concentrations, giving a liquid crystal format, is also based on a strong alignment of collagen [219].

11.8 Sterilisation

As noted previously, although sterilisation is typically at or near the end of a process, its potential impact on the properties of a biomaterial product mean that it should, without doubt, be part of the first plans and strategies in product development. A variety of approaches are possible, with some more suited than others to collagen products, such that the actual number that can be efficiently and effectively used is fairly limited. Evaluation of the pros and cons of different approaches [220–222] should be undertaken in order to select appropriate options. A variety of processes including simple disinfecting steps, can be used in laboratory experiments, but generally it should be recognised that these would not be sufficient for clinical trials or for commercial products, so once a research path is selected a suitable sterilisation approach should be included from early on so that subsequent repetition of work done with a temporary process is not needed. Extensive guidelines on sterilisation processes are available from regulatory authorities and should be consulted at an early stage in a research plan. An end-product sterilisation on the final packaged product is best and should be the target. However, if an end-product sterilisation is not used, then after a prior sterilisation step full sterility must be maintained throughout the remaining steps of the process and packaging. Among the various extra difficulties that this approach introduces, this sterile path should be capable of being scaled up to the desired manufacturing numbers. In many cases, the validation of the sterilisation protocols may be required, and various recommendations are available that should be followed, and can be complemented by other methods [223].

11.8.1 Physical approaches including irradiation

Gamma irradiation

Gamma irradiation is frequently used as a convenient sterilisation process, with irradiation by 25 kGy (2.5Mrads) using a ^{60}Co source [224] being the usual process, although in some cases lower doses have been reported. An advantage is that accurate dose of irradiation is readily achieved, but this can be accompanied by changes in the collagen structure, and the development of some static charge that can make handling the materials difficult.

As well as being useful for fabricated collagen materials, it is now used extensively for tissue grafts and acellular matrix products [225–228]. The main disadvantage for these materials as well as for fabricated collagen-based materials is the damage caused by the irradiation, which can lead to cleavage of the polypeptide chain as well as crosslinking (see chapter 12). This damage has been known for a long time [229–231], and various suggestions have been made for approaches that may mediate the extent of damage. These include inclusion of compounds that are free radical scavengers, such as thiourea or N-acetyl-L-cysteine [232, 233]. Other suggestions have included glucose [234]. Another sugar that has been suggested is ribose [235], which may be assisting through crosslinking the collagen. Certainly, other cross-linking reagents including genipin and carbodiimides can mediate irradiation damage [236, 237].

Gamma irradiation is possibly the method of choice for many collagen-based devices even though there are changes to the collagen. For fabricated, collagen-based devices, irradiation damage does lead to certain observable biological changes. For example, gel contraction, by various cells, is inhibited for irradiated collagen [238, 239]. Also, introduction of cleavage sites and lowering of collagen stability are likely to lead to an increased rate of collagen device turnover [240], but this can be anticipated and controlled.

Electron beam irradiation

Electron beam irradiation provides a sterilisation method that is occasionally reported. Original approaches took long times, but newer technology led to a feasible method [241]. It was found, for example, that treating fabricated collagen-based scaffolds with 15 kGy electron beam irradiation eliminated any bacterial or fungal growth, which did occur with aseptically processed scaffolds, while changes in mechanical properties were not statistically significant [242]. The sterilisation approach did not affect the performance of the scaffold in ligament healing in a porcine model, so, if the equipment is available, then this method can effectively sterilise the scaffolds with only a minimal impact on the *in vivo* function [242]. On the other hand, for more complex structures, such as acellular, tissue-based materials damage to small substructures such as basement membranes and fine elastic fibres could be significant [227].

Pulsed power treatment

Pulsed power methods can potentially sterilise collagen-based materials [243] and also tissue derived materials [244], through short duration pulses at low temperature and at atmospheric pressure. Experiments demonstrated that discharges of 11 kV sterilised samples in 45 s, gave effective sterilisation of connective tissues, although there were notable decreases in elastic modulus, in stiffness and in elastic-strain [244]. Nevertheless, an appropriate balance between sterilisation and degradation can be achieved, and this type of approach had little effect on subsequent cellular growth [243].

11.8.2 Chemical and other approaches

Ethylene oxide

Historically, ethylene oxide sterilisation of collagen-based materials was an important process, in solution and later, more generally, as an efficient gas process [245]. However, it has fallen in popularity, in part because of the decline in the availability of processing facilities and safety issues including inflamability, rather than effectiveness. To work effectively, the process needs some moisture to be present, which is the case for collagen as it has a high binding capacity for moisture. Like other approaches it leads to some modification to the collagen [1]. The reaction with amino groups lowers the stability of the collagen, although the mechanical properties of the sterilised materials were not significantly altered [246].

Supercritical carbon dioxide

The use of supercritical carbon dioxide has emerged more recently as a convenient and efficient sterilisation process [247] that is particularly useful for less robust and biological materials, although inactivation of spores may require a prolonged, potentially more destructive process [248]. It is non-toxic and non-reactive with ease of penetration of materials. It is easily removable by depressurising, leading to the liquid to gas phase transition. It is suitable for collagen-based materials, and can also be used in modifications by addition of H_2O_2 and H_2O_2 in combination with acetic anhydride [248]. Other approaches make it suitable as a terminal sterilisation process that can inactivate spores, while remaining suitable for proteins [249].

Chemical treatments

Various chemical systems have been used for disinfecting and sterilising collagen-based materials and devices. In some cases, this can be used as an end stage process, but in many other cases it is an intermediate step with the following steps needing particular care with isolation protocols. For example, various oxidation processes have been proposed, including peracetic acid followed by keeping the product in glycerol [250] and a combination of copper and H_2O_2 [251] as well as an iodine-based approach [252].

These approaches have avoided the use of aldehydes as these lead to significant modification of collagen. However, for tissue-based devices that are crosslinked using aldehydes, especially glutaraldehyde (GA), aldehydes can be very useful. For example, various cardiovascular devices, including biologic heart valves and biosynthetic vascular prostheses, are processed in 2% GA, and this solution is a good sterilant. For heart valves, some are shipped in GA solution, requiring extensive washing in the theatre prior to implantation, while for other products the GA may be removed in a sterile manner and replaced by aqueous EtOH to maintain sterility. This also needs washing in the theatre prior to surgery, but is less of an issue than GA solution.

Antibiotic treatments
In some instances, antibiotic treatment has been used to disinfect and sanitise products, but the ability to provide a fully sterile product should be considered as limited. This approach may be suitable for certain laboratory trials, but would probably not be satisfactory for commercial products.

11.8.3 Bovine spongiform encephalopathy

A significant issue for the safety of collagen-based medical products came in the mid-1980s with the emergence of 'bovine spongiform encephalopathy' (BSE), also known as 'mad cow disease'. Previously, the human disease 'Creutzfeldt–Jakob Disease', which had been postulated in the early part of the last century [253, 254], had been described as a similar encephalopathy occurrence. This occurred as a rare inherited disease, 'Gerstmann–Straussler–Scheinker-syndrome' or more frequently as a spontaneous new occurrence [255]. There are still concerns about the extent of iatrogenic transmission [256], with dura mater materials having been a significant issue [257]. Similar diseases are 'fatal familial insomnia' [258], 'sporadic fatal insomnia' [259] and 'kuru', which had been described from New Guinea and its transmission was associated with ingestion of diseased tissues, especially neural tissues, during cannibalistic rituals [260]. The transmission of the diseases was attributed to a protein based infectious agent, the prion [261, 262].

Similar diseases had also been described in a number of diverse animal species. The disease 'scrapie' was well-known and was first observed in sheep and goats and was first described in the literature in 1750 [263], but it was almost two centuries later that its transmission was confirmed [264]. Subsequently, a range of other related diseases, such as mink encephalopathy, chronic wasting disease [265] and others have been described [262, 266]. The best known is BSE.

The rapid and massive occurrence of BSE in the United Kingdom was attributed to the possible loss of inactivation with lower temperature rendering processing, but mostly due to the feeding of protein pellets manufactured from abattoir waste that included infected tissues [267]. By 2002 a total of 179 853 infected animals had been identified in the UK, with a maximum of 36 682 in 1992, dropping steadily to 1044 by 2002 [268]. However, stringent regulation had led to there being only one case reported in 2014. The increased surveillance led to the disease being observed in other European countries [269] as well as cases in the USA. To what extent these cases were due to infection or were spontaneous new cases is not clear.

A general solution for the collagen biomaterials industry has been to source collagen for countries that were considered safe, notably Australia and New Zealand. Also, the use of closed herds, with controlled nutrition and health standards along with the use of young animals has been preferred and has proved a good strategy. Nevertheless, spontaneous occurrences are still possible so caution is always needed. Recently, in 2018, a further case in the UK was reported from a well-established closed herd, illustrating the potential for spontaneous, new mutations occurring.

Regulatory authorities look for proof that a collagen material would not be infectious. While there are various methods to eliminate and/or inactivate bacterial and viral contaminants, the same is not true for prion presence. One approach has been to treat collagen for 1 h with 1 M NaOH at room temperature (20 °C), as this approach has been suggested to be effective for prion inactivation [223]. Another has examined the use of various inactivation treatments by solvents including formic acid, trifluoroacetic acid, tetrafluorethanol and hexafluoroisopropanol [270]. Each treatment led to different effects on the collagen, but overall it seemed that formic acid treatment could represent the best approach [270]. Alkali, and other reagents have often been reported for viral inactivation but their utility for prion inactivation is generally not reported. Also, these approaches have often been developed with disinfection of instruments in mind and their effect on a delicate protein preparation is not characterised [271, 272]. So safe sourcing remains an important strategy. An alternative solution is via recombinant production systems.

References

[1] Chvapil M, Kronenthal L and Van Winkle W 1973 *Int. Rev. Connect. Tissue Res.* **6** 1–61
[2] Stenzel K H, Miyata T and Rubin A L 1974 *Annu. Rev. Biophys. Bioeng.* **3** 231–53
[3] Ramshaw J A M, Werkmeister J A and Peters D E 1990 *Current Perspectives on Implantable Devices* vol 2 ed D F Williams (London: JAI Press) pp 151–220
[4] Miyata T, Taira T and Noishiki Y 1992 *Clin. Mater.* **9** 139–48
[5] Daniels J R and Knapp T R 1976 *US Patent* 3949073
[6] Hermann S 1957 *US Patent* 2783212
[7] Pye D J 1960 *US Patent* 2960486.
[8] Wichterle O and Lim D 1960 *Nature* **185** 117
[9] Graham T 1864 *Proc. Roy. Soc.* **13** 335–41
[10] Hardy W B 1899 *J. Physiol.* **24** 158–210.1
[11] Whetham W C D 1904 *The Recent Development of Physical Science* (Philadelphia, PA: P Blakiston's) p 137
[12] Peppas N A and Merrill E W 1976 *J. Polym. Sci. A Polym. Chem.* **14** 441–57
[13] Chai Q, Jiao Y and Yu X 2017 *Gels* **3** 6
[14] Stenzel K H, Dunn M W, Rubin A L and Miyata T 1969 *Science* **164** 1282–83
[15] Dunn M W, Stenzel K H, Rubin A L and Miyata T 1969 *Arch. Ophthalmol.* **82** 840–44
[16] Lai G, Li Y and Li G 2008 *Int. J. Biol. Macromol.* **42** 285–91
[17] Knapp T R, Kaplan E N and Daniels J R 1977 *Plast. Reconstr. Surg.* **60** 398–405
[18] Pohl T 1990 *Meth. Enzymol.* **182** 68–83

[19] Martin R, Farjanel J, Eichenberger D, Colige A, Kessler E, Hulmes D J S and Giraud-Guille M M 2000 *J. Mol. Biol.* **301** 11–7
[20] Wang Y, Silvent J, Robin M, Babonneau F, Meddahi-Pellé A, Nassif N and Giraud-Guille M M 2011 *Soft Matter* **7** 9659–64
[21] Saeidi N, Karmelek K P, Paten J A, Zareian R, DiMasi E and Ruberti J W 2012 *Biomaterials* **33** 7366–74
[22] Shire S J, Shahrokh Z and Liu J 2004 *J. Pharm. Sci.* **93** 1390–402
[23] Gross J and Kirk D 1958 *J. Biol. Chem.* **233** 355–60
[24] Kreger S T, Bell B J, Bailey J, Stites E, Kuske J, Waisner B and Voytik-Harbin S L 2010 *Biopolymers* **93** 690–707
[25] Johnson T D, Lin S Y and Christman K L 2011 *Nanotechnology* **22** 494015
[26] Mahoney M J and Anseth K S 2007 *J. Biomed. Mater. Res. A* **81** 269–78
[27] Mason M N, Arnold C A and Mahoney M J 2009 *Tissue Eng. Part A* **15** 3799–808
[28] Chan B K, Wippich C C, Wu C J, Sivasankar P M and Schmidt G 2012 *Macromol. Biosci.* **12** 1490–501
[29] Kinneberg K R, Nelson A, Stender M E, Aziz A H, Mozdzen L C, Harley B A, Bryant S J and Ferguson V L 2015 *Ann. Biomed. Eng.* **43** 2618–29
[30] Mason B N, Starchenko A, Williams R M, Bonassar L J and Reinhart-King C A 2013 *Acta Biomater.* **9** 4635–44
[31] Kamimura W, Koyama H, Miyata T and Takato T 2014 *J. Biomed. Mater. Res. A* **102** 4309–16
[32] Parmar P A, Skaalure S C, Chow L W, St-Pierre J P, Stoichevska V, Peng Y Y, Werkmeister J A, Ramshaw J A M and Stevens M M 2016 *Biomaterials* **99** 56–71
[33] Braziulis E, Diezi M, Biedermann T, Pontiggia L, Schmucki M, Hartmann-Fritsch F, Luginbühl J, Schiestl C, Meuli M and Reichmann E 2012 *Tissue Eng. Part C Methods* **18** 464–74
[34] Zitnay J L, Reese S P, Tran G, Farhang N, Bowles R D and Weiss J A *Acta Biomater.* **65** 76–87
[35] Gao Y, Kong W, Li B, Ni Y, Yuan T, Guo L, Lin H, Fan H, Fan Y and Zhang X 2018 *Colloids Surf. B Biointerfaces* **2018** 448–56
[36] Gao Y, Li B, Kong W, Yuan L, Guo L, Li C, Fan H, Fan Y and Zhang X 2018 *Int. J. Biol. Macromol.* **118** 2014–20
[37] White L J, Saldin L T, Keane T J, Cramer M C, Shakesheff K M and Badylak S F 2017 *Comprehensive Biomaterials II* vol 2 ed P Ducheyne, K Healy, D E Hutmacher, D W Grainger and C J Kirkpatrick (New York: Elsevier) pp 532–41
[38] Li S T 1991 *US Patent* 5026381
[39] Li S T and Stone K R 1997 *US Patent* 5681353
[40] Nirdlinger M 1904 *US Patent* 753206
[41] Sheppard S E and Hudson J H 1935 *US Patent* 2000042
[42] Schaufler A 2002 *US Patent Application* 2002/0164 322 A1
[43] Sando L, Danon S, Brownlee A G, McCulloch R J, Ramshaw J A M, Elvin C M and Werkmeister J A 2011 *J. Tissue Eng. Regen. Med.* **5** 337–46
[44] Shah C B, Vitaris R F, Fink E D, Orr S, Dowd B, Tranchemontagne A, Ward K, Swaniker H and Patel H A 2010 *US Patent Application* 2010/0260 824 A1
[45] Reichel J 1939 *US Patent* 2166074
[46] Sifferd R H and Schmitt R J 1952 *US Patent* 2610625A

[47] Dagalakis N, Flink J, Stasikelis P, Burke J F and Yannas I V 1980 *J. Biomed. Mater. Res.* **14** 511–28
[48] Nava M M, Draghi L, Giordano C and Pietrabissa R 2016 *J. Appl. Biomater. Funct. Mater.* **14** e223–9
[49] Yannas I V, Lee E, Orgill D P, Skrabut E M and Murphy G F 1989 *Proc. Natl. Acad. Sci. USA* **86** 933–37
[50] Doillon C J, Whyne C F, Brandwein S and Silver F H 1986 *J. Biomed. Mater. Res.* **20** 1219–28
[51] O'Brien F J, Harley B A, Yannas I V and Gibson L J 2005 *Biomaterials* **26** 433–41
[52] Discher D E, Janmey P and Wang Y L 2005 *Science* **310** 1139–43.
[53] Park J S, Chu J S, Tsou A D, Diop R, Tang Z, Wang A and Li S 2011 *Biomaterials* **32** 3921–30
[54] Grier W K, Iyoha E M and Harley B A C 2017 *J. Mech. Behav. Biomed. Mater.* **65** 295–305
[55] Hogrebe N J, Reinhardt J W and Gooch K J 2017 *J. Biomed. Mater. Res A* **105** 640–61
[56] Yannas I V, Tzeranis D S, Harley B A and So P T 2010 *Philos. Trans. A Math. Phys. Eng. Sci.* **368** 2123–39
[57] O'Brien F J, Harley B A, Yannas I V and Gibson L 2004 *Biomaterials* **25** 1077–86
[58] Park S N, Park J C, Kim H O, Song M J and Suh H 2002 *Biomaterials* **23** 1205–12
[59] Haugh M G, Murphy C M and O'Brien F J 2010 *Tissue Eng. Part C Methods* **16** 887–94
[60] Plieva F M, Galaev I Y and Mattiasson B 2007 *J. Sep. Sci.* **30** 1657–71
[61] Bhat S, Lidgren L and Kumar A 2013 *Macromol. Biosci.* **13** 827–37
[62] Heschel I and Rau G 2002 *US Patent* 6447701
[63] Schoof H, Apel J, Heschel I and Rau G 2001 *J. Biomed. Mater. Res.* **58** 352–7
[64] Schoof H, Bruns L, Fischer A, Heschel I and Rau G 2000 *J. Crystal Growth* **15** 122–9
[65] Wu S, Zhu C, He Z, Xue H, Fan Q, Song Y, Francisco J S, Zeng X C and Wang J 2017 *Nat. Commun.* **8** 15154
[66] Divakar P, Yin K and Wegst U G K 2019 *J. Mech. Behav. Biomed. Mater.* **90** 350–64
[67] Kroehne V, Heschel I, Schügner F, Lasrich D, Bartsch J W and Jockusch H 2008 *J. Cell Mol. Med.* **12** 1640–48
[68] Boekema B K H L, Vlig M, Olde-Damink L, Middelkoop E, Eummelen L, Bühren A V and Ulrich M M W 2014 *J. Mater. Sci., Mater. Med.* **25** 423–33
[69] Pennella F, Cerino G, Massai D, Gallo D, Falvo D'Urso Labate G, Schiavi A, Deriu M A, Audenino A and Morbiducci U 2013 *Ann. Biomed. Eng.* **41** 2027–41
[70] Massai D, Pennella F, Gentile P, Gallo D, Ciardelli G, Bignardi C, Audenino A and Morbiducci U 2014 *Biomed. Res. Int.* **2014** 161437
[71] Loh Q L and Choong C 2013 *Tissue Eng. Part B Rev.* **19** 485–502
[72] Shimko D A and Nauman E A 2007 *J. Biomed. Mater. Res. B Appl. Biomater.* **80** 360–9
[73] Al-Munajjed A A, Hien M, Kujat R, Gleeson J P and Hammer J 2008 *J. Mater. Sci., Mater. Med.* **19** 2859–64
[74] Cioffi M, Boschetti F, Raimondi M T and Dubini G 2006 *Biotechnol. Bioeng.* **93** 500–10
[75] Zermatten E, Vetsch J R, Ruffoni D, Hofmann S, Müller R and Steinfeld A 2014 *Ann. Biomed. Eng.* **42** 1085–94
[76] Luo C J, Stoyanov S D, Stride E, Pelan E and Edirisinghe M 2012 *Chem. Soc. Rev.* **41** 4708–35
[77] Aibibu D, Hild M, Wöltje M and Cherif C 2016 *J. Mater. Sci., Mater. Med.* **27** 63
[78] Cross C F, Bevan E J and Beadle C 1892 *GB Patent* 8700 (1892)

[79] Stearn C H 1899 *US Patent* 622087
[80] Thiele E 1902 *US Patent* 710819
[81] Millar A 1898 *US Patent* 611814
[82] Cohoe W P 1915 *US Patent* 1163740
[83] Sharp G A 1935 *US Patent* 1999641
[84] Schulte W 1936 *US Patent* 2058835
[85] Becker O W 1936 *US Patent* 2056595
[86] Schmitt F O 1985 *Annu. Rev. Biophys. Biophys. Chem.* **14** 1–23
[87] Cresswell A 1949 *US Patent* 2475697
[88] Cresswell Awith American Cyanamid 1948 *GB Patent* 606427
[89] Griset E J, Reissmann T L and Nichols J 1963 *US Patent* 3114593
[90] Zahn H 1961 *US Patent* 2988451
[91] Novak L J 1956 *US Patent* 2748774
[92] van Winkle W 1968 *US Patent* 3384073
[93] Kemp P D, Carr R M, Maresh J G, Cavallaro J and Gross J 1995 *S Patent* 5378469
[94] Fofonoff T W and Bell E 1996 *US Patent* 5562946
[95] Cavallaro J F 1998 *US Patent* 5718012
[96] Siriwardane M L, DeRosa K, Collins G and Pfister B J 2014 *Biofabrication* **6** 015012
[97] Kato Y P, Christiansen D L, Hahn R A, Shieh S J, Goldstein J D and Silver F H 1989 *Biomaterials* **10** 38–42
[98] Kato Y P and Silver F H 1990 *Biomaterials* **11** 169–75
[99] Zeugolis D I, Paul R G and Attenburrow G 2008 *J. Appl. Polym. Sci.* **108** 2886–94
[100] Zeugolis D I, Paul R G and Attenburrow G 2008 *J. Biomed. Mater. Res. B Appl. Biomat.* **85** 343–52
[101] Arafat M T, Tronci G, Yin J, Wood D J and Russell S J 2015 *Polymer* **77** 102–12
[102] Meyer M, Baltzer H and Schwikal K 2010 *Mat. Sci. Eng.* C **30** 1266–71
[103] Hoogenkamp H R, Bakker G J, Wolf L, Suurs P, Dunnewind B, Barbut S, Friedl P, van Kuppevelt T H and Daamen W F 2015 *Acta Biomater.* **12** 113–21
[104] Haynl C, Hofmann E, Pawar K, Förster S and Scheibel T 2016 *Nano Lett.* **16** 5917–22
[105] Tucker N, Stanger J J, Staiger M, Razzaq H and Hofman K 2012 *J. Eng. Fibers Fabrics* **7** 63–73
[106] Cooley J F 1900 *GB Patent* 6385
[107] Formhals A 1937 *US Patent* 2077373
[108] Norton C L 1936 *US Patent* 2048651
[109] Taylor G I 1964 *Proc. R. Soc. Lond. A* **280** 383–97
[110] Ding B, Wang X and Yu J 2019 *Electrospinning: Nanofabrication and Applications* (Norwich, NY: William Andrew)
[111] Yoon J, Yang H S, Lee B S and Yu W R 2018 *Adv. Mater.* **30** e1704765
[112] Green T B, King S L and Li L 2015 *US Patent* 9101860
[113] Matthews J A, Wnek G E, Simpson D G and Bowlin G L 2002 *Biomacromolecules* **3** 232–38
[114] Sell S A, McClure M J, Garg K, Wolfe P S and Bowlin G L 2009 *Adv. Drug Deliv. Rev.* **61** 1007–19
[115] Zeugolis D I, Khew S T, Yew E S, Ekaputra A K, Tong Y W, Yung L Y, Hutmacher D W, Sheppard C and Raghunath M 2008 *Biomaterials* **29** 2293–305
[116] Shields K J, Beckman M J, Bowlin G L and Wayne J S 2004 *Tissue Eng.* **10** 1510–17

[117] Chen L, Zhu C, Fan D, Liu B, Ma X, Duan Z and Zhou Y 2011 *J. Biomed. Mater. Res. A* **99** 395–409

[118] Elamparithi A, Punnoose A M and Kuruvilla S 2016 *Artif. Cells Nanomed. Biotechnol.* **44** 1318–25

[119] Buttafoco L, Kolkman N G, Engbers-Buijtenhuijs P, Poot A A, Dijkstra P J, Vermes I and Feijen J 2006 *Biomaterials* **27** 724–34

[120] Shanmugam G, Reddy S M, Natarajan V and Madhan B 2013 *Int. J. Biol. Macromol.* **54** 155–9.

[121] Dong B, Arnoult O, Smith M E and Wnek G E 2009 *Macromol. Rapid Commun.* **30** 539–42

[122] Foltran I, Foresti E, Parma B, Sabatino P and Roveri N 2008 *Macromol. Symp.* **269** 111–18

[123] Yang L, Fitié C F C, van der Werf K O, Bennink M L, Dijkstra P J and Feijen J 2008 *Biomaterials* **29** 955–62

[124] Torres-Giner S, Gimeno-Alcañiz J V, Ocio M J and Lagaron J M 2009 *ACS Appl. Mater. Interfaces* **1** 218–23

[125] Liu T, Teng W K, Chan B P and Chew S Y 2010 *J. Biomed. Mater. Res. A* **95** 276–82

[126] Stachel I, Schwarzenbolz U, Henle T and Meyer M 2010 *Biomacromol.* **11** 698–705

[127] Chen Z, Mo X and Qing F 2007 *Mater. Lett.* **61** 3490–94

[128] Zhou Y, Yao H, Wang J, Wang D, Liu Q and Li Z 2015 *Int. J. Nanomedicine* **10** 3203–15

[129] Chen R, Huang C, Ke Q, He C, Wang H and Mo X 2010 *Colloids Surf. B Biointerfaces* **79** 315–25

[130] McClure M J, Sell S A, Simpson D G, Walpoth B H and Bowlin G L 2011 *J. Vis. Exp.* **47** 2084

[131] Truong Y B, Glattauer V, Briggs K L, Zappe S and Ramshaw J A M 2012 *Biomaterials* **33** 9198–204

[132] Benning L, Gutzweiler L, Tröndle K, Riba J, Zengerle R, Koltay P, Zimmermann S, Stark G B and Finkenzeller G 2017 *J. Biomed. Mater. Res. A* **105** 3231–41

[133] Paten J A, Tilburey G E, Molloy E A, Zareian R, Trainor C V and Ruberti J W 2013 *Biomaterials* **34** 2577–87

[134] Włodarczyk-Biegun M K and Del Campo A 2017 *Biomaterials* **134** 180–201

[135] Mora-Boza A and Lopez-Donaire M L 2018 *Adv. Exp. Med. Biol.* **1058** 221–45

[136] O'Connell C D, Di Bella C, Thompson F, Augustine C, Beirne S, Cornock R, Richards C J, Chung J and Gambhir S *et al* 2016 *Biofabrication* **8** 015019

[137] Hakimi N, Cheng R, Leng L, Sotoudehfar M, Ba P Q, Bakhtyar N, Amini-Nik S, Jeschke M G and Günther A 2018 *Lab Chip* **18** 1440–51

[138] Kim B S, Lee J S, Gao G and Cho D W 2017 *Biofabrication* **9** 025034

[139] Lee H, Han W, Kim H, Ha D H, Jang J, Kim B S and Cho D W 2017 *Biomacromolecules* **18** 1229–37

[140] Ren X, Wang F, Chen C, Gong X, Yin L and Yang L 2016 *BMC Musculoskelet. Disord.* **17** 301

[141] Snyder J E, Hamid Q, Wang C, Chang R, Emami K, Wu H and Sun W 2011 *Biofabrication* **3** 034112

[142] Tan Z, Parisi C, Di Silvio L, Dini D and Forte A E 2017 *Sci. Rep.* **7** 16293

[143] Min D, Lee W, Bae I H, Lee T R, Croce P and Yoo S S 2018 *Exp. Dermatol.* **27** 453–9

[144] Bell A, Kofron M and Nistor V 2015 *Biofabrication* **7** 035007

[145] Diamantides N, Wang L, Pruiksma T, Siemiatkoski J, Dugopolski C, Shortkroff S, Kennedy S and Bonassar L J 2017 *Biofabrication* **9** 034102

[146] Kim Y B, Lee H and Kim G H 2016 *ACS Appl. Mater. Interfaces* **8** 32230–40
[147] Attalla R, Puersten E, Jain N and Selvaganapathy P R 2018 *Biofabrication* **11** 015012
[148] Isaacson A, Swioklo S and Connon C J 2018 *Exp. Eye Res.* **173** 188–93
[149] Izadifar M, Chapman D, Babyn P, Chen X and Kelly M E 2018 *Tissue Eng. Part C Methods* **24** 74–88
[150] Mazzocchi A, Devarasetty M, Huntwork R, Soker S and Skardal A 2018 *Biofabrication* **11** 015003
[151] Kreimendahl F, Köpf M, Thiebes A L, Duarte Campos D F, Blaeser A, Schmitz-Rode T, Apel C, Jockenhoevel S and Fischer H 2017 *Tissue Eng. Part C Methods* **23** 604–15
[152] Yang X, Lu Z, Wu H, Li W, Zheng L and Zhao J 2018 *Mater. Sci. Eng. C Mater. Biol. Appl.* **83** 195–201
[153] Tebb T A, Tsai S W, Glattauer V, White J F, Ramshaw J A M and Werkmeister J A 2006 *Cytotechnology* **52** 99–106
[154] Zhu K, Chen N, Liu X, Mu X, Zhang W, Wang C and Zhang Y S 2018 *Macromol. Biosci.* **18** e1800127
[155] Stratesteffen H, Köpf M, Kreimendahl F, Blaeser A, Jockenhoevel S and Fischer H 2017 *Biofabrication* **9** 045002
[156] Benning L, Gutzweiler L, Tröndle K, Riba J, Zengerle R, Koltay P, Zimmermann S, Stark G B and Finkenzeller G 2018 *J. Biomed. Mater. Res. A* **106** 935–47
[157] Thomas M and Willerth S M 2017 *Front. Bioeng. Biotechnol.* **5** 69
[158] Tarassoli S P, Jessop Z M, Al-Sabah A, Gao N, Whitaker S, Doak S and Whitaker I S 2018 *J. Plast. Reconstr. Aesthet. Surg.* **71** 615–23
[159] Matsumoto K, Nakamura T, Shimizu Y, Ueda H, Sekine T, Yamamoto Y, Kiyotani T and Takimoto Y 1999 *ASAIO J.* **45** 288–92
[160] Li S-T 1993 *US Patent* 5206028
[161] Lubbs E K 1986 *US Patent* 4626286
[162] Edwards G A, Glattauer V, Nash T J, White J F, Brock K A, Werkmeister J A and Ramshaw J A M 1997 *J. Biomed. Mater. Res.* **34** 291–97
[163] Ellender G, Hammond R, Papli R, Mitrangas K, Bateman J F, Glattauer V, Thyer J M, Werkmeister J A and Ramshaw J A M 1992 *Clin. Mater.* **9** 201–9
[164] Chung E J, Jakus A E and Shah R N 2013 *Acta Biomater.* **9** 5153–61
[165] Miyata T, Rubin A L, Stenzel K H and Dunn M W 1979 *US Patent* 4164559
[166] Yannas I V and Sieverding D L 1981 *US Patent* 4252759
[167] Schwarz F, Rothamel D, Herten M, Sager M and Becker J 2006 *Clin. Oral Implants Res.* **17** 403–9
[168] Zerris V A, James K S, Roberts J B, Bell E and Heilman C B 2007 *J. Biomed. Mater. Res. B Appl. Biomater.* **83** 580–8
[169] Davis J 1910 *Johns Hopkins Med. J.* **15** 307–96
[170] Gruskin E, Doll B A, Futrell F W, Schmitz J P and Hollinger J O 2012 *Adv. Drug Deliv. Rev.* **64** 1063–77
[171] Cahn F 1990 *Trends Biotechnol.* **8** 131–36
[172] Malda J and Frondoza C G 2006 *Trends Biotechnol.* **24** 299–304
[173] Glattauer V, White J F, Tsai W B, Tsai C C, Tebb T A, Danon S J, Werkmeister J A and Ramshaw J A M 2010 *J. Biomed. Mater. Res. A* **92** 1301–9
[174] Sehgal P K and Srinivasan A 2009 *Expert Opin. Drug Deliv.* **6** 687–95

[175] Perez R A, Altankov G, Jorge-Herrero E and Ginebra M P 2013 *J. Tissue Eng. Regen. Med.* **7** 353–61
[176] Lima A C, Mano J F, Concheiro A and Alvarez-Lorenzo C 2015 *Stem Cell Rev.* **11** 161–79
[177] Silver F, Sharma V, Berndt D R and Marn L E 1993 *US Patent* 5196185
[178] Berg R A, Silver F H and Pachence J M 1989 *US Patent* 48372285
[179] Dean R C, Silver F H, Berg R A, Phillips P G and Runstadler P W 1989 *US Patent* 4861714
[180] Maffia G J 2003 *US Patent* 6660829
[181] Griffey E S, Livesey S A, Schiff C M and Boerboom L E 2008 *US Patent* 7358284
[182] Zuo H, Peng D, Zheng B, Liu X, Wang Y, Wang L, Zhou X and Liu J 2012 *J. Mater. Sci., Mater. Med.* **23** 2933–44
[183] Nigro J, White J F, Ramshaw J A M, Haylock D N, Nilsson S K and Werkmeister J A 2010 *Biomaterials* **31** 5689–99
[184] Glattauer V, Werkmeister J A and Ramshaw J A M 2007 *Patent Application* WO2006128216-A1
[185] Glattauer V, Tsai W-B, White J F, Nigro J, Tebb T A, Werkmeister J A and Ramshaw J A M 2011 *Regenerative Medicine and Tissue Engineering – Cells and Biomaterials* ed D Eberli (Rijeka: InTech) pp 261–86
[186] Yeager G H and Cowley R A 1948 *Ann. Surg.* **128** 509–19
[187] Pierce E C 1953 *Surgery* **33** 648–57
[188] Eiken O and Norden G 1961 *Acta Chir. Scand.* **121** 90–102
[189] Hufnagel C A 1964 *Ann. Surg.* **159** 828 (discussion)
[190] Assefi I and Parsonnet V 1964 *J. Newark Beth Israel Hosp.* **15** 161–70
[191] Sparks C H 1970 *Ann. Surg.* **172** 787–94
[192] Parsonnet V, Alpert J and Brief D K 1971 *Surgery* **70** 935–39
[193] Sparks C H 1973 *Ann. Surg.* **177** 293–300
[194] Christenson J T and Eklöf B 1979 *Br. J. Surg.* **66** 514–17
[195] Ramshaw J A M, Edwards G A and Werkmeister J A 1995 *Encyclopedic Handbook of Biomaterials and Bioengineering Part B* vol 2 ed D L Wise, D J Trantolo, D E Altobelli, M J Yaszemski, J D Gresser and E R Schwartz (New York: Marcel Dekker) pp 953–78
[196] Campbell J H, Efendy J L and Campbell G R 1999 *Circ. Res.* **85** 1173–78
[197] Voorhees A B, Jaretzki A and Blakemore A H 1952 *Ann. Surg.* **135** 332–36
[198] Sauvage L R 1993 *J. Invest. Surg.* **6** 221–25
[199] Szilagyi D E, France L C, Smith R F and Whitcomb J G 1958 *AMA Arch. Surg.* **77** 538–51
[200] Wesolowski S A 1962 *Evaluation of Tissue and Prosthetic Vascular Grafts* (Springfield, IL: Thomas)
[201] Dardik H, Ibrahim I M, Baier R, Sprayregen S, Levy M and Dardik I I 1976 *J. Am. Med. Assoc.* **236** 2859–62
[202] Dardik H D, Ibrahim I M, Sprayregen S and Dardik I I 1976 *Surgery* **79** 618–24
[203] Karkow W S, Cranley J J, Cranley R D, Hafner C D and Ruoff B A 1986 *J. Vasc. Surg.* **4** 486–92
[204] Parsonnet V, Tiro A C, Brief D K, Alpert J and Brener B J 1978 *Graft Materials in Vascular Surgery* ed H Dardik (Miami, FL: Symposia Specialists) pp 249–62
[205] Christie B, Ketharanathan V and Perloff L J 1980 *J. Surg. Res.* **28** 519–32
[206] Ketharanathan V 1982 *US Patent* 4319363
[207] Moroni L, Boland T, Burdick J A, De Maria C, Derby B, Forgacs G, Groll J, Li Q and Malda J *et al* 2018 *Trends Biotechnol.* **36** 384–402

[208] Werkmeister J A, White J F and Ramshaw J A M 1994 *Med. Prog. Technol.* **20** 231–42
[209] Ramshaw J A M, Casagranda F, White J F and Werkemister J A 1994 *Clin. Mater.* **16** 9–13
[210] White J F, Werkmeister J A, Edwards G A and Ramshaw J A M 1993 *Clin. Mater.* **14** 271–76
[211] Brodsky B and Ramshaw J A M 1994 *Int. J. Biol. Macromol.* **16** 27–30
[212] Pandey S 2006 *Anal. Chim. Acta* **556** 38–45
[213] Meng Z, Zheng X, Tang K, Liu J, Ma Z and Zhao Q 2012 *Int. J. Biol. Macromol.* **51** 440–48
[214] Wang G, Guo J, Zhuang L, Wang Y and Xu B 2015 *Int. J. Biol. Macromol.* **76** 70–9
[215] Tarannum A, Rao J R and Fathima N N 2018 *J. Phys. Chem.* B **122** 1145–51
[216] Tarannum A, Muvva C, Mehta A, Rao J R and Fathima N N 2016 *J. Phys. Chem.* B **120** 6515–24
[217] Ceballos D, Navarro X, Dubey N, Wendelschafer-Crabb G, Kennedy W R and Tranquillo R T 1999 *Exp. Neurol.* **158** 290–300
[218] Nijsure M P, Pastakia M, Spano J, Fenn M B and Kishore V 2017 *J. Biomed. Mater. Res. A* **105** 2429–40
[219] Giraud-Guille M M, Mosser G and Belamie E 2008 *Curr. Opin. Colloid Interface Sci.* **13** 303–13
[220] Grimes M, Pembroke J T and McGloughlin T 2005 *Biomed. Mater. Eng.* **15** 65–71
[221] Wiegand C, Abel M, Ruth P, Wilhelms T, Schulze D, Norgauer J and Hipler U C 2009 *J. Biomed. Mater. Res. B Appl. Biomater.* **90** 710–9
[222] Delgado L M, Pandit A and Zeugolis D I 2014 *Expert Rev. Med. Devices* **11** 305–14
[223] Forest P, Morfin F, Bergeron E, Dore J, Bensa S, Wittmann C, Picot S, Renaud F N R, Freney J and Gagnieu C 2007 *Biomed. Mater. Eng.* **17** 199–208
[224] Artandi C 1977 *Radiation Phys. Chem.* **9** 183–91
[225] Yahia L H, Drouin G and Zukor D 1993 *Biomed. Mater. Eng.* **3** 211–21
[226] Gouk S S, Lim T M, Teoh S H and Sun W Q 2008 *J. Biomed. Mater. Res. B Appl Biomater.* **84** 205–17
[227] Mrázová H, Koller J, Kubišová K, Fujeríková G, Klincová E and Babál P 2015 *Cell Tissue Bank* **17** 255–60
[228] Herbert A, Edwards J H, Jones G L, Ingham E and Fisher J 2017 *J. Biomech.* **57** 157–60
[229] Davidson R J and Cooper D R 1968 *Biochem. J.* **107** 29–34
[230] Cooper D R and Russell A E 1969 *Biochem. J.* **113** 263–69
[231] Cheung D T, Perelman N, Tong D and Nimni M E 1990 *J. Biomed. Mater. Res.* **24** 581–89
[232] Akkus O, Belaney R M and Das P 2005 *J. Orthop. Res.* **23** 838–45
[233] Allaveisi F, Hashemi B and Mortazavi S M 2015 *Cell Tissue Bank.* **16** 97–108
[234] Ohan M P and Dunn M G 2003 *J. Biomed. Mater. Res. A* **67** 1188–95
[235] Attia T, Tupy J, Asker D, Hatton B, Grynpas M and Willett T 2017 *Cell Tissue Bank.* **18** 555–60
[236] Ng K W, Wanivenhaus F, Chen T, Abrams V D, Torzilli P A, Warren R F and Maher S A 2013 *Cell Tissue Bank.* **14** 21–32
[237] Patel J M, Jackson R C, Schneider G L, Ghodbane S A and Dunn M G 2018 *J. Mater. Sci., Mater. Med.* **29** 75
[238] Keller P F, Verin V, Ziegler T, Mermillod B, Popowski Y and Delafontaine P 2001 *J. Investig. Med.* **49** 258–64

[239] Carnevali S, Mio T, Adachi Y, Spurzem J R, Striz I, Romberger D J, Illig M and Rennard S I 2003 *Tissue Cell* **35** 459–69

[240] Chvapil M, Owen J A and Clark D S 1977 *J. Biomed. Mater. Res.* **11** 297–314

[241] Mattern R-H, Pierschbacher M D, Cahn F, Tschopp J F and Malaney T I 2005 *US Patent* 6969523

[242] Proffen B L, Perrone G S, Fleming B C, Sieker J T, Kramer J, Hawes M L, Badger G J and Murray M M 2015 *J. Orthop. Res.* **33** 1015–23

[243] Smith S, Griffiths S, Macgregor S, Beveridge J, Anderson J, van der Walle C and Grant M H 2009 *J. Biomed. Mater. Res. A* **90** 844–51

[244] Heidarkhan Tehrani A, Davari P, Singh S and Oloyede A 2014 *J. Mater. Sci., Mater. Med.* **25** 953–64

[245] Boynton J 1992 *US Patent* 5128101

[246] Olde Damink L H, Dijkstra P J, Van Luyn M J, Van Wachem P B, Nieuwenhuis P and Feijen J 1995 *J. Biomed. Mater. Res.* **29** 149–55

[247] Dillow A K, Dehghani F, Hrkach J S, Foster N R and Langer R 1999 *Proc. Natl. Acad. Sci. USA* **96** 10344–48

[248] Bernhardt A, Wehrl M, Paul B, Hochmuth T, Schumacher M, Schütz K and Gelinsky M 2015 *PLoS One* **10** e0129205

[249] White A, Burns D and Christensen T W 2006 *J. Biotechnol.* **123** 504–15

[250] Lomas R J, Cruse-Sawyer J E, Simpson C, Ingham E, Bojar R and Kearney J N 2003 *Burns* **29** 515–25

[251] Leow-Dyke S F, Rooney P and Kearney J N 2016 *Tissue Eng. Part C Methods* **22** 290–300

[252] Moore M A, McIlroy B K and Phillips R E 1997 *ASAIO J.* **43** 23–30

[253] Creutzfeldt H G 1920 *Z. Neurol. Psych.* **57** 1–18

[254] Jakob A M 1921 *Z. Neurol. Psych.* **64** 147–228

[255] Gerstmann J, Straussler E and Scheinker I 1936 *Z. Neurol. Psych.* **154** 736–76

[256] Brown P, Preece M, Brandel J P, Sato T, McShane L, Zerr I, Fletcher A, Will R G and Pocchiari M *et al* 2000 *Neurology* **55** 1075–81

[257] Ae R, Hamaguchi T, Nakamura Y, Yamada M, Tsukamoto T, Mizusawa H, Belay E D and Schonberger L B 2018 *MMWR Morb. Mortal Wkly. Rep.* **67** 274–78

[258] Lugaresi E, Medori R, Montagna P, Baruzzi A, Cortelli P, Lugaresi A, Tinuper P, Zucconi M and Gambetti P 1986 *New Engl. J. Med.* **315** 997–1003

[259] Manetto V, Medori R, Cortelli P, Montagna P, Tinuper P, Baruzzi A, Rancurel G, Hauw J J and Vanderhaeghen J J *et al* 1992 *Neurology* **42** 312–9

[260] Zigas V and Gajdusek D C 1957 *Med. J. Aust.* **44** 745–54

[261] Prusiner S B 1982 *Science* **216** 136–44

[262] Schneider K, Fangerau H, Michaelsen B and Raab W H-M 2008 *Brain Res. Bull.* **77** 343–55

[263] Leopoldt J G 1750 *Nützliche und auf die Erfahrung gegründete Einleitung zu der Land-Wirthschafft* vol 5 (Sorau: Johann Gottlieb Rothen)

[264] Cuillé J and Chelle P-L 1936 *Compt. Rend. Hebdomadaires des séances de l'Académie des Sciences* **203** 1552–54

[265] Gilch S, Chitoor N, Taguchi Y, Stuart M, Jewell J E and Schätzl H M 2011 *Top. Curr. Chem.* **305** 51–77

[266] Tee B L, Longoria Ibarrola E M and Geschwind M D 2018 *Neurol. Clin.* **36** 865–97

[267] Southwood T R E, Epstein M A, Martin W B and Walton J 1989 *Report of the Working Party on Bovine Spongiform Encephalopathy* (London: Ministry of Agriculture, Fisheries and Food, Department of Health, HMSO)

[268] Smith P G and Bradley R 2003 *Br. Med. Bull.* **66** 185–98

[269] Supervie V and Costagliola D 2004 *Vet. Res.* **35** 349–62

[270] Doillon C J, Drouin R, Côte M F, Dallaire N, Pageau J F and Laroche G 1997 *J. Biomed. Mater. Res.* **37** 212–21

[271] Beekes M, Lemmer K, Thomzig A, Joncic M, Tintelnot K and Mielke M 2010 *J. Gen. Virol.* **91** 580–9

[272] Wagenführ K and Beekes M 2012 *Prion* **6** 1–6

IOP Publishing

John A M Ramshaw and Veronica Glattauer

Chapter 12

Chemical modifications

Chemical modifications, particularly to give a crosslinked product, are a core technology for collagen-based biomedical device manufacture. Crosslinking can provide strength, durability, control of turnover and immunological limitation, while single site modifications can control the properties of collagen. A wide range of strategies have been developed and evaluated.

It is essential for a wide variety of applications for collagen as a biomedical material that the collagen matrix, either as tissue or a purified and reconstituted material, is stabilised using chemical or physical methods [1]. Reactions that lead to single point modifications are generally destabilising to various extents, while crosslinking reactions add to stability.

The chemical reactions of proteins, globular and fibrous, are well documented, in a wide range of papers and in many reviews and books [1–5]. The reactions of collagens are no different to those of other proteins, other than being modulated by the specific compositions and the accessibility of all non-Gly residues, albeit varying between the Xaa and Yaa positions (see chapter 2). However, although a broad range of crosslinking methods are available, those that are suited to biomedical products are much more limited.

A wide variety of modifications have been studied, all aiming to enhance the stability, durability and strength, while minimising, preferably eliminating any immunogenicity or cytotoxicity, as well as controlling the rate of turnover. The field is continually evolving. Some strategies that are well tested and well accepted, such as chromic stabilisation of sutures [6], would possibly not be approved as new products today by regulatory authorities. Another well-established process, glutaraldehyde (GA) stabilisation, which has been used effectively for over 50 years, still leads to concerns due to its potential disadvantages of cytotoxicity [7] and leading to calcification [8]. This has encouraged development of other approaches. In some

doi:10.1088/978-0-7503-2096-2ch12

cases, such as the use of epoxides, the original enthusiasm may have subsided, whereas others, such as the use of sugars and plant products, may still be in an ascendency. Whatever approach is used, the final product must have appropriate biological and physical properties, balancing desirable and undesirable effects. And, reagents that are unsuited for one application, such as providing long-term durability, may be ideal for another, such as short-term control of turnover.

Are product aesthetics important? For some, this is potentially the case. In this regard, strategies that add material to products can change the colour. For example, aldehydes typically cause yellowing while plant products can cause brown through to inky blue colorations.

Various crosslinking strategies have emerged. Often, a single approach is employed, whereby homodimer or heterodimer reagents can be used. More recent approaches have also considered multifunctional, dendrimer reagents [9]. In some cases, additives are included that are incorporated as bridges in the crosslinking [10]. A further opportunity arises with strategies that can enable material delivery and subsequent *in situ* crosslinking, frequently using a two-step approach where single point modifications are introduced and then these crosslinked in a second step. The other approach that has been used, and may deserve further attention, is the use of a combination of strategies. For example, pre-stabilisation with one reagent could be used to optimise tissue mechanics or maintain shape, while the second reagent could provide durability or control of immunogenicity.

The strategies developed for chemical modifications often utilise specific amino acid side chains. Thus, chemical approaches frequently target the ε-amino group of Lys residues, but can also use the side chain carboxyl groups of Asp or Glu residues. Additional copies of these acidic side chains occur in deamidated, anionic collagens. The frequency and distribution of amino acids may affect their availability and suitability for reactions. Some suitable side chains, such as Cys and His occur very rarely, while others, such as Arg, are quite common but have not yet been fully exploited (see chapter 2). Some crosslinking methods have emphasised that no additional chemical residues are retained, through formation of 'zero-length' crosslinks [1].

12.1 Chemical crosslinking methods

12.1.1 Aldehyde-based crosslinks

Glutaraldehyde

The emerging ready availability of GA in the 1950s led to its use in the tanning industry [11] and subsequently for the preparation of collagen-based biomaterials [12], notably heart valves, for which there are many reviews (e.g., [13]).

In solution, GA consists of a mixture of chemical forms in rapid equilibrium, with the linear dialdehyde being only 4% and the monohydrate, $CHO\text{-}CH_2\text{-}CH_2\text{-}CH_2\text{-}CH(OH)_2$ being only 8% at 25 °C [14]. The most abundant species is probably the linear dihydrate that comprises just over half of the solution [14]. While the amounts of the various components do not vary greatly with concentration or pH, the concentration of free aldehyde increases rapidly with temperature [14]. The optimum pH for reaction is at neutrality (figure 12.1), and drops off below pH 4. Higher

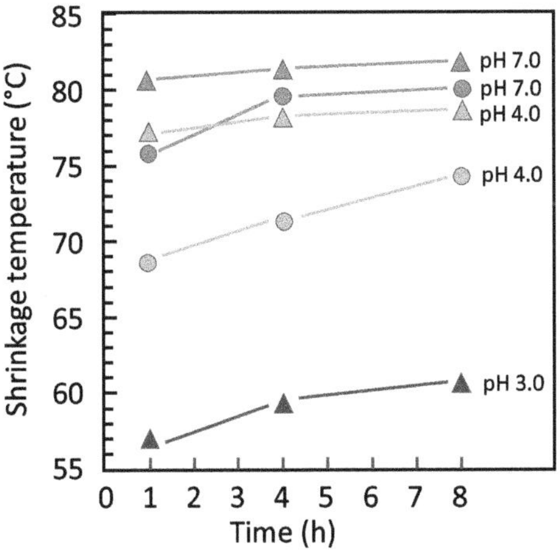

Figure 12.1. Changes in the shrinkage temperature of rat tail tendon collagen after treatment with GA under various conditions, based on the data of [15]. • represents 0.1% GA, while ▲ 1.0% GA, in Na phosphate buffer, pH 3.0 (blue), pH 4.0 (green) and pH 7.0 (green).

shrinkage temperatures can be obtained by using higher GA concentrations, but there is little increase above 2% w/v GA. The solution complexity means that the mechanism of collagen stabilisation, while probably initially based on a reaction of an aldehyde with a Lys ε-amino group to give an aldimine product (figure 12.2), is substantially more complex [13].

A significant issue in using GA is its propensity to form polymers, probably arising from a cyclic hemiacetal form in solution [16]. The formation of polymers may tend to coat a material and prevent full penetration of reagent [17], and also may depolymerise later to give cytotoxic products [18, 19] and accelerate any calcification that is happening. Greater problems may exist with devices that are shipped in GA to maintain sterility.

Various strategies have been suggested to minimise the issues arising from GA polymer formation. One has suggested using lower concentrations of GA [20], with extended times at moderate (0.5%) GA concentration being used [3]. Post treatment washing is used on some occasions, and free aldehyde groups eliminated by a glycine wash [21]. Another strategy is to react at a lower pH, with addition of NaCl to control swelling, as less polymeric material accumulates at low pH [7, 22]. The amount of monomer is considerably greater at higher temperatures [14], so fixation at 50 °C has been proposed [23]. Clearly, this approach is only suitable for tissue-based devices as soluble, reconstituted materials would denature at the elevated temperature. Finally, more recently, GA vapour has been used for the stabilisation of electrospun gelatin materials (see [24]). GA vapour is also suitable for stabilising collagen sponges [25]. In experiments where the concentration of GA solution providing the vapour and the time of exposure were varied (figure 12.3), the crosslinking occurred fairly rapidly and a maximum extent of crosslinking could be obtained in 24 h over a 20% w/v GA solution. The use of crosslinking by GA acetals in EtOH solution may also limit polymer formation [26].

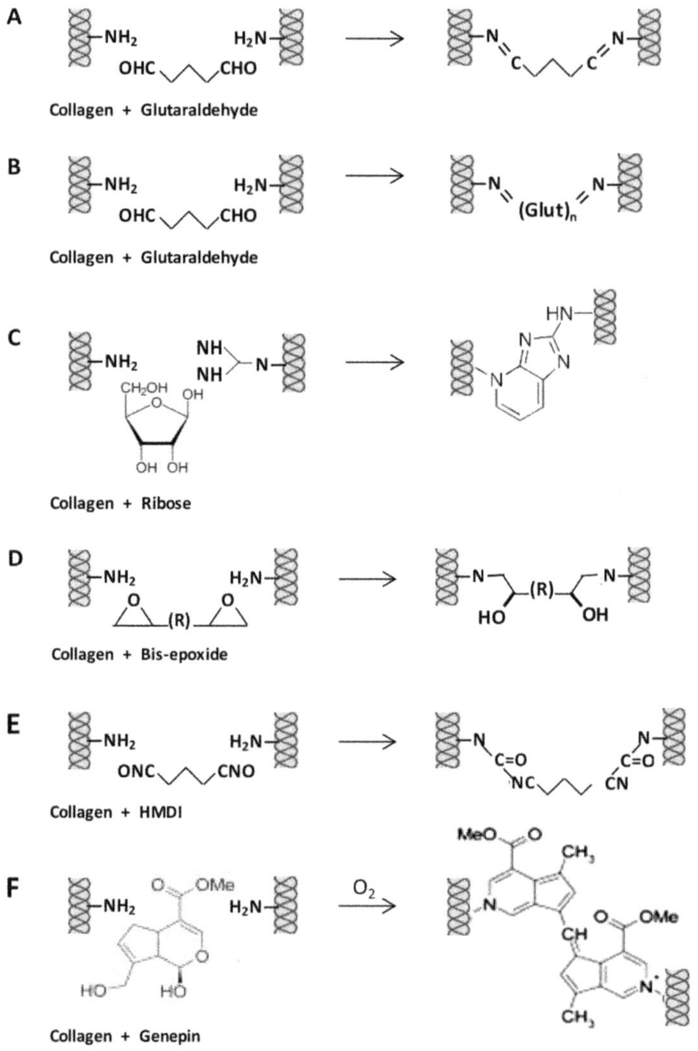

Figure 12.2. Examples of chemical reactions that can be used in collagen crosslinking. (A) GA, (B) GA as a polymer, (C) ribose, (D) bis-epoxide, (E) diisocyanate, (F) genipin.

Another set of variations on the GA fixation procedure involve the use of organic solvents. In one case, it was shown that using organic solvents to dissolve the GA led to significant improvements in the products, with, in some cases, a better match of some of the fixed tissue mechanical properties to those of native tissue [27]. In other examples, washing or storage of fixed products in EtOH led to a reduction in the extent of calcification [22].

Formaldehyde

Formaldehyde is an effective crosslinking agent for proteins, even though it has only a single aldehyde functional group [28]. The side chain ε-amino groups of Lys

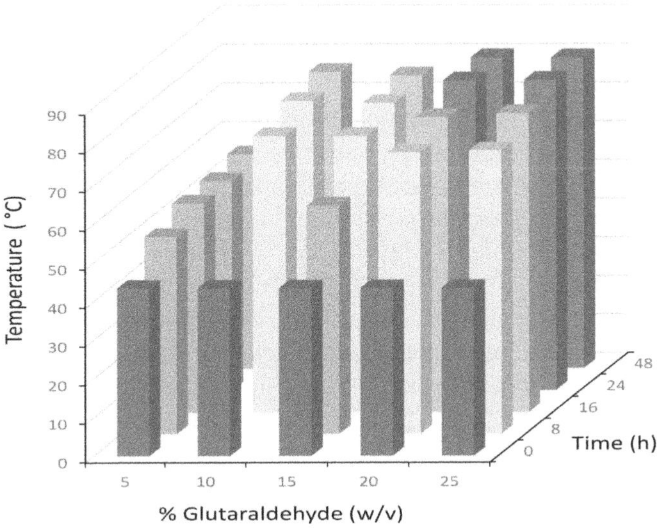

Figure 12.3. Variations in the extent of vapour-phase crosslinking of collagen sponges, determined by DSC, arising from the time and concentration of GA during the procedure. (From [24], with permission from Hindawi Publishing.)

residues are the initial points of reaction, but reactions with other side chains also occur as the complex crosslinking network is established [29]. The reagent was examined early on for processing of biological heart valves. While initial results seemed promising [30], subsequently a lack of durability and concerns over mechanical stability and cytotoxic leaching led to it being discontinued for this type of device [31]. Despite this, the reagent may still be useful for devices that are for short-term, topical applications [32].

Other bis-aldehydes

Various aliphatic dialdehydes, including dialdehyde starch and alginate, have been examined as collagen crosslinking reagents. The key differences that emerged compared to GA was that they differ in solution and do not form into a complex mixture nor form into polymers [33], and generally react much more slowly, and less efficiently than GA [34]. Thus, they could be suitable for a short-term implant, but not for devices requiring long-term durability. As well as dialdehyde starch, other dialdehyde polysaccharides have been examined, including chitosan dialdehyde [35] and also alginate dialdehyde, which has been used to develop composite collagen-based materials [36–38], as well as dialdehyde gum [38].

12.1.2 Other crosslink approaches

Sugar-based crosslinks

A natural set of crosslinks, the advanced glycosylation end-product (AGE) system, that is based on sugars, especially glucose has been substantially characterised and a variety of specific crosslinking entities identified (chapter 5) (figures 12.2 and 5.4).

This non-enzymatic process that uses reducing sugars is significant in ageing, and particularly in tissue-based complications in diabetes. Collagen turnover is generally very slow and so the AGE derived crosslinks accumulate and lead to detrimental effects, such as arterial wall stiffening and other tissue changes [39].

Under normal, neutral salt conditions (e.g., physiological buffered saline, PBS) the reaction rate of glucose modification of collagen is very slow, and is much too slow for effective use in biomaterial production. Glucose typically exists in a ring structure with extremely little of a linear form with a free aldehyde being present in the equilibrium solution [40]. The available aldehyde concentration is much greater for ribose, which has been used successfully in biomaterial production [41] and commercial materials. Ribose crosslinked collagen shows an increase in tensile stress [42], and increased resistance to pepsin, although extensively crosslinked material became more susceptible to pepsin [43]. Smaller compounds, such as threose [44] and glyoxals [45, 46] lead rapidly to AGE-like modifications and crosslinks.

Epoxides

A wide range of polyglycidyl ether (polyepoxy) compounds were proposed as useful crosslinking agents for collagens as they gave collagen tissue-based products that remained pliable, hydrophilic and that did not discolour (yellow) like GA treated tissues [47]. The range of potential reagents included bifunctional and multi-functional options, often using sorbitol and glycerol frameworks [48, 49] which react primarily (about 80%) with the ε-amino group of Lys residues [50]. The reaction is favoured by a high pH, increased temperature, and higher reagent concentrations [51]. The 1,4-butanediol diglycidyl ether format has been a frequent choice of reagent [52].

These reagents have a range of advantages. For example, the compounds are water-soluble and do not readily self-polymerise, so excess reagent is readily removed by washing. As a consequence, a more uniform distribution of crosslinking is found in fixed tissue [53]. Importantly, they exhibit a lower cytotoxicity than, for example, glutaraldehyde or formaldehyde [54], and do not lead to calcification [47]. The materials can be haemocompatible [50] and also can have preferable mechanical properties [55].

This group of reagents has been used successfully for the stabilisation of a wide range of tissues, including heart valves, pericardium, blood vessels and tendons (see [1]). Nevertheless, the rate of reaction is much slower than that of GA and is not limited by a diffusion rate [51]. Hence, some authors have suggested several days for treatment, so this approach to stabilisation may not be ideal for the processing schedules preferred for large manufacturing processes.

Isocyanates

Isocyanates are homobifunctional reagents that react to give crosslinks between the ε-amine groups of Lys residues. The most commonly used reagent has been hexamethylene diisocyanate (HMDI), which was used in the 1960s to make a clinically approved modified gelatin as plasma expander [56]. Subsequently, HMDI was applied to collagen-based biomaterial production [7]. The reagent may be best used in an organic solvent [3]. A wide range of potential alternative compounds exist, some of which have been assessed for leather production, but would probably not provide biomedical devices advantages above HMDI [57].

Studies with collagen showed that HMDI was an effective crosslinking reagent. The reagent did not have polymer forms and so excess reagent could be washed out readily and therefore did not leave residues. *In vivo* studies on crosslinked collagen sponges showed minimal tissue reaction, while cellular invasion of the sponge occurred gradually giving full infiltration through the sponge material. Hence, HMDI provides a better response profile than equivalent material that had been treated with GA [7]. This system has also been reported to have a lower propensity to lead to calcification than GA treated materials, perhaps arising from the use of an organic solvent [3].

Plant derived compounds

Plant derived chemicals, vegetable tannins, have been used for collagen stabilisation in leather production for more than 2 millennia into prehistoric times. In particular, materials such as oak bark, which is rich in complex polyphenolic tannic acid compounds (tannins) (from the Old High German word *tanna* for oak or fir tree), and simpler structures such as gallic acid have been preferred. While such compounds occur naturally in foods and have been approved as *generally approved as safe* in certain food applications, they have found only limited use in collagen stabilisation for biomedical applications [58]. An early report [6] indicated that tannic acid treated materials resorbed fairly rapidly, possible too rapidly to be useful. Recently, there has been renewed interest in using this crosslink, including applications for collagen-based wound dressings [59], for stabilising pericardium [60], and as a component of a collagen ink for bioprinting [61]. Other polyphenolic compounds have been suggested for crosslinking collagen-based biomaterials, including gallic acid, catechin, epigallocatechin and oleuropein [62–65], often based on prior experience with these agents in the leather industry.

Some crosslinking agents have been derived from plants that are suggested for herbal remedies and nutritional supplements. Most notably, these include genipin that is extracted from fruits of Cape Jasmin (*Gardenia jasminoides*) [66] as well as nordihydroguaiaretic acid from the Creosote Bush (*Larrea tridentata*) [67, 68], curcumin from turmeric (*Curcuma longa*) [69] and the polyphenolic quercetin, which is found in a range of plants [70, 71]. No doubt a further range of alternatives could emerge [72].

Genipin has emerged as the plant derived crosslinking agent for collagen creating most interest. Generally, it is used as a monomer, but polymeric formats have also been considered [73]. Prior to use in crosslinking, the reagent has been shown to be neither cytotoxic nor genotoxic [74, 75]. Nevertheless, some questions were raised on cytotoxicity with this appearing to be cell and dose dependent [76]. A mechanism for the crosslinking reaction (figure 12.2) involves a single point reaction at Lys residues and dimerisation of these modified residues. The reaction with collagen is most rapid at neutral pH, although not as rapid as GA, and is effective at 25 °C, with no improvement at higher temperatures up to 45 °C [77] and needs to be used as a fresh

Table 12.1. Examples of some other, less commonly used crosslinking systems for collagen.

Compound	Examples of applications	References
Azide-alkyne cycloaddition	Cornea	[81]
Bis(nitrophenyl) esters	General	[82]
Bis-imidates	Dermal	[83–86]
Chloro-*s*-triazines	General	[87]
Chrome salts	Sutures, wound dressings	[6, 88]
3,4-dihydroxyphenylalanine	Gels	[89]
Glycerol	Pericardium, dermis	[90]
β-nitro alcohols	Ophthalmic	[91]
1,3-phenylenediacetic acid	Ophthalmic	[92]
Riboflavin	Keratoconus	[93, 94]
Silane	Bone repair	[95]

solution as it is not stable over time [77]. The crosslinking reaction leads to an increase in the T_d of tissue, and while this is less than for GA, there is a different chemical entity introduced and the crosslink density is comparable [78]. Studies on pericardium samples showed that the crosslinking was stable, superior to GA, and that no toxic material leached from the stabilised tissues, unlike GA [79].

Furfural (2-furancarboxaldehyde), which is derived from plants pentosans, will react with collagen leading to increased denaturation temperature and improved mechanical properties, while not being cytotoxic [80], with a mechanism related to ribose crosslinking.

Overall, a characteristic of the plant derived crosslinking compounds is that they impart colour, sometimes quite intense, on the product.

Other approaches

A variety of other crosslinking and stabilising agents have also been used to prepare collagen-based medical devices (table 12.1), but these methods have often not received significant on-going attention. In some cases, these have not been examined in detail or have not been shown to have utility for biomedical applications.

12.1.3 Introducing zero-length crosslinks

In the preceding sections, the crosslinking approaches lead to the incorporation of additional chemical moieties within the crosslinking bonds that are being formed (figure 12.2). On the other hand, reagents that are termed zero-length crosslinkers lead to crosslink formation without the introduction and incorporation of any new compounds. This is achieved through catalysing formation of bonds between amino acids from adjacent molecules (figure 12.4). This approach is seen as advantageous as it does not introduce any additional compounds that could lead to cytotoxicity or other detrimental effects.

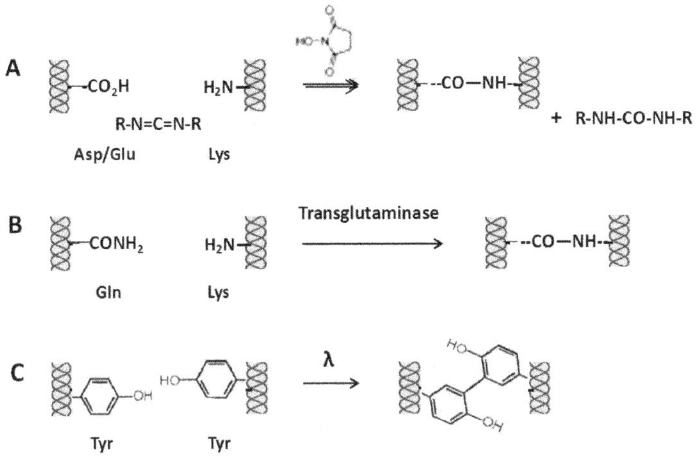

Figure 12.4. Examples of reactions that can be used to produce 'zero-length' collagen crosslinks: (A) carbodiimide, with N-hydroxysuccinimide catalysis, (B) transglutaminase, (C) dityrosine formation.

Carbodiimides

The development of water-soluble carbodiimides led to an initial report on formation of peptide bonds between amino acids [96]. These carbodiimides were then introduced for aqueous, protein crosslinking using gelatin [97]. The gelatin was rapidly crosslinked, within 30 min, and it was noted that unlike other methods available then, it did not leave a residue in the treated material [97]. The reagent reacts with a carboxyl group, from either an aspartic or glutamic acid side chain, to form an *o*-acylurea that then reacts with an adjacent amino group, from a lysine side chain, to form a peptide bond crosslink while releasing the urea derivative of the reagent (figure 12.4).

Subsequently, water-soluble carbodiimides have emerged as important reagents, for peptide chemistry and for protein crosslinking, with 1-ethyl-3(3-dimethylamino-propyl)carbodiimide (EDC) emerging as the most commonly used reagent. This included collagen crosslinking, initially for surface modifications [98], but subsequently for reconstituted collagen materials [99, 100] and for tissues [101, 102]. With skin tissue, it was shown that the crosslinking increased the tissue thermal stability from 56 °C to 73 °C. However, this could be increased to 86 °C if N-hydroxysuccinimide (NHS) was added to assist the reaction (figure 12.4) [102]. The use of NHS, or alternatively use of *N*-hydroxysulfosuccinimide [103], enhances the crosslinking by reacting through the *o*-acylurea to form an activated ester that is more stable in solution. This intermediate can then react with an amine to form the crosslink. EDC-NHS crosslinked dermal collagen was stable to proteolysis [104], while crosslinked pericardium had a resistance to CNBr digestion similar to GA treated tissue [105]. Mechanical testing indicated that the crosslinked pericardium was more extensible and more elastic than glutaraldehyde treated material [105]. When extruded collagen fibres were crosslinked using EDC-NHS, an increase in fibrillar organisation was observed by atomic force microscopy and spectroscopy [106]. Although developed as an aqueous system, it is also possible to add a co-solvent, and addition or substitution with EtOH has been used [107, 108]. As a special case, cyanamide could be regarded as the smallest carbodiimide [105], but is not particularly effective in crosslinking pericardial tissue [105]. Its use may, therefore, be limited to combination strategies.

Transglutaminase

Transglutaminase (protein-glutamine:amine γ-glutamyl transferase) provides an enzymatic method that enables the crosslinking of proteins via Lys and Gln side chains through formation of *N*-ε-(γ-L-glutamyl)-L-lysine bonds (figure 12.4). Enzymes with this activity are widespread, including microbial (mTG) as well as animal tissue (tTG) sources. In mammals, tTG is Ca^{2+} dependent and plays important physiological roles including stabilising the tissue at the dermal-epidermal junction in wound healing [109].

Both mTG [110–112] and tTG [112] are clearly effective in crosslinking gelatin, and have been used to prepare potential materials and tissue mimics [113]. For collagen, however, the results are equivocal. Good crosslinking is again achieved when the collagen is partially or fully denatured, for example, by urea [114] or by heat [115, 116]. For native collagen, the extent of crosslinking seems limited [117, 118], probably as a result of inadequate accessibility of reactive Lys and/or Gln residues in the triple-helical conformation [119]. Measurements of viscosity [117], thermal stability [118] and mechanical and biochemical properties [120] show only limited crosslinking, while other studies suggest that there is no crosslinking unless some denatured collagen is present [115]. It has been suggested that crosslinking could be enhanced by inclusion of other proteins, instead of gelatin [121], although this would need validation that no adverse immune effects arise in applications. In this respect, it is important that the enzyme used is well purified to be free of extraneous polypeptides [122], especially for mTG as preparations initially intended for the food industry are often stabilised by a protein addition, such as by milk powder. Similarly, it is likely that enzyme residues may remain after crosslinking [123]. Despite these concerns, transglutaminases have been used successfully to crosslink collagen for biomedical applications [120, 124, 125].

Acyl azides

In this method, carboxyl groups from Asp or Glu residues were transformed in a three-step procedure, using hydrazine, to give acyl azide intermediates. These acyl azides can then react with amino group from Lys residues, giving an isopeptide linkage between the side chains [126]. Pericardium crosslinked by this method gave a good thermal stability of 83 °C, comparable to GA, when tissue swelling was minimised by addition of 1 M NaCl [126]. Samples prepared using this method, for example, collagen sponges, were durable *in vivo* and did not calcify, even after three months implant time in a rat subcutaneous model [127].

Although this initial method was effective with various tissues [126, 128], the multi-step process was not convenient. This problem was resolved through the use of diphenylphorylazide [129]. This reagent forms the acyl azide directly without intermediate steps, although it requires dimethyl formamide or a similar solvent, and gave similar results to the more complex procedure [129]. Tissue and manufactured collagen materials stabilised by this method were not cytotoxic [130, 131] while membranes used clinically for treatment of periodontal disease performed well and showed no adverse reactions [132]. Despite this utility, the method does not seem to have been widely adopted.

Dityrosine crosslinks

At one time, it was suggested that dityrosine crosslinks may be present as native crosslinks for collagen [133], but this view has not been sustained. Other contemporary studies looked at introducing dityrosine crosslinks through oxidative processes (figure 12.4), either chemically or enzymatically [133–135]. More recently, light activated processes have been examined [136]. Systems such as the use of Ru^{II} [137], which works for gelatin [138] and Fe^{II} in a Fenton type reaction [139, 140] could be employed. However, collagens typically have a very low Tyr content (table 2.1), so it is possible that other bonds are formed as well as dityrosine. The Tyr content could be augmented, for example, by reaction with tyramine or by the Bolton-Hunter reagent [141]. Tyr residues can also be added or substituted in recombinant collagens [142].

12.2 Physical crosslinking

The methods examined for physical crosslinking of collagen have involved heat and irradiation, and these can be used in combination with chemical approaches [143].

12.2.1 Direct, non-catalysed reactions

Dehydrothermal crosslinking

Early studies examined heating of collagen. For example, it was reported that dehydrated and dried collagen, on heating to 105 °C up to 180 °C, did not lose its fibrous structure or become brittle, which was quite distinct from heating of wet tissue [144]. It was much later that more detailed studies were reported, and, as with several other crosslinking systems, dehydrothermal crosslinking was initially characterised in studies on gelatin [145]. Both gelatin and collagen (figure 2.3) have loosely bound and some more tightly bound water molecules [146]. Removal of water, by using and then maintaining high vacuum, allows the sample to be heated without hydrolysis, or denaturation of collagen. The heat and vacuum together then lead to an amide condensation reaction, in which an inter-molecular amide (peptide bond) linkages involving lysine and aspartic acid or glutamic acid residues form with elimination of a water molecule. Thus, the bond formed is the same as that which forms in carbodiimide crosslinking (figure 12.4). Subsequently, dehydrothermal crosslinking was shown to stabilise native collagen [147] and it was shown that in the absence of water the denaturation of the collagen triple-helix did not occur.

A wide range of potential conditions have been reported for collagen crosslinking by different research groups [148–154]. Temperatures between 60 °C to 180 °C have been reported for times from 1 up to 5 days. The aim is to get a balance between crosslinking and degradation; a low level of fragmentation of the collagen has been reported, even under mild conditions [155], and seen to increase with increasing temperature [149]. However, vacuum conditions have not always been reported, and can be variable unless special attention is paid to system quality. Since the elimination of water is critical to the prevention of degradation and to the crosslinking, variations in the vacuum quality can lead to product differences.

Dehydrothermal crosslinking has been suggested for several biomaterial manufacturing approaches. These include stabilisation and giving tensile strength to reconstituted fibril-based tendon prostheses [156]. However, this approach may be more suited to applications where tensile properties are not critical, such as for wound and burns dressings or tissue engineering scaffolds [150, 151, 153, 157–159].

Irradiation induced crosslinking

UV irradiation is effective for crosslinking collagen solutions, but is restricted to only thin samples of tissue or reconstituted materials as the transmittance is limited and the crosslinking diminishes with increasing opacity and depth. It is a rapid process and easily controlled. Thus, UV irradiation of a solution causes an increase in viscosity, and gels can be formed when oxygen is absent [160]. The extent of crosslinking is greater for native collagen with telopeptides, probably as the Tyr content is higher than for atelocollagen [160]. At the same time, irradiation leads to degradation of the collagen, accompanied by a loss of thermal stability; electrophoresis methods can show the degradation too [161]. The method can also be applied to translucent tissues, such as cornea [162]. For reconstituted collagen fibril materials, a short period of UV irradiation leads to an increase in tensile strength and in shrinkage temperature, although bond cleavage and partial fragmentation occur with increasing time, leading to a greater protease susceptibility [155, 163]. Similarly, it can be used for appropriate configurations of sponges [164]. The triple-helix provides some protection from irradiation damage relative to an unfolded gelatin structure [165].

The introduction of gamma irradiation for sterilisation (see chapter 11) provided a method which could also be used for crosslinking, so long as a balance between degradation and crosslinking can be achieved. It was recognised early on that collagen degradation was an issue to be considered [166, 167]. Clearly, if higher levels of irradiation are used it could be employed concurrently with the sterilisation of the device. However, the extent of irradiation typically used for sterilisation leads to significant damage to the collagen, with significant numbers of peptide bonds being cleaved [168], so lower levels could be preferable for a crosslinking strategy. Gamma-ray irradiation of collagen solutions leads to a lowering of solubility, an increase in susceptibility to trypsin digestion, as well as changes to the collagen chain composition that can be seen by SDS-PAGE [169]. The changes to the chains will depend on the extent of radiation, but will include scission events, but with some of the activated polypeptide chains that are produced also leading to crosslinking. It has also been noted that the extent of damage can depend on whether the sample is wet or dry, but this cannot always be selected depending on the nature of the device.

Gamma-ray irradiation can be used to crosslink soluble collagen to form clear insoluble gels, for example, for ophthalmic applications [48]. Gamma irradiation can also be used to form collagen hydrogels, with the properties depending on the pH of the collagen used [170]. Selected hydrogels have also been prepared as substrates for cell proliferation [171] and stem cell studies [172].

If collagen mixtures are treated by gamma irradiation, for example, mixtures containing polyethylene glycol, dextran, or other polymers, then the material characteristics can be dramatically changed by the formation of a polymer network and possibly reduced collagen crosslinking [173, 174]. On the other hand, inclusion of lower molecular weight compounds can also be beneficial as some can be used as radioprotectants, for example, by free radical scavenging [175, 176]. Inclusion of prior crosslinking has also been used for various collagen materials, including tendon xenografts [177] and dermal equivalents [178].

Other methods

Electron beam irradiation has been proposed for the crosslinking of native [179] and reconstituted [180] collagen materials. Better crosslinking could be obtained with partial denaturation [180], while in both cases, scission of the collagen chains was seen [179, 180], with the extent of this change potentially limiting its use in crosslinking, and also as a sterilisation method [181].

12.2.2 Catalysed reactions

Dye-mediated photo-crosslinking

Dye-mediated photo-oxidation was proposed in the 1960s for collagen biomaterial stabilisation [182]. It received increased attention, especially during the 1990s [183–188], especially for stabilising pericardial tissue for heart valve applications. However, despite showing some clinical potential [189], it has fallen from favour due to design issues, although the method has been examined as a method for *in situ* collagen crosslinking [190] and most recently its potential for stabilisation of pericardial patches has been re-examined [191].

Dye-mediated photo-oxidation, using visible light, in general modifies several amino acids, including His, Tyr, Trp and Met, with the extent of modification dependent on pH [192]. For collagen, at neutral pH, His is significantly lost during irradiation, in a time dependent manner. Only minor changes occur to Met and Tyr, while Trp is absent in interstitial collagens [193]. As the amount of His decreases, new, higher molecular weight CNBr fragments appear [193], and the collagen becomes less readily extracted [186, 194]. Accompanying these changes, an increased resistance to pepsin solubilisation of the tissue was found, also confirming crosslink formation [186, 194]. It was of interest, however, that little change was observed in T_S [193, 194], suggesting that the new crosslinks are thermally labile.

Various dyes could be used for dye-mediated photo-crosslinking [182]. From these, methylene blue has emerged as the dye of choice, as it has the advantage of being readily removed from stabilised tissue by washing whereas some other dyes, e.g., Rose Bengal, remain bound to the collagen tissue [193, 194].

The potential range of crosslinks that form, presumably incorporating the His residues, have not yet been fully defined, although distinct, newly formed crosslinks have been observed [186] using a diphenylboronic acid labelling method [195]. Nevertheless, *in vivo* trials have shown that stabilised pericardial tissue is stable [189, 196] and does not lead to extensive calcification [197], while it was non-cytotoxic and passed a wide range of general tests for biocompatibility [196, 197].

12.3 Assessing the effectiveness of crosslinking

Various approaches have been used to examine the nature, extent and effectiveness of crosslinking. The applicability of different methods depends on the nature of the crosslinking method and the format of the collagen material. In all cases, it is possible that although a crosslink is intended, some single point modifications could also be present. This needs to be considered in interpretation of analytical data. Some examples are given in the following sections.

12.3.1 Physical methods

Thermal methods

A frequently used method for assessing the effectiveness and extent of crosslinking is measurement of the denaturation (shrinkage) temperature (T_s) of the material. For tissue-based materials this is determined by an isometric tension approach [198, 199], in which the sample is increased in temperature while kept at constant length immersed in phosphate buffered saline, or another appropriate solution. In some protocols, the sample is initially held at a minimum tension [198] while in another a high initial tension is used [200]. A sharp increase in force arises when the sample shrinks. This later approach may be better for study of rate of relaxation after shrinkage. Increasing the extent of crosslinking increases the T_s that is observed. However, T_s is not a value that can necessarily be compared between different crosslinking systems, as variations in the thermal stability of different crosslink systems can exist. While crosslinks typically increase the T_s, for irradiated samples, such as after gamma irradiation, the T_s may fall [201]. If the T_s is measured using a modified tensile testing apparatus, it is possible to measure the tensile properties of the denatured tissue after shrinkage. If it is assumed that the material is now equivalent to a random rubber, then the equations developed by Flory and colleagues enable an estimate to be obtained for the average molecular weight of the collagen chain between crosslinks [202].

For samples, such as gels and many reconstituted collagen materials, that cannot be readily secured for isometric tension measurements, calorimetric studies using differential scanning calorimetry can be performed; the method can also be used on tissue samples as an alternative to isometric tension measurement [203, 204].

Mechanical methods

Uniaxial tensile measurements, using classical approaches [205–207], have been used to examine native tissues and the changes that occur with crosslinking [208–210]. Increases in crosslinking extent lead to increases in the tensile strength, although other mechanical parameters may not follow an appropriate trend. The availability now of biaxial testing systems is particularly useful for studying tissues used in biomedical devices, such as pericardium, as these materials generally show anisotropic properties and biaxial testing allows a better understanding of their properties [211, 212].

Examination of the tensile properties of collagen sponges and particularly collagen gels is more problematic. For stronger gels it may be possible if the sample can be adequately grasped; this can be aided by incorporating netting fabric in the ends of the gel sample when it is made, and using sandpaper or a similar product in the jaws that hold the sample. Therefore, for gel samples, particularly for soft gel samples compression testing is a useful approach, with initial compression strength increasing with fibril density [213, 214]. Nano indentation is another method that is applicable to certain sample types. The swelling of gels under changing solvent conditions can also be used to examine the extent of crosslinking [215]. Compression and swelling data, obtained under appropriate conditions can also be analysed by Flory-Rehner theory [202, 216].

Rheological properties have also been used to examine crosslinking effects. For example, for collagen solutions and dispersions viscosity can be measured at defined solution conditions [217, 218], with the increasing molecular weights arising from crosslinking increasing viscosity. For collagen gels, systems such as parallel plate rheometry/dynamic mechanical analysis can be used [219, 220].

Spectroscopic methods

Although spectroscopic methods are not routine, they can be useful in specific instances. Certain crosslinking reactions, for example, using sugars or genipin, lead to the development of fluorescence accompanying the crosslink formation. Increases in fluorescence intensity are readily observed [220] and for ribose crosslinking can be quantitated by emission measurement at 445 nm with excitation at 370 nm [221].

Crosslinking generally leads to disruptions in collagen structure or organisation. This disruption, for example, caused by sugar-based reactions and by chain scission by gamma irradiation, can be assessed by Fourier transform infrared spectroscopy (FTIR), for example, by examining changes to the area ratio of the 1678/1692 cm^{-1} sub-bands within the amide I peak [201, 222]. Glycated collagen can also be examined by other methods, including Raman spectroscopy [223].

Other methods

Various other methods will be found to examine the effects of crosslinks. These will match particular needs for the material, such as oxygen or solute permeability and may only be used infrequently. Such tests include measurement of the impedance of the material over a wide range of frequencies providing data on the dielectric constant [224], as well as measurement of hydraulic permeability [214].

12.3.2 Chemical and biological methods

Chemical methods

Chemical assessment of the extent of crosslinking can examine either the number of crosslinks formed, or the proportion of unmodified reactive sites. Choice of analysis depends on the nature and stability of the crosslink, especially to acid hydrolysis. For example, the bonds formed during glutaraldehyde crosslinking are stable to acid hydrolysis [32]. Other bonds, such as the peptide bond crosslink formed by carbodiimide, are lost during acid hydrolysis.

For acid stable crosslinking, any unreacted Lys residues can be measured through colorimetric methods, including trinitrobenzenesulfonic acid (TNBS) [225]. Conversely, free carboxyl groups on intact denatured protein can also be quantitated [226]. Alternatively, hydrolysates can be examined, for example, by amino acid hydrolysis. The limitation to such approaches is that they do not differentiate the extent to which modified amino acids are actually involved in crosslink formation, as opposed to being non-crosslinked single residue modifications. In some cases, the involvement of a specific amino acid in crosslinking has been inferred, and amino acid analysis can be used to follow the loss of the amino acid, for example, His in photo-crosslinking [193]. On the other hand, if the nature of the crosslink has been established, then amino acid analysis may potentially be used to identify it, or a modified or stabilised form, as with natural crosslinks (see chapter 5). Alternatively, a system that specifically separates crosslinks, such as interaction with diphenylboronic acid, and allows potential identification and quantitation could be used [195].

Biochemical methods

The most frequently used biochemical approach for examining the extent of crosslinking is to examine the protease sensitivity of the product. Various enzymes have been used, including trypsin and bacterial collagenase [163].

Biological examination

For all collagen-based devices, the final outcome in clinical application is the test for performance, including the effectiveness of any crosslinking which may affect durability and other characteristics. Prior to implantation, initially in appropriate animal models, a variety of biological tests are possible based on cell biology methods. These, for example, include examining cell infiltration, viability and phenotypic changes in cell culture [227, 228]. For materials where surface properties are important, for example, blood contact materials, the ease of cell coverage, for example, by endothelial cells, can be assessed [229]. For collagen gels, crosslinking can limit the extent of gel shrinkage by included cells, but quantitation of crosslinks is not readily assessed by the approach [230].

One effect that has been observed for certain crosslinking strategies, notably GA, is the propensity to induce calcification of an implant, especially with biological heart valve prostheses (see chapter 13). This issue is best addressed in an animal model, with a rat subcutaneous model being used extensively as a preliminary test [231].

12.4 Site-specific chemical modifications

Many of the functional, chemical groups that are frequently used in crosslinking reagents can also be used in single, site-selective modification reagents [232]. These include aldehydes, epoxides, imidates and isocyanates that react with Lys side chains in a specific manner [2]. On the other hand, the physical crosslinking approaches are rarely used for single site modification as they frequently lack the specificity required. The following sections will, therefore, focus on further functional groups that are typically not utilised or have low efficiencies in crosslinking strategies.

12.4.1 Additional reactions with amino groups

Acyl chlorides

Chemical entities that are available as acid chlorides will react readily with Lys side chains, allowing a wide variety of modifications such as acetylation, benzoylation, sulfoylation and pyrophosphatation [57]. An interesting option is the attachment of monoalkyl hydrocarbon chains. These alkyl additions can stabilise collagen, enhanced by increasing chain length [233], while providing an approach to allow development of hydrophobic collagen films [234].

Anhydrides

Anhydrides have proved to be reagents that are frequently used in collagen modification, reacting with Lys side chains. Acetic anhydride is the reagent of choice for acetylation of collagen [235] and was shown to destabilise the collagen [236]. A similar modification was achieved by adding succinyl groups through using succinic anhydride, again reducing the stability of the collagen [237]. An emerging modification is addition of norbornene entities, which can be achieved through reaction with carbic anhydride (endo-cis-5-norbornene-2,3-dicarboxylic anhydride) [238]. The norbornene functional groups are particularly useful for subsequent thiol-ene reactions, as the strained ring structure is more reactive than unstrained structures. These, and many other anhydride reactions, give stable products that are not reversible. In some instances, reversible modification could be valuable. This can be achieved by reaction with maleic anhydride [239] or citraconic anhydride [240]. Both reactions take place above neutral pH, and can be reversed by acid treatment, with removal of maleyl groups in days and citraconyl groups in hours. Other peptide chemistry approaches, such as fluorenylmethyloxycarbonyl (FMOC) modification [241] could also be used for reversible modification of Lys residues.

Methacrylation of Lys side chains, using methacrylic anhydride was introduced as a method that allowed photo-crosslinking of collagen chains [242]. Subsequently, this approach has been used extensively with gelatin to produce scaffolds for tissue engineering [243–246]. More recently, use of this anhydride reaction for methacrylation of collagen, and for tissues, has started to gain renewed attention [215, 247, 248].

Methacrylation, or other modifications, can be made on collagen prior to product manufacture so that once the material has been fabricated it can then be readily crosslinked, for example, for thin sheets obtained by electrospinning [249].

Alternative approaches for adding methacryl groups to collagen is through reaction of bi-functional methacryl-*N*-hydroxysuccinimide derivatives [250] or by use of acrylamide-isocyanate reagents [251].

Figure 12.5. The reaction of iminothiolane (Traut's reagent) with collagen to introduce thiol functions.

Thiolation

Cys residues are rarely found in native collagens but can be readily introduced into recombinant proteins where they can be used for crosslinking or introduction of site-specific modifications [142, 252]. For native collagens, thiols groups can be introduced by chemical modification of Lys residues.

Probably the most regularly used reagent for thiolation of proteins is iminothiolane (Traut's reagent), which reacts with amino groups on Lys residues (figure 12.5) [253]. However, other systems, including γ-thiobutyrolactone [254] have been used. Quantitation of the thiolation can be measured using Ellman's reagent [255].

Introduction of sulfhydryl groups on collagens allows other opportunities to be explored. These include the introduction of crosslinking through oxidation [142]. Others, such as thiol-ene reactions, are discussed in the following section.

12.4.2 Reactions with other functional groups

Glutamine or lysine

While transglutaminase may have limited effectiveness for crosslinking, it does provide a route for single site modification of a collagen substrate [256]. Gln containing (or terminated) peptides or appropriate substrates can react with Lys residues. Alternatively, Lys containing compounds could react with Gln residues. This latter approach may be preferable as Gln residues are often less frequent in collagens than Lys residues.

Glutamine or asparagine

The amide groups on Asn and Gln residues are susceptible to alkaline hydrolysis [237]. This releases the amides as ammonia and leaves Asp and Glu side chains. This leads to a lowering of the isoelectric point (pI) of collagen, to around a pI of 5 with full deamidation, and gives a product termed anionic collagen [257].

Glutamic acid or aspartic acid

The carboxylic acid side chains of Asp and Glu can be modified by methylation using methanol in acidic conditions [237], with the extent of modification increasing with time of incubation. A further reaction for these acidic residues is activation to produce an N-hydroxysuccinimide modification, which can then react with amino groups, either from another collagen or from an introduced molecule.

Sulfhydryl groups

While Cys residues are rare in collagen, they can be introduced by designed changes in recombinant collagens [142, 252], giving limited, specific locations for modifications. Alternatively, other collagens can be modified by Traut's reagent (as above), but this lacks the site specificity. Once sulfhydryl groups are present, they can be readily modified by a wide variety of reactions [258], for example, with chemical entities terminated by an acyl halide group [252]. Alternative reagents include those with maleimide [259] or sulfone [260] or vinyl sulfone groups [261].

Recently, thiol-ene coupling reactions have gained increased attention (particularly with gelatin so far) [262–264] with norbornene often being used as the -ene component [263]. The reaction is often initiated by UV light [248, 262], but other milder, visible light systems have emerged. These include use of catalysts, including Ru^{II} complexes [265], 9-mesityl-10-methylacridium tetrafluoroborate [266] or dyes [267, 268].

Tyrosine groups

Tyrosine groups are also rare in collagens, but can be introduced if desired by reaction with tyramine or with Bolton-Hunter reagent [138]. Catalysed photo-crosslinking could then be used to introduce single residue modifications [269], but can also be used for crosslinking [142]. Introduced Tyr residues can also be changed to dihydroxyphenylalanine residues through enzyme modification [270], for example, using the readily available mushroom tyrosinase.

References

[1] Ramshaw J A M, Glattauer V and Werkmeister J A 2000 *Handbook of Biomaterials Engineering* ed D L Wise (New York: Marcel Dekker) pp 717–38
[2] Wong S S 1991 *Chemistry of protein conjugation and cross-linking* (Boca Raton, FL: CRC)
[3] Khor E 1997 *Biomaterials* **18** 95–105
[4] Dunn R M 2012 *Plast. Reconstr. Surg.* **130** 18S–26S
[5] Oryan A, Kamali A, Moshiri A, Baharvand H and Daemi H 2018 *Int. J. Biol. Macromol.* **107** 678–88
[6] Lister J 1881 *Brit. Med. J.* **1881** 219–21
[7] Chvapil M, Speer D, Mora W and Eskelson C 1983 *J. Surg. Res.* **35** 402–9
[8] Golomb G, Schoen F J, Smith M S, Linden J, Dixon M and Levy R J 1987 *Am. J. Pathol.* **127** 122–30
[9] Duan X and Sheardown H 2005 *J. Biomed. Mater. Res. A* **75** 510–8
[10] McKegney M, Taggart I and Grant M H 2001 *J. Mater. Sci., Mater. Med.* **12** 833–44
[11] Fein M L and Filachione E M 1957 *J. Am. Leather Chem. Assoc.* **52** 17–23
[12] Carpentier A, Lemaigre G, Robert L, Carpentier S and Dubost C 1969 *J. Thorac. Cardiovasc. Surg.* **58** 467–83
[13] Nimni M E, Cheung D T, Strates B, Kodama M and Sheikh K 1988 *Collagen* vol 3 ed M E Nimni (Boca Raton, FL: CRC) pp 1–38
[14] Holloway C E and Dean F H 1975 *J. Pharm. Sci.* **64** 1078–9
[15] Peters D E, Stephens L J and Ramshaw J A M 1990 *Das Leder* **41** 129–33
[16] Hardy P M, Nicholls A C and Rydon H W 1969 *J. Chem. Soc., Chem. Commun.* **1969** 565–6
[17] Fisher J, Gorham S D, Howie A M and Wheatley D J 1987 *Life Supp. Syst.* **5** 189–93
[18] Speer D P, Chvapil M, Volz R G and Holmes M D 1979 *Clin. Orthop.* **144** 326–35
[19] Speer D P, Chvapil M, Eskelson C D and Ulreich J 1980 *J. Biomed. Mater. Res.* **14** 753–64

[20] Cheung D T, Perelman N, Ko E C and Nimni M E 1985 *Connect. Tissue Res.* **13** 109–15
[21] Gendler E, Gendler S and Nimni M E 1984 *J. Biomed. Mater. Res.* **18** 727–36
[22] Casagranda F, Werkmeister J A, Ramshaw J A M and Ellender G 1994 *J. Mat. Sci.:Mat. Med.* **5** 332–7
[23] Ruijgrok J M, de Wijn J R and Boon M E 1994 *Clin. Mater.* **17** 23–7
[24] Peng Y Y, Glattauer V and Ramshaw J A M 2017 *Int. J. Biomaterials* **2017** ID 8947823
[25] Lickorish D, Ramshaw J A M, Werkmeister J A, Glattauer V and Howlett C R 2004 *J. Biomed. Mater. Res. A* **68** 19–27
[26] Yoshioka S A and Goissis G 2008 *J. Mater. Sci., Mater. Med.* **19** 1215–23
[27] Gratzer P F, Periera C A and Lee J M 1996 *J. Biomed. Mater. Res.* **31** 533–43
[28] Fraenkel-Conrat H, Cooper M and Olcott H S 1945 *J. Am. Chem. Soc.* **67** 950–4
[29] Kunkel G R, Mehrabian M and Martinson H G 1981 *Mol. Cell. Biochem.* **34** 3–13
[30] O'Brien M F 1967 *J. Thorac. Cardiovasc. Surg.* **53** 392–7
[31] Buch W S, Kosek J C and Angell W W 1970 *J. Thorac. Cardiovasc. Surg.* **60** 673–7
[32] Barker H, Oliver R, Grant R and Stephen L 1980 *Biochim. Biophys. Acta* **632** 589–97
[33] Bowes J H and Cater C W 1968 *Biochim. Biophys. Acta* **168** 341–52
[34] Hardy P M, Nichols A C and Rydon H N 1972 *J. Chem. Soc. Perkin II* **1972** 2270–8
[35] Liu X, Dan N, Dan W and Gong J 2016 *Int. J. Biol. Macromol.* **82** 989–97
[36] Du T, Chen Z, Li H, Tang X, Li Z, Guan J, Liu C, Du Z and Wu J 2016 *Int. J. Biol. Macromol.* **82** 580–8
[37] Zhu S, Yu X, Xiong S, Liu R, Gu Z, You J, Yin T and Hu Y 2017 *Mater. Sci. Eng. C Mater. Biol. Appl.* **78** 727–37
[38] Ding W, Wang Y N, Zhou J and Shi B 2018 *Carbohydr. Polym.* **201** 549–56
[39] Sims T J, Rasmussen L M, Oxlund H and Bailey A J 1996 *Diabetologia* **39** 946–51
[40] Dworkin J P and Miller S L 2000 *Carbohydrate Res.* **329** 359–65
[41] Brodsky B, Berg R A, Avigad G, Eikenberry E, Jain M and Tanaka S 1990 *US Patent* 4971954
[42] Bai P, Phua K, Hardt T, Cernadas M and Brodsky B 1992 *Connect. Tissue Res.* **28** 1–12
[43] Tian S F, Toda S, Higashino H and Matsumara S 1996 *J. Biochem. (Tokyo)* **120** 1153–62
[44] Kinnunen J, Kokkonen H T, Kovanen V, Hauta-Kasari M, Vahimaa P, Lammi M J, Töyräs J and Jurvelin J S 2012 *J. Biomed. Opt.* **17** 97003
[45] Nagaraj R H, Shipanova I N and Faust F M 1996 *J. Biol. Chem.* **271** 19338–45
[46] Degenhardt T P, Thorpe S R and Baynes J W 1998 *Cell. Mol. Biol.* **44** 1139–45
[47] Imamura E, Sawatani O, Koyanagi H, Noishiki Y and Miyata T 1989 *J. Cardiac Surg.* **4** 50–7
[48] Miyata T 1992 *Clin. Mater.* **9** 139–48
[49] Tu R, Shen S H, Lin D, Hata C, Thyagarajan K, Noishiki Y and Quijano R C 1994 *J. Biomed. Mater. Res.* **28** 677–84
[50] Wang E, Thyagarajan K, Tu R, Lin D, Hata C, Shen S H and Quijano R C 1993 *Int. J. Artif. Organs* **16** 530–6
[51] Tu R, Lu C-L, Thyagarajan K, Wang E, Nguyen H, Shen S, Hata C and Quijano R C 1993 *J. Biomed. Mater. Res.* **27** 3–9
[52] Zeeman R, Dijkstra P J, van Wachem P B, van Luyn M J, Hendriks M, Cahalan P T and Feijen J 1999 *J. Biomed. Mater. Res.* **46** 424–33
[53] Shen S-H, Sung H W, Tu R, Hata C, Lin D, Noishiki Y and Quijano R C 1994 *J. Appl. Biomater.* **5** 159–62

[54] Nishi C, Nakajima N and Ikada Y 1995 *J. Biomed. Mater. Res.* **29** 829–34

[55] Sung H-W, Cheng W-H, Chiu I-S, Hsu H-L and Liu S-A 1996 *J. Biomed. Mater. Res. (Appl. Biomaterials)* **33** 177–86

[56] Moeller J and Sykudes A 1962 *Dtch. Med. Wschr.* **87** 726–9

[57] Heidemann E 1993 *Fundamentals of Leather Manufacture* (Darmstadt: Eduard Roether)

[58] Heijmen F H, du Pont J S, Middelkoop E, Kreis R W and Hoekstra M J 1997 *Biomaterials* **18** 749–54

[59] Natarajan V, Krithica N, Madhan B and Sehgal P K 2013 *J. Biomed. Mater. Res. B Appl. Biomater.* **101** 560–7

[60] Cwalina B, Turek A, Nozynski J, Jastrzebska M and Nawrat Z 2005 *Int. J. Artif. Organs* **28** 648–53

[61] Lee J, Yeo M, Kim W, Koo Y and Kim G H 2018 *Int. J. Biol. Macromol.* **110** 497–503

[62] Goo H C, Hwang Y S, Choi Y R, Cho H N and Suh H 2003 *Biomaterials* **24** 5099–113

[63] Madhan B, Subramanian V, Rao J R, Nair B U and Ramasami T 2005 *Int. J. Biol. Macromol.* **37** 47–53

[64] Reddy R R, Phani Kumar B V, Shanmugam G, Madhan B and Mandal A B 2015 *J. Phys. Chem.* B **119** 14076–85

[65] Antunes A P M, Attenburrow G, Covington A D and Ding J 2008 *J. Leather Sci.* **2** 1–12

[66] Sung H W, Chang Y, Chiu C T, Chen C N and Liang H C 1999 *J. Biomed. Mater. Res.* **47** 116–26

[67] Koob T J, Willis T A, Qiu Y S and Hernandez D J 2001 *J. Biomed. Mater. Res.* **56** 40–8

[68] Lü X, Zhai W, Zhou Y, Zhou Y, Zhang H and Chang J 2010 *J. Mater. Sci., Mater. Med.* **21** 473–80

[69] Gopinath D, Ahmed M R, Gomathi K, Chitra K, Sehgal P K and Jayakumar R 2004 *Biomaterials* **25** 1911–7

[70] Gomathi K, Gopinath D, Rafiuddin Ahmed M and Jayakumar R 2003 *Biomaterials* **24** 2767–72

[71] Greco K V, Francis L, Huang H, Ploeg R, Boccaccini A R and Ansari T 2018 *J. Tissue Eng. Regen. Med.* **12** e1716–24

[72] Balalaie A, Rezvani M B and Mohammadi Basir M 2018 *Dent. Mater. J.* **37** 173–82

[73] Mu C, Zhang K, Lin W and Li D 2013 *J. Biomed. Mater. Res. A* **101** 385–93

[74] Sung H W, Huang R N, Huang L L H and Tsai C C 1999 *J. Biomater. Sci. Polym. Ed.* **10** 63–78

[75] Tsai C C, Huang R N, Sung H W and Liang H C 2000 *J. Biomed. Mater. Res.* **52** 58–65

[76] Wang C, Lau T T, Loh W L, Su K and Wang D A 2011 *J. Biomed. Mater. Res. B Appl. Biomater.* **97** 58–65

[77] Sung H W, Chang Y, Liang I L, Chang W H and Chen Y C 2000 *J. Biomed. Mater. Res.* **52** 77–87

[78] Sung H W, Chang Y, Chiu C T, Chen C N and Liang H C 1999 *Biomaterials* **20** 1759–72

[79] Sung H W, Liang I L, Chen C N, Huang R N and Liang H F 2001 *J. Biomed. Mater. Res.* **55** 538–46

[80] Lakra R, Kiran M S, Usha R, Mohan R, Sundaresan R and Korrapati P S 2014 *Int. J. Biol. Macromol.* **65** 252–7

[81] Lee H J, Fernandes-Cunha G M, Na K S, Hull S M and Myung D 2018 *Adv. Healthc. Mater.* **7** e1800560

[82] Zahn H, Schade F and Siepmann E 1963 *Leder* **14** 299–304

[83] Tzaphlidou M 1983 *J. Histochem. Cytochem.* **31** 1274–8

[84] Rhee W M, Berg R A, Rosenblatt J S, Schroeder J A, Braga L J, Smestad T L and Freeman A 1997 *US Patent* 5643464

[85] Charulatha V and Rajaram A 1997 *J. Biomed. Mater. Res.* **36** 478–86

[86] Lakra R, Kiran M S and Sai K P 2015 *Biomed. Mater.* **10** 065015

[87] Zerlotti E 1967 *Nature* **214** 1304–6

[88] Sinha R N, Verma P K and Maden P 1972 *Ind. J. Plastic Surg.* **5** 56–60

[89] Gade J N, Fellman J H and Bentley J P 1991 *J. Biomed. Mater. Res.* **25** 799–811

[90] Liao K, Seifter E, Gong G, Yellin E L and Frater R W M 1991 *ASAIO Trans.* **37** M175–6

[91] Paik D C, Wen Q, Airiani S, Braunstein R E and Trokel S L 2008 *Exp. Eye Res.* **87** 279–85

[92] Tronci G, Doyle A, Russell S J and Wood D J 2013 *J. Mater. Chem.* B **1** 5478–88

[93] Brennan-Pierce E P, MacAskill I, Price R B and Lee J M 2014 *Biomed. Mater. Eng.* **24** 1659–71

[94] Subasinghe S K, Ogbuehi K C and Dias G J 2018 *Graefes Arch. Clin. Exp. Ophthalmol.* **256** 1363–84

[95] Raghavan R N, Muthukumar T, Somanathan N and Sastry T P 2013 *Mater. Sci. Eng. C Mater. Biol. Appl.* **33** 1983–8

[96] Sheehan J C and Hlavka J J 1956 *J. Org. Chem.* **21** 439–41

[97] Sheehan J C and Hlavka J J 1957 *J. Am. Chem. Soc.* **79** 4528–9

[98] Venkataramani E S, Senatore F F, Lokapur A K, Shankar H and Feola M 1986 *Res. Commun. Chem. Pathol. Pharmacol.* **54** 421–4

[99] Kato Y P, Christiansen D L, Hahn R A, Shieh S J, Goldstein J D and Silver F H 1989 *Biomaterials* **10** 38–42

[100] Marchand R, Woerly S, Bertrand L and Valdes N 1993 *Brain Res. Bull.* **30** 415–22

[101] van Wachem P B, van Luyn M J, Olde Damink L H, Dijkstra P J, Feijen J and Nieuwenhuis P 1994 *Int. J. Artif. Organs* **17** 230–9

[102] Olde Damink L H, Dijkstra P J, van Luyn M J, van Wachem P B, Nieuwenhuis P and Feijen J 1996 *Biomaterials* **17** 765–73

[103] Staros J V, Wright R W and Swingle D M 1986 *Anal. Biochem.* **156** 220–2

[104] Olde Damink L H, Dijkstra P J, van Luyn M J, van Wachem P B, Nieuwenhuis P and Feijen J 1996 *Biomaterials* **17** 679–84

[105] Lee J M, Edwards H H, Pereira C A and Samii S I 1996 *J. Mater. Sci., Mater. Med.* **7** 531–41

[106] Shepherd D V, Shepherd J H, Ghose S, Kew S J, Cameron R E and Best S M 2015 *APL Mater.* **3** pii: 014902

[107] Barnes C P, Pemble C W, Brand D D, Simpson D G and Bowlin G L 2007 *Tissue Eng.* **13** 1593–605

[108] Nam K, Kimura T and Kishida A 2008 *Macromol. Biosci.* **8** 32–7

[109] Raghunath M, Hopfner B, Aeschlimann D, Luthi U, Meuli M, Altermatt S, Gobet R, Bruckner-Tuderman L and Steinmann B 1996 *J. Clin. Invest.* **98** 1174–84

[110] McDermott M K, Chen T, Williams C M, Markley K M and Payne G F 2004 *Biomacromolecules* **5** 1270–9

[111] Broderick E P, O'Halloran D M, Rochev Y A, Griffin M, Collighan R J and Pandit A S 2005 *J. Biomed. Mater. Res. B Appl. Biomater.* **72** 37–42

[112] Bertoni F, Barbani N, Giusti P and Ciardelli G 2006 *Biotechnol. Lett.* **28** 697–702

[113] Fang J Y, Tan S J, Yang Z, Tayag C and Han B 2014 *PLoS One* **9** e105616

[114] Nomura Y, Toki S, Ishii Y and Shirai K 2001 *Biosci. Biotechnol., Biochem.* **65** 982–5
[115] Stachel I, Schwarzenbolz U, Henle T and Meyer M 2010 *Biomacromol.* **11** 698–705
[116] Cheng S, Wang W, Li Y, Gao G, Zhang K, Zhou J and Wu Z 2019 *Food Chem.* **271** 527–35
[117] Chen R N, Ho H O and Sheu M T 2005 *Biomaterials* **26** 4229–35
[118] Zeugolis D I, Panengad P P, Yew E S, Sheppard C, Phan T T and Raghunath M 2010 *J. Biomed. Mater. Res. A* **92** 1310–20
[119] Jelenska M M, Fesus L and Kopec M 1980 *Biochim. Biophys. Acta* **616** 167–78
[120] Chau D Y, Brown S V, Mather M L, Hutter V, Tint N L, Dua H S, Rose F R and Ghaemmaghami A M 2012 *Biomed. Mater.* **7** 045011
[121] Wu X, Liu Y, Liu A and Wang W 2017 *Int. J. Biol. Macromol.* **98** 292–301
[122] O Halloran D M, Collighan R J, Griffin M and Pandit A S 2006 *Tissue Eng.* **12** 1467–74
[123] Schloegl W, Klein A, Fürst R, Leicht U, Volkmer E, Schieker M, Jus S, Guebitz G M and Stachel I *et al* 2012 *Eur. J. Pharm. Biopharm.* **80** 282–8
[124] Fortunati D, Chau D Y, Wang Z, Collighan R J and Griffin M 2014 *Amino Acids* **46** 1751–61
[125] Zhao L, Li X, Zhao J, Ma S, Ma X, Fan D, Zhu C and Liu Y 2016 *Mater. Sci. Eng. C Mater. Biol. Appl.* **68** 317–26
[126] Petite H, Rault I, Huc A, Menasche P and Herbage D 1990 *J. Biomed. Mater. Res.* **24** 179–87
[127] Anselme K, Petite H and Herbage D 1992 *Matrix* **12** 264–73
[128] Simmons D M and Kearney J N 1993 *Biotechnol. Appl. Biochem.* **17** 23–9
[129] Petite H, Frei V, Huc A and Herbage D 1994 *J. Biomed. Mater. Res.* **28** 159–65
[130] Petite H, Duval J L, Frei V, Abdul-Malak N, Sigot-Luizard M F and Herbage D 1995 *Biomaterials* **16** 1003–8
[131] Marinucci L, Lilli C, Guerra M, Belcastro S, Becchetti E, Stabellini G, Calvi E M and Locci P 2003 *J. Biomed. Mater. Res. A* **67** 504–9
[132] Benqué E, Zahedi S, Brocard D, Oscaby F, Justumus P and Brunel G 1997 *J. Clin. Periodont.* **24** 544–9
[133] Waykole P and Heidemann E 1976 *Connect. Tissue Res.* **4** 219–22
[134] LaBella F, Waykole P and Queen G 1968 *Biochem. Biophys. Res. Commun.* **30** 333–8
[135] Tenovuo J and Paunio K 1979 *Arch. Oral Biol.* **24** 591–4
[136] Vashi A V, Werkmeister J A, Vuocolo T, Elvin C M and Ramshaw J A M 2012 *J. Biomed. Mater. Res. A* **100** 2239–43
[137] Fancy D A and Kodadek T 1999 *Proc. Natl Acad. Sci. USA* **96** 6020–4
[138] Elvin C M, Vuocolo T, Brownlee A G, Sando L, Huson M G, Liyou N E, Stockwell P R, Lyons R E and Kim M *et al* 2010 *Biomaterials* **31** 8323–31
[139] Fenton H J H 1894 *J. Chem. Soc. Trans.* **65** 899–911
[140] Qin G, Rivkin A, Lapidot S, Hu X, Preis I, Arinus S B, Dgany O, Shoseyov O and Kaplan D L 2011 *Biomaterials* **32** 9231–43
[141] Bolton A E and Hunter W M 1973 *Biochem. J.* **133** 529–39
[142] Stoichevska V, An B, Peng Y Y, Yigit S, Vashi A V, Kaplan D L, Werkmeister J A, Dumsday G J and Ramshaw J A M 2016 *J. Biomed. Mater. Res. A* **104** 2369–76
[143] Hu Y, Liu L, Dan W, Dan N, Gu Z and Yu X 2013 *Int. J. Biol. Macromol.* **55** 221–30
[144] Ssadikow W S 1910 *Kolloidchem. Beihefte* **1** 119–220 Cited by: Alexander J 1923 *Glue and Gelatine* (New York: American Chemical Society) p 116
[145] Yannas I V and Tobolsky A V 1967 *Nature* **215** 509–10

[146] Bella J, Brodsky B and Berman H M 1995 *Structure* **3** 893–906

[147] Silver F H, Yannas I V and Salzman E W 1979 *J. Biomed. Mater. Res.* **13** 701–16

[148] Weadock K, Olson R M and Silver F H 1983 *Biomater. Med. Devices Artif. Organs* **11** 293–318

[149] Gorham S D, Light N D, Diamond A M, Willins M J, Bailey A J, Wess T J and Leslie N J 1992 *Int. J. Biol. Macromol.* **14** 129–38

[150] Wang M C, Pins G D and Silver F H 1994 *Biomaterials* **15** 507–12

[151] Geesin J C, Brown L J, Liu Z and Berg R A 1996 *J. Biomed. Mater. Res.* **33** 1–8

[152] Haugh M G, Jaasma M J and O'Brien F J 2009 *J. Biomed. Mater. Res. A* **89** 363–9

[153] Madaghiele M, Calò E, Salvatore L, Bonfrate V, Pedone D, Frigione M and Sannino A 2016 *J. Biomed. Mater. Res. A* **104** 186–94

[154] Nakada A, Shigeno K, Sato T, Hatayama T, Wakatsuki M and Nakamura T 2017 *J. Biomed. Mater. Res. B Appl. Biomater.* **105** 2301–7

[155] Weadock K S, Miller E J, Bellincampi L D, Zawadsky J P and Dunn M G *J. Biomed. Mater. Res.* **29** 1373–9

[156] Goldstein J D, Tria A J, Zawadsky J P, Kato Y P, Christiansen D and Silver F H 1989 *J. Bone Joint Surg. Am.* **71** 1183–91

[157] Koide M, Osaki K, Konishi J, Oyamada K, Katakura T, Takahashi A and Yoshizato K 1993 *J. Biomed. Mater. Res.* **27** 79–87

[158] Ueda H, Nakamura T, Yamamoto M, Nagata N, Fukuda S, Tabata Y and Shimizu Y 2003 *J. Control. Release* **88** 55–64

[159] Sun L, Li B, Yao D, Song W and Hou H 2018 *J. Mech. Behav. Biomed. Mater.* **80** 51–8

[160] Miyata T, Sohde T, Rubin A L and Stenzel K H 1971 *Biochim. Biophys. Acta* **229** 672–80

[161] Jariashvili K, Madhan B, Brodsky B, Kuchava A, Namicheishvili L and Metreveli N 2012 *Biopolymers* **97** 189–98

[162] Lin J T and Cheng D C 2017 *PLoS One* **12** e0175002

[163] Weadock K S, Miller E J, Keuffel E L and Dunn M G 1996 *J. Biomed. Mater. Res.* **32** 221–6

[164] Shibata H, Shioya N and Kuroyanagi Y 1997 *J. Biomed. Sci. Polymer Edn.* **8** 601–21

[165] Rabotyagova O S, Cebe P and Kaplan D L 2008 *Mater. Sci. Eng. C Mater. Biol. Appl.* **28** 1420–9

[166] Davidson R J and Cooper D R 1968 *Biochem. J.* **107** 29–34

[167] Cooper D R and Russell A E 1969 *Biochem. J.* **113** 263–9

[168] Cheung D T, Perelman N, Tong D and Nimni M E 1990 *J. Biomed. Mater. Res.* **24** 581–9

[169] Liu B, Harrel R, Davis R H, Dresden M H and Spira M 1989 *J. Biomed. Mater. Res.* **23** 833–44

[170] Inoue N, Bessho M, Furuta M, Kojima T, Okuda S and Hara M 2006 *J. Biomater. Sci. Polym. Ed.* **17** 837–58

[171] Zhang X, Xu L, Huang X, Wei S and Zhai M 2012 *J. Biomed. Mater. Res. A* **100** 2960–9

[172] Takitoh T, Bessho M, Hirose M, Ohgushi H, Mori H and Hara M 2015 *J. Biosci. Bioeng.* **119** 217–25

[173] Edwards G A, Glattauer V, Nash T J, White J F, Brock K A, Werkmeister J A and Ramshaw J A M 1997 *J. Biomed. Mater. Res.* **34** 291–7

[174] Zhang Y, Zhang X, Xu L, Wei S and Zhai M 2015 *J. Biomater. Sci. Polym. Ed.* **26** 162–80

[175] Seto A, Gatt C J and Dunn M G 2008 *Clin. Orthop. Relat. Res.* **466** 1788–95

[176] Seto A, Gatt C J and Dunn M G 2009 *Clin. Orthop. Relat. Res.* **467** 2994–3001

[177] Roe S C, Milthorpe B K, True K, Rogers G J and Schindhelm K 1992 *Clin. Mater.* **9** 149–54
[178] Nguyen T D, Cornillet-Stoupy J, Gillery P and Maquart F X 1991 *Experientia* **47** 725–8
[179] Fujisato T, Tomihata K, Tabata Y, Iwamoto Y, Burczak K and Ikada Y 1999 *J. Biomater. Sci. Polym. Ed.* **10** 1171–81
[180] Jiang B, Wu Z, Zhao H, Tang F, Lu J, Wei Q and Zhang X 2006 *Biomaterials* **27** 15–23
[181] Mrázová H, Koller J, Kubišová K, Fujeríková G, Klincová E and Babál P 2016 *Cell Tissue Bank.* **17** 255–60
[182] Kuntz E 1964 *US Patent* 3152976
[183] Mechanic G L 1992 *US Patent* 5147514
[184] Bengtsson L A, Phillips R and Haegerstrand A N 1995 *Ann. Thorac. Surg.* **60** S365–8
[185] Bianco R W, Phillips R, Mrachek J and Witson J 1996 *J. Heart Valve Dis.* **5** 317–22
[186] Moore M A, Bohachevsky I K, Cheung D T, Boyan B D, Chen W M, Bickers R R and McIlroy B K 1994 *J. Biomed. Mater. Res.* **28** 611–8
[187] Moore M A 1997 *J. Heart Valve Dis.* **6** 521–6
[188] McIlroy B K, Robinson M D, Chen W M and Moore M A 1997 *J. Heart Valve Dis.* **6** 416–23
[189] Schoen F J 1998 *J. Heart Valve Dis.* **7** 174–9
[190] Johnson T S, O'Neill A C, Motarjem P M, Amann C, Nguyen T, Randolph M A, Winograd J M, Kochevar I E and Redmond R W 2007 *J. Surg. Res.* **143** 224–9
[191] Baird C W, Myers P O, Piekarski B, Borisuk M, Majeed A, Emani S M, Sanders S P, Nathan M and Del Nido P J 2017 *Interact. Cardiovasc. Thorac. Surg.* **24** 240–4
[192] Verweij H, Dubbleman T M A R and van Steveninck J 1981 *Biochim. Biophys. Acta* **647** 87–94
[193] Ramshaw J A M, Stephens L J and Tulloch P A 1994 *Biochim. Biophys. Acta* **1206** 225–30
[194] Moore M A, Chen W M, Phillips R E, Bohachevsky I K and McIlroy B K 1996 *J. Biomed. Mater. Res.* **32** 209–14
[195] Graham L and Gallop P M 1994 *Anal. Biochem.* **217** 298–305
[196] Moore M A and Phillips R E 1997 *J. Heart Valve Dis.* **6** 307–15
[197] Moore M A and Adams A K 2001 *J. Biomed. Mater. Res.* **56** 24–30
[198] Rigby B J 1961 *Biochim. Biophys. Acta* **47** 529–34
[199] Rigby B J 1967 *Biochim. Biophys. Acta* **140** 548–51
[200] Pereira C A, Lee J M and Haberer S A 1990 *J. Biomed. Mater. Res.* **24** 345–61
[201] Shah N B, Wolkers W F, Morrissey M, Sun W Q and Bischof J C 2009 *Tissue Eng. Part C Methods* **15** 33–40
[202] Flory P J and Rehner J 1943 *J. Chem. Phys.* **11** 521–6
[203] Sun W Q and Leung P 2008 *Acta Biomater.* **4** 817–26
[204] Zhu K, Slusarewicz P and Hedman T 2011 *J. Orthop. Res.* **29** 8–13
[205] Hindman H and Burr G S 1949 *Trans. ASME* **71** 789–96
[206] Han P 1992 *Tensile Testing* (Materials Park, OH: ASM International)
[207] Roeder R K 2013 *Characterization of Biomaterials* ed A Bandyopadhyay and S Bose (Amsterdam: Elsevier) pp 49–104
[208] Broom N D 1977 *J. Biomechanics* **10** 707–24
[209] Olde Damink L H H, Dijkstra P J, van Luyn M J A, van Wachem P B, Nieuwenhuis P and Feijen J 1995 *J. Mater. Sci., Mater. Med.* **6** 460–72
[210] Gratzer P F and Lee J M 2001 *J. Biomed. Mater. Res.* **58** 172–9

[211] Sacks M S, Chuong C J and More R 1994 *ASAIO J.* **40** M632–7

[212] Langdon S E, Chernecky R, Pereira C A, Abdulla D and Lee J M 1999 *Biomaterials* **20** 137–53

[213] Lee C R, Grodzinsky A J and Spector M 2001 *Biomaterials* **22** 3145–54

[214] Serpooshan V, Julien M, Nguyen O, Wang H, Li A, Muja N, Henderson J E and Nazhat S N 2010 *Acta Biomater.* **6** 3978–87

[215] Tronci G, Grant C A, Thomson N H, Russell S J and Wood D J 2015 *J. R. Soc. Interface* **12** 20141079

[216] Van der Sman R G 2015 *Food Hydrocolloids* **48** 94–101

[217] Cerny L C, Stasiw D M and Tinelli W C 1973 *Biorheology* **10** 565–75

[218] Friess W and Schlapp M 2001 *Eur. J. Pharm. Biopharm.* **51** 259–65

[219] Sheu M T, Huang J C, Yeh G C and Ho H O 2001 *Biomaterials* **22** 1713–9

[220] Sundararaghavan H G, Monteiro G A, Lapin N A, Chabal Y J, Miksan J R and Shreiber D I 2008 *J. Biomed. Mater. Res. A* **87** 308–20

[221] Tanaka S, Avigad G, Eikenberry E F and Brodsky B 1988 *J. Biol. Chem.* **263** 17650–7

[222] Schmidt F N, Zimmermann E A, Campbell G M, Sroga G E, Püschel K, Amling M, Tang S Y, Vashishth D and Busse B 2017 *Bone* **97** 243–51

[223] Guilbert M, Said G, Happillon T, Untereiner V, Garnotel R, Jeannesson P and Sockalingum G D 2013 *Biochim. Biophys. Acta* **1830** 3525–31

[224] Figueiró S D, Macêdo A A, Melo M R, Freitas A L, Moreira R A, de Oliveira R S, Góes J C and Sombra A S 2006 *Biophys. Chem.* **120** 154–9

[225] Habeeb A F S A 1966 *Anal. Biochem.* **14** 328–36

[226] Everaerts F, Torrianni M, Hendriks M and Feijen J 2007 *J. Biomed. Mater. Res. A* **83** 1176–83

[227] Garcia Y, Collighan R, Griffin M and Pandit A 2007 *J. Mater. Sci., Mater. Med.* **18** 1991–2001

[228] Vicens-Zygmunt V, Estany S, Colom A, Montes-Worboys A, Machahua C, Sanabria A J, Llatjos R, Escobar I and Manresa F *et al* 2015 *Respir. Res.* **16** 82

[229] Wissink M J, van Luyn M J, Beernink R, Dijk F, Poot A A, Engbers G H, Beugeling T, van Aken W G and Feijen J 2000 *Thromb. Haemost.* **84** 325–31

[230] Keller P F, Verin V, Ziegler T, Mermillod B, Popowski Y and Delafontaine P 2001 *J. Investig. Med.* **49** 258–64

[231] Levy R J, Schoen F J, Levy J T, Nelson A C, Howard S L and Oshry L J 1983 *Am. J. Pathol.* **113** 143–55

[232] Hu Q Y, Berti F and Adamo R 2016 *Chem. Soc. Rev.* **45** 1691–719

[233] Forns P, Lauer-Fields J L, Gao S and Fields G B 2000 *Biopolymers* **54** 531–46

[234] Forest P O, Karoum R and Gagnieu C H 2007 *J. Biomed. Mater. Res. A* **80** 758–67

[235] Bose S M and Joseph K T 1958 *Arch. Biochem. Biophys.* **74** 46–56

[236] Fujimoto D 1970 *Biochim. Biophys. Acta* **200** 172–4

[237] Rauterberg J and Kühn K 1968 *Hoppe Seylers Z. Physiol. Chem.* **349** 611–22

[238] Mũnoz Z, Shih H and Lin C-C 2014 *Biomat. Sci.* **2** 1063–72

[239] Butler P J, Harris J I, Hartley B S and Leberman R 1967 *Biochem. J.* **103** 78P–9P

[240] Dixon H B F and Perham R N 1968 *Biochem. J.* **109** 312–4

[241] Chang C D and Meienhofer J 1978 *Int. J. Pept. Protein Res.* **11** 246–9

[242] Brinkman W T, Nagapudi K, Thomas B S and Chaikof E L 2003 *Biomacromolecules* **4** 890–5

[243] Benton J A, DeForest C A, Vivekanandan V and Anseth K S 2009 *Tissue Eng. Part A* **15** 3221–30
[244] Nichol J W, Koshy S T, Bae H, Hwang C M, Yamanlar S and Khademhosseini A 2010 *Biomaterials* **31** 5536–44
[245] Yue K, Trujillo-de Santiago G, Alvarez M M, Tamayol A, Annabi N and Khademhosseini A 2015 *Biomaterials* **73** 254–71
[246] Loessner D, Meinert C, Kaemmerer E, Martine L C, Yue K, Levett P A, Klein T J, Melchels F P, Khademhosseini A and Hutmacher D W 2016 *Nat. Protoc.* **11** 727–46
[247] Drzewiecki K E, Parmar A S, Gaudet I D, Branch J R, Pike D H, Nanda V and Shreiber D I 2014 *Langmuir* **30** 11204–11
[248] Li L, Lu C, Wang L, Chen M, White J F, Hao X, McLean K M, Chen H and Hughes T C 2018 *ACS Appl. Mater. Interfaces* **10** 13283–92
[249] Song X, Dong P, Gravesande J, Cheng B and Xing J 2018 *Int. J. Biol. Macromol.* **120** 2086–93
[250] Cosgriff-Hernandez E M, Hahn M S, Russell B, Wilems T, Munoz-Pinto D, Browning M B, Rivera J and Höök M 2010 *Acta Biomater.* **6** 3969–77
[251] Browning M B, Guiza V, Russell B, Rivera J, Cereceres S, Höök M, Hahn M S and Cosgriff-Hernandez E M 2014 *Tissue Eng. Part A* **20** 3130–41
[252] Stoichevska V, Peng Y Y, Vashi A V, Werkmeister J A, Dumsday G J and Ramshaw J A M 2017 *J. Biomed. Mater. Res. A* **105** 806–13
[253] Jue R, Lambert J M, Pierce L R and Traut R R 1978 *Biochemistry* **17** 5399–406
[254] Yamauchi K, Takeuchi N, Kurimoto A and Tanabe T 2001 *Biomaterials* **22** 855–63
[255] Ellman G L 1959 *Arch. Biochem. Biophys.* **82** 70–7
[256] Jones M E and Messersmith P B 2007 *Biomaterials* **28** 5215–24
[257] Bet M R, Goissis G and Lacerda C A 2001 *Biomacromolecules* **2** 1074–9
[258] Chalker J M, Bernardes G J, Lin Y A and Davis B G 2009 *Chem. Asian J.* **4** 630–40
[259] Smyth D G, Blumenfeld O O and Konigsberg W 1964 *Biochem. J.* **91** 589–95
[260] Badescu G, Bryant P, Swierkosz J, Khayrzad F, Pawlisz E, Farys M, Cong Y, Muroni M and Rumpf N *et al* 2014 *Bioconjug. Chem.* **25** 460–9
[261] Morales-Sanfrutos J, Lopez-Jaramillo J, Ortega-Muñoz M, Megia-Fernandez A, Perez-Balderas F, Hernandez-Mateo F and Santoyo-Gonzalez F *Org. Biomol. Chem.* **8** 667–75
[262] Russo L, Battocchio C, Secchi V, Magnano E, Nappini S, Taraballi F, Gabrielli L, Comelli F and Papagni A *et al* 2014 *Langmuir* **30** 1336–42
[263] Truong V X, Hun M L, Li F, Chidgey A P and Forsythe J S 2016 *Biomater Sci.* **4** 1123–31
[264] Parrish J, Lim K S, Baer K, Hooper G J and Woodfield T B F 2018 *Lab Chip* **18** 2757–75
[265] Tyson E L, Ament M S and Yoon T P 2012 *J. Org. Chem.* **78** 2046–50
[266] Zhao G, Kaur S and Wang T 2017 *Org. Lett.* **19** 3291–4
[267] Shih H and Lin C-C 2013 *Macromol. Rapid Commun.* **34** 269–73
[268] Batchelor R R, Kwandou G, Spicer P T and Stenzel M H 2017 *Polym. Chem.* **8** 980–4
[269] Endrizzi B J, Huang G, Kiser P F and Stewart R J 2006 *Langmuir* **22** 11305–10
[270] Udenfriend S B 1966 *Pharm. Rev.* **18** 43–51

Part IV

Applications of collagen in medical products

Chapter 13

Applications for intact tissue collagen

Collagen is used in a range of different formats in biomedical applications. One of these formats is where the structure of an intact tissue is used and processed to make it suitable for the particular clinical application. This chapter discusses the range of options for collagen formats. It then continues to discuss a selection of the tissue-based products and uses that have emerged.

Collagens are used in biomedical applications in a wide range of formats and formulations. It is hard to make distinctions between these different systems as they typically form a spectrum of different approaches to solving clinical problems (figure 13.1). In this section, concerning 'applications of collagen in medical products', the information has been arranged in a quasi-chronological order. Thus, this initial chapter covers tissue-based materials where the structure is maintained during processing, starting with sutures. This tissue-based type of material became significant in the late 1960s as various forms of bioprosthetic heart

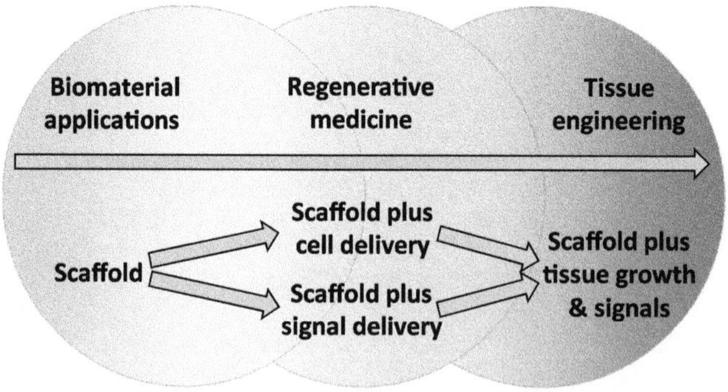

Figure 13.1. An overview of the continuum of options in the use of collagen as a biomedical material.

doi:10.1088/978-0-7503-2096-2ch13

valves emerged. More recently, the development of acellular matrix (ACM) as a processing strategy for intact tissues has seen a resurgence in development of novel tissue-based materials (see this chapter). Although the development of materials based on processed and purified collagens had been established in the 1940s using comminuted tissue, in part due to military requirements (see [1]), it was in the 1970s that significant interest in this type of material, and materials derived from soluble collagen gained greater interest [2] (see chapter 14). Afterwards, in the late 1970s and through the 1980s, the development of biosynthetic materials emerged, providing an effective biologically-based solution for arterial replacement [3] (see chapter 15). Subsequently, the identification and characterisation of growth factors [4, 5] and other bioactive molecules, and their availability by recombinant production, led to the development of coatings (see chapter 17) and composite materials (see chapter 18), while cells seeding also led to other composite formats. Nowadays, these coated and composite materials are often referred to as materials for regenerative medicine (see chapter 16). Also, principally from around 1990 onwards, the combination of a collagen supporting structure along with signals and cells led to new materials that were termed as tissue engineering constructs (see chapter 16). Undoubtedly, the present direction for research into collagen-based materials is to develop complex, multi-component devices with enhanced properties and functions.

In this chapter, we discuss devices based on tissue collagen where the intrinsic structure of the tissue has been maintained, and where the processing is to maintain this structure rather than as a basis for formation of more complex composite structures.

A range of tissue-based products have been described for a range of medical applications (table 13.1) that are distinct from products based on purified and reconstituted collagen (see chapter 14). Typically, products for a clinical application are found in only one of these formats, and only occasionally have both strategies been examined for a particular indication.

Table 13.1. Examples of applications of tissue collagen-based biomaterials in a range of medical areas.

Medical areas and applications		Pioneering examples	Recent examples and reviews
Cardiovascular surgery	Vessel replacement	[6]	
	Heart valves	[7–9]	[10–12]
	Myocardial repair		[13]
Dermatology	Dermal replacement and dressing	[14–16]	[17–19]
General surgery	Hernia repair	[20]	[21–23]
	Adhesion barriers	[24]	[25]
Neurosurgery	Dural repair		[26, 27]
Ophthalmology	Corneal repair and shields	[28]	[29]
Orthopaedics	Tendon	[30, 31]	

When a tissue-based strategy has been selected, it is based on several advantages. These can include the original shape and size of the material, and its strength, durability and persistence, augmented generally by chemical crosslinking. The products typically have a low immunogenicity, further masked by crosslinking, while some, possibly many, of the potential cell-matrix interactions are preserved. The one disadvantage that has been observed, especially for cardiovascular applications such as heart valves, has been the propensity for non-specific calcification to occur, often reducing the lifespan of the device.

The number of different tissues that have been used is relatively limited. In some cases, the tissue has found application in several different medical disciplines. In the discussion that follows, therefore, the focus has been on some of the tissues that have been used, rather than to specific disciplines, based on direct use of processed/stabilised tissue or the use of ACM. These are discussed by reference to just a few of the numerous papers that are available.

13.1 Stabilised tissues

13.1.1 Intestine

If we follow a chronological context, then clearly the 'catgut' suture based on animal intestine is the first tissue-based material to consider. It has been suggested that intestine-based sutures may have been used around 3750 BCE in Egypt [32]. Certainly, by 150 CE, the physician Galen was using this type of collagen suture [32, 33]. There seems to be little further in the literature for many centuries. Medieval drawings and woodcuts show sutures ready to use or being used (see, for example, [34], while in 1656 letters patent were granted by Louis XIV to the Guild who made these sutures, along with racquet and musical instrument strings [32]. A sampler cloth, possibly made in the 18th century, showing numerous surgical suture examples and methods is on display at the museum of the Institute for Medical History in Zurich, Switzerland.

The work of Joseph Lister (1st Baron Lister, OM, PC, PRS, 1827–1912) had a significant impact on sutures as we now know them. For example, he introduced the use of chrome salts to produce a chromic suture with extended resorption, an addition to plain gut sutures [35]. While wine may have been used from millennia before to sanitise sutures [32], Lister clearly identified the need for sterility in surgery [36] and this led, over the next 40 years, to many dozens of approaches for sterilisation of sutures and to eliminate issues such as 'catgut tetanus' [37]. Once this problem had been resolved, relatively few changes to the material seem to have occurred. One, however, was the introduction of eyeless needled sutures, by George Merson (the Mersuture) and by others (e.g., [38]) enabling only a single strand to be taken through tissue rather than a more damaging doubled strand. Subsequently, changes included moving from packaging in liquid in a glass tube (figure 13.2) to packing after plasticising in a foil envelope (e.g., [39]) that was also suited to the emerging use of ^{60}Co gamma sterilisation.

Finally, although intestine has been the main tissue source, other tissues have been considered, notably tendon. Among these, kangaroo tendon was proposed [40]

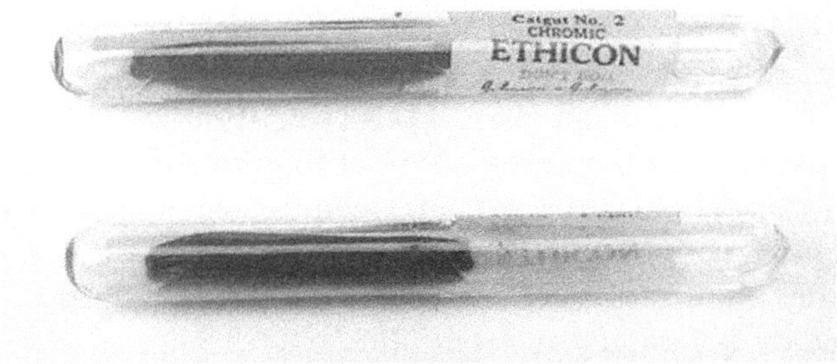

Figure 13.2. Chromic sutures packaged in solution in glass.

and became commercially available during the interwar period. However, the use of collagen-based sutures has not grown in the last decades due to the emergence of a range of polymeric suture materials [41].

A tissue-based material that has emerged strongly in recent years is small intestine submucosa (SIS), normally from a porcine source. The material shows a very low immunogenicity, and products that are fully acellular and free of DNA perform best with least inflammation [42]. Similarly, different processing approaches can lead to different performance results, with, for example, a more open structure obtained by lyophilisation giving better neovascularisation [43], while perforation can also be used to enhance porosity [44]. In some uses, SIS is used without crosslinking, whereas in other examples a crosslinked product is made, for example, using 1-ethyl-3-(3-dimethylaminopropyl)carbodiimide (EDC) in the presence of N-hydroxysuccinimide (NHS), with the crosslinking giving longer-term durability when it is needed [45].

SIS material has been used for a variety of clinical indications. It has been approved for use, for example, in hernia repair, ureteral reconstruction and stress incontinence treatments [22], with other applications proposed and being examined. Initial studies reported the success of the material in various small animal abdominal repair protocols [46, 47], including when an analgesic was delivered with the material [48]. Other applications that have been described include use in tendon augmentation [31], bladder augmentation [49] and peripheral nerve repair.

SIS is also used in other formats where the tissue structure has been lost, including as a gel [50] and as reconstituted crosslinked sponge [51].

13.1.2 Amnion

In the few years prior to the Great War, several groups reported on the potential to use amnion membrane for wound repair, particularly in burns treatment [14–16]. Although not realised at this time, this tissue has the advantage that it does not express various HLA or other antigens and so typically does not experience immunological rejection and could be seen as an immuno-privileged tissue [52–54]. This contrasts with skin. The use of autologous skin had been examined much earlier. In the recent era, this was seen as starting with Reverdin [55], and followed by others (see [56, 57]). However, this advance was some 3000 years after the Hindus had examined skin grafting (see [57]). Later skin from cadavers was examined for skin grafting. However, it became generally accepted that non-autologous skin could not be used for grafting (see [58]) excepting between monozygotic twins [59; others cited in 58].

The initial studies using amnion [14–16] did not attract much immediate attention, and the next studies appeared more than two decades later, with use in tissue reconstruction followed by a move from untreated amnion to a processed and dehydrated form, amnioplastin [60], and later to other formats (see [61]), including those using stabilisation of the collagen. Various stabilisation methods have been proposed, including glutaraldehyde (GA) [62, 63], carbodiimide [64], genipin [65] and irradiation [62].

A range of studies have shown that amnion is a suitable substrate for the growth of a wide range of cells, including stem cells [66], without any adverse effects [67], suggesting a broad applicability to a range of medical applications. In practice, however, the main uses for amnion have been in ophthalmic and dermal applications. For ophthalmic use, amnion materials have been reported mainly in treatment of severe chemical and thermal injuries [29, 68–71] as well as ulceration and other disorders [72], generally with successful outcomes, although some less than positive results have been reported [73]. For wound management, the pliability of amnion allows it to readily conform to irregular surfaces, giving it a clear advantage [74]. The membrane is able to prevent heat and water loss from burns, with water loss being recognised as a major concern, while also providing a barrier to minimise bacterial contamination and providing clear relief from pain [17, 75]. In clinical use, amnion was reported to be potentially superior to allograft and xenograft skin materials that were available at the time, but to lack vascularisation [75, 76]. Recent years have seen a major growth in reported studies along with reviews of the use of amnion [77].

Some other applications for amnion that have been considered include its use in bladder reconstruction [78], as a membrane, either to control post-operative adhesions [25] or for guided bone regeneration [79], as a pericardial substitute [80] and for tracheal repair [81]. Its use in these and other applications will no doubt increase significantly now that a choice of different commercially processed, off-the-shelf products is now available (e.g., [17, 77]).

13.1.3 Pericardium

Processed pericardium has proved a very versatile material, being assessed for a variety of applications, with cardiac applications being foremost. Indeed, probably the most extensively used tissue collagen-based device is the bioprosthetic heart valve, which has now been used successfully for more than five decades. Bioprosthetic heart valves broadly fall into two types; those made from stabilised and shaped pericardium or those made from processed animal heart valves (see the following section). The requirements for heart valves, whether bioprosthetic or mechanical are stringent as failure can be a significant problem, even catastrophic. Yet they must be very durable to endure constant motion, as with a natural valve which may go though perhaps more than 2.5 billion beats in a lifetime.

Early intracardiac operations were necessarily simple ones which did not interrupt the circulation [82, 83], prior to the development of the heart-lung machine technology [84, 85] and the numerous subsequent improvements in this technology [86–88]. Pericardium patches were initially used as an adjunct in valve repair [83]. The pericardium-based valve was developed and evaluated in the next decade, 1950s, onwards by various groups [7, 8, 89], and has become a frequently used device. Currently, there are some hundreds of thousands of heart valve replacements performed every year [90]. Surgical preferences depend on various factors, including the age and life status of the patient, and change over time as new research results emerge. As a rough approximation, bioprosthetic valves represent about half of these patients and pericardial valves around half of the bioprosthetic devices, although these rough values vary between countries and institutions and have been changing over time [91].

Various design modifications and improvement have regularly occurred or been proposed [92]. These include the design and properties of the supporting stent/sewing ring, where changing from a rigid structure to a more flexible structure provided clinical benefits [92, 93]. The principle stabilisation agent has been GA, which has been associated with calcification of the device during implant [94] (see below). As such, other fixation strategies for pericardium for valves have been proposed, including using lower GA amounts [95] and alternative reagents such as polyglycidyl ethers [96] and carbodiimides [97]. These also include the source species for the tissue. While bovine has been the predominant material [98, 99], other suggestions have included porcine [100], horse [101], donkey [102] and yak [96]. The physical and mechanical properties of the tissue vary between species. For example, bovine tissue is thinner than donkey [102] and porcine tissues [100]. The mechanical properties of these tissues have been examined by many groups, often using different methods. However, comparative tables of these properties have been established [103]. The principle concerns on this device continue to be tissue durability and calcification (discussed below).

The emerging use for pericardium in cardiac surgery is its application in transcatheter valves [104, 105]. In this application, the physical and mechanical properties of the material are important, particularly as it needs to survive the packing in the device and its subsequent deployment. Crimping is used to allow delivery diameters as low as 6 mm for a 23mm deployed device, and can be managed so as to prevent major damage, although further design improvements could be valuable [106]. It has been suggested that pericardium with a reduced thickness may be more suitable, such as porcine or calf, with foetal calf being too thin and lacking in mechanical properties [11]. Data on performance of this type of device is still relatively limited, and may not yet allow a fair comparison [107], but it is now a preferred option for certain patients, such as those for whom normal valve replacement has a prohibitive risk [107]. Data accumulation in registries will assist in future evaluation [108].

Another cardiac application for pericardium that has been proposed is as a delivery vehicle for cells, especially stem cells, to damaged cardiac muscle after myocardial infarction [13, 109, 110] with good cell support *in vitro* [13] and promising animal studies [110].

Pericardium has also been proposed for other clinical indications. These include repair of cranial defects [111] and for certain pelvic organ prolapse indications [112, 113].

13.1.4 Heart valve

Undoubtedly, the concept of utilising a xenogeneic heart valve as a replacement for a diseased or damaged natural valve has bought benefits, notably quality of life and in longer lifetimes, to many hundreds of thousands of patients. Conceptually, it seems simple, but the practicalities of bringing a successful device to patients are considerably more complex. Initial use of natural valves was in the mid-1950s, through homografts (see [114]) and the procedure was later improved through inclusion of a supporting stent [115]. Availability of fresh tissue was always going to be a problem. Hence, in the mid-1960s the concept of a xenograft emerged [9], again using a supporting frame, and was followed by a commercial valve (Hancock) in 1968 (see [114]). The initial fixation approach using formaldehyde proved to be inadequate [116], but was replaced successfully by GA. The first commercial implants were in 1969 and 1970. GA continues to be the fixation approach of choice, although it is seen to have problems, most notably its propensity to cause calcification (see below). These animal valves are generally of porcine origin, but other species have been proposed, including a marsupial source [117].

There is always discussion on whether to use a bioprosthetic or a mechanical heart valve, with patient welfare being the driving factor [91, 118]. And if a bioprosthetic valve is the appropriate choice, then should a pericardium-based or processed (porcine) valve be the choice? Overall, the failure point of valves is typically discussed as being in the range of 10 to 15 years, although this may be an underestimate, especially for older (>65 yo) patients [12] and longer times up to 24 years have been reported [119]. Various studies have been reported, based on the data from a single institution [120], or detailed compilations for a single device [121] or a compilation from many institutions based on literature surveys [12]. The number of cases studied ranges from 2.5, 12.5 up to 42 thousand implants respectively. These data suggested that failure was predominantly through leaflet tear for porcine valves and fibrosis/calcification for pericardial valves [120]. They also suggested that the mean times to valve failure were similar between most devices excepting that one manufacturers' pericardium-based device had a significantly lower risk [12]. This difference in design performance suggests that further progress can be made. Another, tissue processing area where progress may emerge is in the use of lyophilised tissue [122, 123].

13.1.5 Issues with calcification of tissue-based devices

Calcification is distinct from mineralisation in that it is unplanned, even though the nature of the mineral deposit and aspects of its deposition may be the same or very similar to the natural mineralisation process.

Calcification of heart valves has been recognised for a long time, both in natural cardiac tissues [124, 125] and particularly in bioprosthetic devices [126–128]. A major consequence of the mineral deposits, especially at the junction of the leaflet to the annular tissue, is the slow destruction of the collagenous tissues leading to tears and failure in the bioprosthetic tissue [128], albeit that collagen fibre damage may occur in bioprosthetic heart valves independent of calcification [129].

The mechanism and circumstances by which the calcification occurs, both in native and in bioprosthetic cardiac tissues, have not been clear and remain uncertain with many factors probably combining [130, 131]. For bioprosthetic tissue a number of suspects have been proposed and strategies developed to address these potential causes. The leading suspect was GA (e.g., [132]), especially when valves were shipped in GA solution. Hence alternative stabilisation strategies were suggested. These include modifications to the GA procedure [133] and also include use of different strategies, such as di- and polyglycidyl ethers [96, 134], dye-mediated photo-oxidation [135] and carbodiimides [136] as well as combination approaches, for example, using prior treatment with formaldehyde [137] and a genipin-GA combination [138]. Approaches that potentially retain most of the natural tissue 'lubricants', the proteoglycans, with their glycosaminoglycan chains, and hyaluronic acid, may also be important [139, 140]. The distinct alternative was to consider unfixed, cryopreserved or acellular valves [141], while unfixed tissue engineered devices are now being proposed (see chapter 16).

Another approach was that instead of replacing GA, strategies were developed that aimed to quench unreacted aldehyde groups after stabilisation, including with amino acids, such as glutamic acid [142]. An extension of this approach was to use diamines [143], or to chemically incorporate potential bioactive compounds, such as 3-(4-hydroxy-3,5-di-tert-butylphenyl) propyl amine, within the tissue [144]. This compound was selected to mitigate against any oxidative changes that are indicated by the presence of oxidised Tyr moieties [145], with oxidative changes recently being seen as having previously unrecognised importance [145, 146].

In all these approaches, the structural and mechanical status of the tissue during the stabilisation may alter the subsequent performance. Studies using GA stabilisation with both static and dynamic loading have indicated that dynamic protocols led to more proteolytically sensitive tissue [147], while higher loading led to stiffer tissue [148].

Other strategies have looked at pre- and post-stabilisation treatments. One approach was to include bisphosphonates [149, 150], such as ethane-1-hydroxy-1,1-bisphosphonate, within the tissue device. This class of compounds inhibit calcium phosphate mineral formation and deposition, and the systemic toxicity is limited by specific inclusion in the device. Pre-treatment of tissue with 1,4-dioxane was found to increase the aggregation of collagen fibres [151] while pre-treatment with EtOH was very positive, perhaps due to removal of cholesterol and phospholipid [152]. Subsequently, other variations on the use of EtOH, including with neomycin treatment to minimise glycosaminoglycan loss as an alternative to GA [153] and with triglycidyl amine, also as an alternative to GA, have been considered [154]. Post-stabilisation treatment, with long chain alcohols, such as

1,2-octanediol, have also been proposed, as they also reduce the phospholipid content of the product [155].

Most recently a strategy has developed that is based on tissue modification by removal of α-Gal (Galα1,3-Galβ1–4GlcNAc-R) epitopes from the tissue, either *in vitro* through treatment with α-galactosidase [156] or *in vivo* through the development of galactosyltransferase knockout animals [10, 138]. This strategy arises from the α-Gal epitopes being known to be involved in the stimulation of the immune system and hence probably to degeneration of biological heart valves [157]. Development of porcine valves lacking α-Gal epitopes appears promising [10, 158] and may provide a distinct benefit in the durability of these valves when long-term data becomes available. Further to the α-Gal epitopes, other possible epitopes of concern include those with N-glycolylneuraminic acid, which could be tackled using similar strategies [159].

13.1.6 Other tissues

Several other tissues have been proposed for adapting to medical applications, but have not always obtained broad acceptance. These include, for example, sheep dermis, which was crosslinked by various methods and examined *in vivo* in a rat subcutaneous model. Of the samples that were examined, the one processed by EDC with NHS showed the most promise [160]. Another example was the proposal of using a processed and stabilised bovine ureter as a small diameter vascular prosthesis. Examination in a canine femoral artery model indicated good performance, with adequate mural integrity and haemocompatibility [161].

13.2 Acellular matrix

In recent years, after the methodology for production of ACM emerged (see chapter 8), a large number of products have become available. Thus, there are at least 45 named products from over 35 companies, with more companies becoming involved with further products in the pipeline [162]. These commercial products have been derived mainly from dermis, although a range of other tissues have been employed (see table 13.2) [162].

Table 13.2. Examples of tissues that have been used in commercial acellular matrix production (see [162]).

Human pulmonary valve	Porcine dermis
Human aortic valves	Porcine mesothelium
Human demineralised bone	Porcine small intestine submucosa
Human dermis	Porcine urinary bladder matrix
Human fascia lata	
	Equine pericardium
Bovine pericardium	
Bovine bone	Ovine forestomach submucosa
Bovine dermis/foetal dermis	

Table 13.3. Examples of clinical applications suggested for the range of commercial acellular matrix products.

Abdominal surgery	Orthopaedic surgery
– hernia repair	– tendon repair
– abdominal wall repair	– tendon augmentation
	– spine fusion procedures
Cardiovascular surgery	– extender in the spine
– valve replacements	– rotator cuff tear repair
– leaflet defect repairs	– rotator cuff reinforcement
– septal defect repairs	– bone graft extender in the spine
– pericardial repair and reconstruction	– bone void filler
– vascular repairs	– bone defect repair
– cardiac tissue repair	– lateral ankle stabilisation
– arterial repair	– osseous gaps and bone defects
– atrial and ventricle repair	
– repair of outflow tracts	Pelvic region surgery
	– pelvic organ prolapse repair
Dentistry and craniofacial surgery	– pelvic floor reconstruction
– gingival augmentation	– rectal prolapse repair
– ridge augmentation	– pelvic soft tissue repair
– intraosseous defect repair	– urethral sling implantation
– endodontal augmentation	– cystocele repair
– periodontal regeneration	– enterocele repair
– mandibular reconstruction	
– maxillofacial surgery	Plastic surgery
– duraplasty	– breast repair
– osseous defects in the cranium	– breast reconstruction
– sinus lifting or augmentation	– oncologic soft tissue defect repair
Dermal repair	
– full-/split-thickness skin regeneration	
– repair of soft tissues	
– burns/scars treatment	
– acute wound repair	
– chronic wound repair	

These various products have been proposed for a wide range of potential clinical applications (table 13.3) (see [162]).

In these instances (table 13.3), the clinical uses are predominantly based on using the ACM alone, where it supports host-based regeneration, based on the intrinsic biochemistry and topology of the ACM. However, the addition of signals, such as incorporation of growth factors, can be employed to facilitate more rapid healing. It has been broadly suggested that maintaining the intrinsic structure and topology of the ACM is of benefit in getting quality new tissue formation [163]. In some cases, the ACM is converted into an acellular gel, where these structural features are removed, but the biochemistry retained [164]. ACM has more recently been examined extensively as a potential supporting structure (scaffold) for tissue engineering, when both cells and signals (often as growth factors) are included in the system (see chapter 16). In many cases, a more limited approach is used allowing what is termed a regenerative medicine approach (see chapter 16).

Some clinical applications for ACM have been successful and are routinely used. Some examples include dermal treatments, abdominal wall repair and breast reconstruction. ACM materials seem particularly useful in repair of severe damage, and especially in heavily infected and otherwise compromised situations.

Dermal repair by ACM has now been used for a considerable time [165, 166], and a significant amount of data on its use has now accumulated [167, 168]. The products may be particularly useful in slow to heal locations, such as on the heel and other extremities [169, 170]. In many cases the material is used as supplied, but should it be revitalised? This could provide advantages if it is consistent with the time-line and location of the specific therapy [171].

Given the success of ACM in dermal applications, it has also been introduced in many other clinical indications (table 13.3), including for abdominal wall/hernia repair [143, 172, 173]. A range of products, over 10, are available from different sources and with differing processing for ventral hernia repair [174]. Cost analysis suggests that its primary use is much greater in cost than for the traditional mesh approaches [175]. However, in cases where there is significant infection due to the nature of the event, or after prior surgery [176, 177], or after organ transplant surgery [178], the use of ACM may be a preferable approach and lead to a lower whole-of-episode cost in these cases.

A further example of the use of ACM that has emerged is in breast reconstruction, especially for cases where revisions are required [179–181]. Data are now emerging that suggest the utility of ACM materials [182, 183], so increases in use may emerge. It has also been suggested that an ACM layer could be used to lower the possibility of capsular contraction around conventional, silicone shell breast implants [184], similar to the previously suggested use of a collagen sponge [185].

References

[1] Schmitt F O 1985 *Annu. Rev. Biophys. Biophys. Chem.* **14** 1–23

[2] Chvapil M, Kronenthal L and Van Winkle W 1973 *Int. Rev. Connect. Tissue Res.* **6** 1–61

[3] Christie B, Ketharanathan V and Perloff L J 1980 *J. Surg. Res.* **28** 519–32

[4] Metcalf D 1981 *Tissue Growth Factors* ed R Baserga (NewYork: Springer) pp 343–84

[5] Sparrow L G, Metcalf D, Hunkapiller M W, Hood L E and Burgess A W 1985 *Proc. Natl Acad. Sci. USA* **82** 292–96

[6] Dardik H, Miller N, Dardik A, Ibrahim I M, Sussman B, Berry S M, Wolodiger F, Kahn M and Dardik I 1988 *J. Vasc. Surg.* **7** 336–46

[7] Bakst A A, Boley S J, Joseph E and Loewe L 1958 *J. Thorac. Surg.* **35** 492–502

[8] Sauvage L R, Wood S J, Bill A H and Logan G A 1962 *J. Thorac. Cardiovasc. Surg.* **44** 67–72

[9] O'Brien M F and Clarebrough J K 1966 *Med. J. Aust.* **2** 228–30

[10] McGregor C, Byrne G, Rahmani B, Chisari E, Kyriakopoulou K and Burriesci G 2016 *Acta Biomater.* **41** 204–9

[11] Caballero A, Sulejmani F, Martin C, Pham T and Sun W 2017 *J. Mech. Behav. Biomed. Mater.* **75** 486–94
[12] Wang M, Furnary A P, Li H F and Grunkemeier G L 2017 *Ann. Thorac. Surg.* **104** 1080–87
[13] Vashi A V, White J F, McLean K M, Neethling W M L, Rhodes D I, Ramshaw J A M and Werkmeister J A 2015 *J. Biomed. Mater. Res. A* **103** 1999–2005
[14] Davis J 1910 *Johns Hopkins Med. J.* **15** 307–96
[15] Sabella N 1913 *Med. Rec. NY* **83** 478–80
[16] Stern M 1913 *J. Am. Med. Assoc.* **60** 973–74
[17] Sheikh E S, Sheikh E S and Fetterolf D E 2014 *Int. Wound J.* **11** 711–17
[18] Chattopadhyay S and Raines R T 2014 *Biopolymers* **101** 821–33
[19] Chaudhari A A, Vig K, Baganizi D R, Sahu R, Dixit S, Dennis V, Singh S R and Pillai S R 2016 *Int. J. Mol. Sci.* **17** E1974
[20] James N L, Poole-Warren L A, Schindhelm K, Milthorpe B K, Mitchell R M and Howlett C R 1991 *Biomaterials* **16** 801–9
[21] Limpert J N, Desai A R, Kumpf A L, Fallucco M A and Aridge D L 2009 *Am. J. Surg.* **198** e60–5
[22] Luo J-C, Chen W, Chen X-H, Qin T-W, Huang Y-C, Xie H-Q, Li X-Q, Qian Z-Y and Yang Z-M 2011 *Biomaterials* **32** 706–13
[23] Warwick A M, Velineni R, Smart N J and Daniels I R 2016 *Hernia* **20** 321–5
[24] Young R L, Cota M D, Zund G, Mason B A and Wheeler J M 1991 *Fertil. Steril.* **55** 624–8
[25] Tao H and Fan H 2009 *Eur. Spine. J.* **18** 1202–12
[26] He S K, Guo J H, Wang Z L, Zhang Y, Tu Y H, Wu S Z, Huang F G and Xie H Q 2017 *Mater. Sci. Eng. C Mater. Biol. Appl.* **73** 267–74
[27] Seo Y, Kim J W, Dho Y S, Chowdhury T, Kim S and Park C K 2018 *J. Clin. Neurosci.* **58** 187–91
[28] Ruffini J J, Aquavella J V and LoCascio J A 1989 *Ophthal. Surg.* **20** 21–5
[29] Westekemper H, Figueiredo F C, Siah W F, Wagner N, Steuhl K P and Meller D 2017 *Br. J. Ophthalmol.* **101** 103–7
[30] Roe S C, Milthorpe B K, True K, Rogers G J and Schindhelm K 1992 *Clin. Mater.* **9** 149–54
[31] Badylak S F, Tullius R, Kokini K, Shelbourne K D, Klootwyk T, Voytik S L, Kraine M R and Simmons C 1995 *J. Biomed. Mater. Res.* **29** 977–85
[32] Holder E J 1949 *Postgrad. Med. J.* **25** 427–33
[33] Mackenzie D 1973 *Med. Hist.* **17** 158–68
[34] Dietz U A, Kuhfuss I, Debus E S and Thiede A 2006 *World J. Surg.* **30** 141–48
[35] Lister J 1881 *Brit. Med. J.* **1881** 219–21
[36] Lister J 1867 *Brit. Med. J.* **2** 246–48
[37] Dietz U A, Kehl F, Hamelmann W and Weisser C 2007 *World J. Surg.* **31** 2275–83
[38] Morton H D 1925 *US Patent* 1558037
[39] Zoller H F and Neumeyer J L 1961 *US Patent* 2993589
[40] Hutchinson J 1913 *Proc. Roy. Soc. Med. (Surg. Sect.)* **6** 272–4
[41] Pillai C K and Sharma C P 2010 *J. Biomater. Appl.* **25** 291–366
[42] Zheng M H, Chen J, Kirilak Y, Willers C, Xu J and Wood D 2005 *J. Biomed. Mater. Res. B Appl. Biomater.* **73** 61–7
[43] Janis A D, Johnson C C, Ernst D M and Brightman A O 2012 *J. Biomater. Appl.* **26** 1013–33

[44] Poulose B K, Scholz S, Moore D E, Schmidt C R, Grogan E L, Lao O B, Nanney L, Davidson J and Holzman M D 2005 *J. Surg. Res.* **123** 262–7
[45] Mayer S, Decaluwe H, Ruol M, Manodoro S, Kramer M, Till H and Deprest J 2015 *PLoS One* **10** e0132021
[46] Zhang J, Wang G Y, Xiao Y P, Fan L Y and Wang Q 2011 *Biomaterials* **32** 7086–95
[47] Lee A J, Lee S H, Chung W H, Kim D H, Chung D J, Do S H and Kim H Y 2013 *J. Vet. Sci.* **14** 175–84
[48] Suckow M A, Wolter W R, Fecteau C, Labadie-Suckow S M and Johnson C 2012 *J. Biomater. Appl.* **27** 231–37
[49] Kropp B P, Rippy M K, Badylak S F, Adams M C, Keating M A, Rink R C and Thor K B 1996 *J. Urol.* **155** 2098–104
[50] Crapo P M and Wang Y 2010 *Acta Biomater.* **6** 2091–96
[51] Kim M S, Hong K D, Shin H W, Kim S H, Kim S H, Lee M S, Jang W Y, Khang G and Lee H B 2005 *Int. J. Biol. Macromol.* **36** 54–60
[52] Dua H S and Azuara-Blanco A 1999 *Br. J. Ophthalmol.* **83** 748–52
[53] Adinolfi M, Akle C A, McColl I, Fensom A H, Tansley L, Connolly P, Hsi B L, Faulk W P, Travers P and Bodmer W F 1982 *Nature* **295** 325–27
[54] Kubo M, Sonoda Y, Muramatsu R and Usui M 2001 *Invest. Ophthalmol. Vis. Sci.* **42** 1539–46
[55] Reverdin J L 1869 *Bull. Imp. Soc. Chir. Paris* **10** 511–15
[56] Ehrenfried A 1909 *Boston Med. Surg. J.* **161** 911–17
[57] Hauben D J, Baruchin A and Mahler A 1982 *Ann. Plast. Surg.* **9** 242–45
[58] Gibson T and Medawar P B 1943 *J. Anat.* **77** 299–309
[59] Brown J B 1937 *Surgery* **1** 558–63
[60] Chao Y C, Humphreys S and Penfield W 1940 *Br. Med. J.* **1** 517
[61] Shortt A J, Secker G A, Lomas R J, Wilshaw S P, Kearney J N, Tuft S J and Daniels J T 2009 *Biomaterials* **30** 1056–65
[62] Fujisato T, Tomihata K, Tabata Y, Iwamoto Y, Burczak K and Ikada Y 1999 *J. Biomater. Sci. Polym. Ed.* **10** 1171–81
[63] Spoerl E, Wollensak G, Reber F and Pillunat L 2004 *Ophthalmic Res.* **36** 71–7
[64] Lai J Y 2015 *Mater. Sci. Eng. C Mater. Biol. Appl.* **51** 28–36
[65] Gobinathan S, Zainol S S, Azizi S F, Iman N M, Muniandy R, Hasmad H N, Yusof M R B, Husain S, Abd Aziz H and Lokanathan Y 2018 *J. Biomater. Sci. Polym. Ed.* **29** 2051–67
[66] Shakouri-Motlagh A, O'Connor A J, Kalionis B and Heath D E 2019 *J. Biomed. Mater. Res. A* **107** 232–42
[67] Wilshaw S P, Kearney J, Fisher J and Ingham E 2008 *Tissue Eng. Part A* **14** 463–72
[68] Shimazaki J, Yang H Y and Tsubota K 1997 *Ophthalmology* **104** 2068–76
[69] Sridhar M S, Bansal A K, Sangwan V S and Rao G N 2000 *Am. J. Ophthalmol.* **130** 134–37
[70] John T 2003 *Ophthalmol. Clin. North Am.* **16** 43–65
[71] Tejwani S, Kolari R S, Sangwan V S and Rao G N 2007 *Cornea* **26** 21–6
[72] Suri K, Kosker M, Raber I M, Hammersmith K M, Nagra P K, Ayres B D, Halfpenny C P and Rapuano C J 2013 *Eye Contact Lens* **39** 341–47
[73] Joseph A, Dua H S and King A J 2001 *Br. J. Ophthalmol.* **85** 1065–69
[74] Matthews R N, Bennett J P and Faulk W P 1981 *Br. J. Plast. Surg.* **34** 76–8
[75] Bose B 1979 *Ann. R. Coll. Surg. Engl.* **61** 444–47

[76] Colocho G, Graham W P, Greene A E, Matheson D W and Lynch D 1974 *Arch. Surg.* **109** 370–73
[77] McKenna B and Summers N J 2018 *Clin. Podiatr. Med. Surg.* **35** 1–9
[78] Fishman I J, Flores F N, Scott F B, Spjut H J and Morrow B 1987 *J. Urol.* **138** 1291–94
[79] Li W, Ma G, Brazile B, Li N, Dai W, Butler J R, Claude A A, Wertheim J A, Liao J and Wang B 2015 *Langmuir* **31** 8642–53
[80] Francisco J C, Correa Cunha R, Cardoso M A, Baggio Simeoni R, Mogharbel B F, Picharski G L, Silva Moreira Dziedzic D, Guarita-Souza L C and Carvalho K A 2016 *Transplant. Proc.* **48** 2845–49
[81] Jorge L F, Francisco J C, Bergonse N, Baena C, Carvalho K A T, Abdelwahid E, Neto J R F, Moreira L F P and Guarita-Souza L C 2018 *J. Tissue Eng. Regen. Med.* **12** e1525–30
[82] Samways D W 1898 *Lancet* **151** 927
[83] Templeton J Y and Gibbon J H 1949 *Ann. Surg.* **129** 161–76
[84] Stokes T H and Gibbon J H 1950 *Surg. Gynec. Obst.* **91** 138–56
[85] Dennis C, Spreng D S, Nelson G E, Karlson K E, Nelson R M, Thomas J V, Eder W P and Varco R L 1951 *Ann. Surg.* **134** 709–21
[86] Romaine-Davis A 1992 *John Gibbon and His Heart-Lung Machine* (Philadelphia, PA: University of Pennsylvania Press)
[87] Stoney W S 2009 *Circulation* **119** 2844–53
[88] Jauhar S 2018 *Heart: A history* (New York: Farrar, Straus and Giroux)
[89] Bakst A A and Loewe L 1957 *Surg. Forum* **8** 383–86
[90] Manji R A, Lee W and Cooper D K C 2015 *Int. J. Surg.* **23** 280–84
[91] Isaacs A J, Shuhaiber J, Salemi A, Isom O W and Sedrakyan A 2015 *J. Thorac. Cardiovasc. Surg.* **149** e3
[92] Vesely I 2003 *Cardiovasc. Pathol.* **12** 277–86
[93] Reis R L, Hancock W D, Yarbrough J W, Glancy D L and Morrow A G 1971 *J. Thorac. Cardiovasc. Surg.* **62** 683–89
[94] Goissis G, Giglioti Ade F and Braile D M 2011 *Artif. Organs* **35** 484–89
[95] Umashankar P R, Sabareeswaran A and Shenoy S J 2017 *J. Biomed. Mater. Res. B Appl. Biomater.* **105** 2145–52
[96] Tang Z and Yue Y 1995 *ASAIO J.* **41** 72–8
[97] Tam H, Zhang W, Feaver K R, Parchment N, Sacks M S and Vyavahare N 2015 *Biomaterials* **66** 83–91
[98] Hiester E D and Sacks M S 1998 *J. Biomed. Mater. Res.* **39** 215–21
[99] Mao J, Wang Y, Philippe E, Cianciulli T, Vesely I, How D, Bourget J M, Germain L, Zhang Z and Guidoin R 2017 *Morphologie* **101** 77–87
[100] Fentie I H, Allen D J, Schenck M H and Didio L J 1986 *J. Submicrosc. Cytol.* **18** 53–65
[101] Rassoli A, Fatouraee N and Guidoin R 2018 *Artif. Organs* **42** 630–39
[102] Chen S, Xu L, Liu Y, Li Q, Wang D, Wang X and Liu T 2013 *Artif. Organs* **37** 248–55
[103] Aguiari P, Fiorese M, Iop L, Gerosa G and Bagno A 2016 *Interact. Cardiovasc. Thorac. Surg.* **22** 72–84
[104] Vesely I 2010 *J. Heart Valve Dis.* **19** 543–58
[105] Emmert M Y and Hoerstrup S P 2017 *Eur. Heart J.* **38** 2710–13
[106] Guidoin R, Zegdi R, Lin J, Mao J, Rochette-Drouin O, How D, Guan X, Bruneval P and Wang L *et al* 2016 *Morphologie* **100** 234–44
[107] Arora S, Ramm C J, Misenheimer J A and Vavalle J P 2017 *Cardiovasc. Diagn. Ther.* **7** 1–3

[108] Arora S, Ramm C J, Strassle P D, Vaidya S R, Caranasos T G and Vavalle J P 2017 *Am. J. Cardiol.* **120** 331–36

[109] Spinali K L and Schmuck E G 2018 *Adv. Exp. Med. Biol.* **1098** 115–30

[110] Kameli S M, Khorramirouz R, Eftekharzadeh S, Fendereski K, Daryabari S S, Tavangar S M and Kajbafzadeh A M 2018 *J. Biomed. Mater. Res. A* **106** 2670–78

[111] Hirata H H, Munhoz M A, Plepis A M, Martins V C, Santos G R, Galdeano E A and Cunha M R 2015 *Injury* **46** 1215–22

[112] Guerette N L, Peterson T V, Aguirre O A, Vandrie D M, Biller D H and Davila G W 2009 *Obstet. Gynecol.* **114** 59–65

[113] Milani R, Frigerio M, Beretta C, Cola A, Spelzini F and Manodoro S 2017 *Minerva Ginecol.* **69** 311–13

[114] Angell W W and Angell J D 1980 *Prog. Cardiovasc. Dis.* **23** 141–66

[115] Angell W W, Iben A B and Shumway N E 1968 *Arch. Surg.* **97** 826–30

[116] Buch W S, Kosek J C, Angell W W and Shumway S E 1970 *J. Thorac. Cardiovasc. Surg.* **60** 673–82

[117] Neethling W M L, van Riet S, van den Heever J J, Hough J and Hodge A J 2001 *J. Card. Surg.* **16** 392–99

[118] Reineke D, Gisler F, Englberger L and Carrel T 2016 *Expert Rev. Cardiovasc. Ther.* **14** 423–30

[119] Bassetto S, Menardi A C, Alves Junior L, Rodrigues A J and Évora P R 2011 *Rev. Bras. Cir. Cardiovasc.* **26** 653–57

[120] Grunkemeier G L, Furnary A P, Wu Y, Wang L and Starr A 2012 *J. Thorac. Cardiovasc. Surg.* **144** 1381–86

[121] Johnston D R, Soltesz E G, Vakil N, Rajeswaran J, Roselli E E, Sabik J F, Smedira N G, Svensson L G, Lytle B W and Blackstone E H 2015 *Ann. Thorac. Surg.* **99** 1239–47

[122] Leirner A A, Tattini V and Pitombo R N 2009 *Artif. Organs* **33** 221–29

[123] Wang S, Goecke T, Meixner C, Haverich A, Hilfiker A and Wolkers W F 2012 *Tissue Eng. Part C Methods* **18** 517–25

[124] Rudis-Jicinsky J 1902 *Tr. Am. Roentgen Ray Soc.* **3** 57–65

[125] Grayson C E 1948 *Calif. Med.* **68** 121–25

[126] Horowitz M S, Goodman D J, Hancock E W and Popp R L 1976 *J. Thorac. Cardiovasc. Surg.* **71** 450–57

[127] Lamberti J J, Wainer B H, Fisher K A, Karunaratne H B and Al-Sadir J 1979 *Ann. Thorac. Surg.* **28** 28–32

[128] Schoen F J and Levy R J 2009 *Biomaterials* **30** 4445–46

[129] Sacks M S and Schoen F J 2002 *J. Biomed. Mater. Res.* **62** 359–71

[130] Hutcheson J D, Goettsch C, Rogers M A and Aikawa E 2015 *Semin. Cell Dev. Biol.* **46** 68–77

[131] Hutcheson J D, Blaser M C and Aikawa E 2017 *Circ. Res.* **120** 270–73

[132] Grimm M, Eybl E, Grabenwöger M, Spreitzer H, Jäger W, Grimm G, Böck P, Müller M M and Wolner E 1992 *Surgery* **111** 74–8

[133] Neethling W M L, Glancy R and Hodge A J 2010 *J. Heart Valve Dis.* **19** 778–85

[134] Pereira C A, Lee J M and Haberer S A 1990 *J. Biomed. Mater. Res.* **24** 345–61

[135] Moore M A and Phillips R E 1997 *J. Heart Valve Dis.* **6** 307–15

[136] Leong J, Munnelly A, Liberio B, Cochrane L and Vyavahare N 2013 *J. Biomater. Appl.* **27** 948–60

[137] McClurg W M, Lawford P V, Hughes H and Rogers S 1996 *J. Heart Valve Dis.* **5** 343–47
[138] Lim H G, Choi S Y, Yoon E J, Kim S H and Kim Y J 2013 *Tissue Eng. Part A* **19** 2339–48
[139] Mercuri J J, Lovekamp J J, Simionescu D T and Vyavahare N R 2007 *Biomaterials* **28** 496–503
[140] Tripi D R and Vyavahare N R 2014 *J. Biomater. Appl.* **28** 757–66
[141] Courtman D W, Pereira C A, Omar S, Langdon S E, Lee J M and Wilson G J 1995 *J. Biomed. Mater. Res.* **29** 1507–16
[142] Grimm M, Grabenwöger M, Eybl E, Moritz A, Böck P, Müller M M and Wolner E 1992 *J. Card. Surg.* **7** 58–64
[143] Buinewicz B and Rosen B 2004 *Ann. Plast. Surg.* **52** 188–94
[144] Christian A J, Alferiev I S, Connolly J M, Ischiropoulos H and Levy R J 2015 *J. Biomed. Mater. Res. A* **103** 2441–48
[145] Christian A J, Lin H, Alferiev I S, Connolly J M, Ferrari G, Hazen S L, Ischiropoulos H and Levy R J 2014 *Biomaterials* **35** 2097–102
[146] Lee S, Levy R J, Christian A J, Hazen S L, Frick N E, Lai E K, Grau J B, Bavaria J E and Ferrari G 2017 *J. Am. Heart Assoc.* **6** e005648
[147] Ellsmere J C, Khanna R A and Lee J M 1999 *Biomaterials* **20** 1143–150
[148] Langdon S E, Chernecky R, Pereira C A, Abdulla D and Lee J M 1999 *Biomaterials* **20** 137–53
[149] Golomb G, Dixon M, Smith M S, Schoen F J and Levy R J 1986 *ASAIO Trans.* **32** 587–90
[150] Vyavahare N R, Chen W, Joshi R R, Lee C H, Hirsch D, Levy J, Schoen F J and Levy R J 1997 *Cardiovasc. Pathol.* **6** 219–29
[151] Forti F L, Goissis G and Plepis A M 2006 *J. Biomater. Appl.* **20** 267–85
[152] Vyavahare N, Hirsch D, Lerner E, Baskin J Z, Schoen F J, Bianco R, Kruth H S, Zand R and Levy R J 1997 *Circulation* **95** 479–88
[153] Raghavan D, Shah S R and Vyavahare N R 2010 *J. Biomed. Mater. Res. B Appl. Biomater.* **92** 168–77
[154] Connolly J M, Bakay M A, Alferiev I S, Gorman R C, Gorman J H, Kruth H S, Ashworth P E, Kutty J K and Schoen F J *et al* 2011 *Ann. Thorac. Surg.* **92** 858–65
[155] Pathak C P, Adams A K, Simpson T, Phillips R E and Moore M A 2004 *J. Biomed. Mater. Res. A* **69** 140–44
[156] Nam J, Choi S Y, Sung S C, Lim H G, Park S S, Kim S H and Kim Y J 2012 *Korean J. Thorac. Cardiovasc. Surg.* **45** 380–89
[157] McGregor C G, Kogelberg H, Vlasin M and Byrne G W 2013 *J. Heart Valve Dis.* **22** 383–90
[158] Naso F, Stefanelli U, Buratto E, Lazzari G, Perota A, Galli C and Gandaglia A 2017 *Tissue Eng. Part A* **23** 1181–95
[159] Lee W, Long C, Ramsoondar J, Ayares D, Cooper D K, Manji R A and Hara H 2016 *Xenotransplantation* **23** 370–80
[160] van Wachem P B, van Luyn M J, Olde Damink L H, Dijkstra P J, Feijen J and Nieuwenhuis P 1994 *J. Biomed. Mater. Res.* **28** 353–63
[161] Burns P, Edwards G A, Roberts G R, Ketharanathan V and Hatami-Monazah H 1989 *ASAIO Trans.* **35** 214–18
[162] Parmaksiz M, Dogan A, Odabas S, Elçin A E and Elçin Y M 2016 *Biomed. Mater.* **11** 022003

[163] Livesey S A, del Campo A A, Nag A, Nichols K B and Coleman C 1994 *US Patent* 5336616
[164] Hernandez M J and Christman K L 2017 *JACC Basic Transl. Sci.* **2** 212–26
[165] Wainwright D J 1995 *Burns* **21** 243–48
[166] Ruszczak Z 2003 *Adv. Drug Deliv. Rev.* **55** 1595–611
[167] Kirsner R S, Bohn G, Driver V R, Mills J L, Nanney L B, Williams M L and Wu S C 2015 *Int. Wound J.* **12** 646–54
[168] Rosenbaum A J, Banerjee S, Rezak K M and Uhl R L 2018 *J. Am. Acad. Orthop. Surg.* **26** 833–43
[169] Guo X, Mu D and Gao F 2017 *Int. J. Surg.* **40** 1–7
[170] Khoo R and Jansen S 2018 *Int. Wound J.* **15** 205–11
[171] Eweida A M and Marei M K 2015 *Biomed. Res. Int.* **2015** 839694
[172] Parker D M, Armstrong P J, Frizzi J D and North J H 2006 *Curr. Surg.* **63** 255–58
[173] Silverman R P 2011 *Aesthet. Surg. J.* **31** 24S–9S
[174] Brown P 2009 *AORN J.* **90** 513–20
[175] Fischer J P, Basta M N, Krishnan N M, Wink J D and Kovach S J 2016 *Plast. Reconstr. Surg.* **137** 647–59
[176] Albino F P, Patel K M, Nahabedian M Y, Attinger C E and Bhanot P 2015 *Ann. Plast. Surg.* **75** 629–33
[177] Gowda A U, Chang S M, Chopra K, Matthews J A, Sabino J, Stromberg J A, Zahiri H R, Pinczewski J and Holton L H *et al* 2016 *Int. Wound J.* **13** 972–76
[178] Gowda A U, McNichols C H, Asokan I, Matthews J A, Buckingham E B, Sabino J, Maddox J S, Slezak S, Rasko Y and Singh D P 2016 *Ann. Plast. Surg.* **77** 674–77
[179] Slavin S A and Lin S J 2012 *Plast. Reconstr. Surg.* **130** 70S–85S
[180] Maxwell G P and Gabriel A 2014 *Plast. Reconstr. Surg.* **134** 932–38
[181] Salibian A A, Frey J D, Choi M and Karp N S 2016 *Plast. Reconstr. Surg. Glob. Open* **4** e1139
[182] Sobti N and Liao E C 2016 *Plast. Reconstr. Surg.* **138** 959–67
[183] Sobti N, Ji E, Brown R L, Cetrulo C L, Colwell A S, Winograd J M, Austen W G and Liao E C 2018 *Plast. Reconstr. Surg.* **141** 541–49
[184] Wan D and Rohrich R J 2016 *Plast. Reconstr. Surg.* **137** 826–41
[185] Wallace D G, Rosenblatt J and Ksander G A 1992 *J. Biomed. Mater. Res.* **26** 1517–34

IOP Publishing

Biophysical and Chemical Properties of Collagen: Biomedical Applications

John A M Ramshaw and Veronica Glattauer

Chapter 14

Applications for purified collagen

Purified collagen, either in the form of comminuted fibres or as solubilised collagen, has been used for many decades for the fabrication of collagen-based medical devices. These have proved successful in clinical applications and have paved the way toward new, improved device formats. This chapter briefly discusses some examples of clinical and emerging uses for purified collagen in different applications.

The previous chapter, chapter 13, discussed the difficulties in categorising collagen-based devices as of one type or another with a wide range of new materials containing additional components beyond collagen. In this chapter we explore applications where collagen has been significantly processed, to the extent where its macroscopic structure has been lost. Frequently, these products are based on collagen that has been made soluble and then reformed, but also includes products where small segments of individual fibrils have been isolated and reformed into a structure.

These purified collagen products have been used for a number of clinical indications (table 14.1) that are generally distinct to those where tissue-based devices have proved valuable (table 13.1). In a few cases, however, there is overlap between these different formats. One of the main advantages of the purified collagen materials is that they typically have a low immunogenicity, arising from the loss of telopeptides. Also, these products can have controlled turnover, with additional crosslinking slowing the rate of resorption or dispersion of an implant. Generally, the cell-matrix interactions are broadly retained, unless high levels of crosslinking start to interfere with the binding sites. Depending on the fabrication approach, porosity can be controlled, allowing these materials to be made sufficiently porous to allow cell migration, or alternatively with reduced porosity that prevents cell migration.

Some examples of the types of clinical indications where these products have been proposed or have reached clinical use are shown (table 14.1). However, analogous to the previous chapter, the discussion of a limited number of the applications follows

doi:10.1088/978-0-7503-2096-2ch14

Table 14.1. Examples of applications of purified collagen-based biomaterials in a range of medical areas.

Medical areas and applications		Pioneering examples	Recent examples and reviews
Cardiovascular	Myocardial infarction		[1, 2]
Dentistry	Periodontal attachment	[3, 4]	[5, 6]
	Alveolar ridge augmentation	[7]	[8]
Dermatology	Tissue augmentation	[9]	[10, 11]
	Tissue replacement	[12, 13]	[14]
	Dressings	[15, 16]	[17, 18]
Esophageal/laryngeal surgery	Augmentation	[19]	[20]
General surgery	Hernia repair	[21]	[22]
	Adhesion barrier	[23]	[24]
	Haemostasis	[25]	[26]
	Tissue adhesive	[27]	[28]
Neurosurgery	Nerve repair	[29, 30]	[31, 32]
	Spinal cord treatment		[33]
	Brain treatment		[34]
	Dural repair		[35, 36]
Ophthalmology	Corneal implant	[37, 38]	[39]
	Vitreous replacement	[40]	[41]
Orthopaedics	Bone repair	[42, 43]	[44, 45]
	Articular cartilage repair	[46]	[47, 48]
	Ligament reconstruction	[49]	[50]
	Tendon repair		[50, 51]
	Meniscus repair	[52, 53]	[54, 55]
Otology	Tympanic membrane repair	[56, 57]	[58]
Urology	Ureter replacement	[59]	[60]
	Renal repair	[61]	
	Dialysis membrane	[62]	
	Urinary incontinence	[63]	

the type of material rather than a clinical use. In some cases, products based on recombinant collagens (chapter 9) are emerging where the recombinant protein substitutes for a soluble collagen derived from tissue by conventional methods (chapter 8). Recombinant collagen can provide a uniform, defined material, which is biochemically pure, and virus and disease free.

In most cases these purified collagen materials (table 14.1) are made from type I collagen. However, if the product has been obtained from dermal tissue, it may also contain some type III collagen, perhaps 5% to 10%, unless the collagen has been further purified by NaCl fractionation (see chapter 8). The performance of the materials may depend on whether any denaturation of the collagen has occurred during fabrication and sterilisation as denatured chains can change the phenotype and binding of fibroblasts [64]. This is possibly due to the emergence of cryptic binding sites, such as Arg-Gly-Asp, which would normally be unavailable for binding in the constrained triple helical structure of native collagens. In some cases, the use of other collagen types has been proposed, including type II collagen and type VI collagen for articular cartilage repair [65–67]. Also, type IV collagen has been identified as forming a gel [68, 69] and could be fabricated as a film suitable for tympanic membrane repair [70]. The species of origin is generally bovine, but a range of other species have been proposed with marine collagens attracting significant attention [71, 72]. Some products using recombinant human collagens have been proposed [73], while others are now marketed [74]. Bacterial collagen products are also likely to appear [75, 76], particularly those that are engineered to include novel design principles [77].

14.1 Gels and hydrogels

A dilute solution, 2 mg ml^{-1}, of soluble collagen along with cell nutrients was used to encourage healing of periodontal repair in dogs [3]. More frequently, however, collagen solutions have been used in a concentrated form, e.g., >35 mg ml^{-1}, where the material forms a viscous gel. (The distinction, as used in this text, between gels (un-crosslinked) and hydrogels (crosslinked) was discussed in chapter 11.) These collagen gels are able to be dispensed through a fine gauge needle. If the solution is in physiological saline there is a strong tendency for the collagen to self-assemble into a fibrous gel, such as can be obtained on heating to 37 °C [78]. It is also possible to deliver this material through a fine gauge needle, as the sheer thinning that occurs during delivery allows steady flow to be maintained [79, 80]. If the collagen solution is kept in physiological phosphate, of equivalent ionic strength to physiological saline, then the gel is retained and the likelihood of fibre formation is reduced.

14.1.1 Tissue augmentation

In the late 1970s, collagen gels became a significant product for dermal augmentation [9, 81] and were the dominant product into the 1990s [82]. Previously, synthetic polymer materials had been discussed but most showed significant foreign body reactions and only a few, non-resorbable materials existed [83]. Numerous studies on the performance of these implants lead to a significant increase in the understanding of the response to a collagen implant, and especially of the immunological implications (see chapter 7) (see [84, 85]). Immunological issues have proved the main concern with the use of collagen fillers. Pre-treatment skin tests revealed that around 3% of potential patients displayed an adverse response [86, 87] and subsequently did not proceed to treatment. However, those patients that did proceed to implantation then led to a further 1% to 2% with hypersensitivity symptoms [86, 88], despite the previous skin test. Once implanted, the collagen may lead to a mild inflammatory response with potentially some deposition of new connective tissue. Fibroblast invasion may occur within a few weeks, but is slower than found in animal models. Over time there may be deposits of new collagenous tissue or else the implant disperses and is lost [86, 88].

A range of variations of collagen fillers have emerged. The initial change was to move from a clear gel to a fibril gel, followed by introduction of low levels of glutaraldehyde crosslinking, with the aim of extending the lifetime of the implant before resorption or dispersion [89–91]. Subsequently, a product concept emerged that combined the temporary collagen with permanent polymethylmethacrylate (PMMA) beads [92]. The PMMA beads, at around 20% of the mixture, needed to be between 30–40 microns to allow injection through a 27 gauge needle [92]. Further research to eliminate any electrostatic charge on the PMMA particle was needed to remove the granulomatous reactions observed after injection of previous formulations [93]. Subsequent studies showed that the collagen was resorbed after about one month and was replaced by new host tissue by three months giving extended management of wrinkles [94]. The concept works well over extended periods, with the granulomas found in about 1.7% of patients, and these could mostly be corrected by treatment [95]. More recently, a 'rapid polymerising collagen' material has been developed using a porcine collagen gel that contains EDTA to prevent fibril formation prior to injection whence the change of ionic strength leads to fibril formation [96].

The demand for dermal fillers has grown significantly in recent years [97], driven by aesthetic concerns and a desire in some to try and minimise the inevitable natural aging of the face in a simple manner without surgical intervention [98]. A wide range of temporary, degradable materials such as collagen have been joined by semi-permanent and permanent products [97, 99]. Some materials, such as silicone are banned in Europe and the USA, but may unfortunately be used by some and also used elsewhere [100]. Hyaluronic acid (HyA) emerged significantly as the main, resorbable competitor to collagen in the late 1990s [82], with the material becoming readily available from bacterial rather than tissue sources. This alternative, HyA, was allowed to be used without the need for pre-testing for sensitivities, a commercial advantage, but it too has shown some adverse events [101–104], sometimes after a reasonable delay [105] and dependent on the extent of crosslinking that had been included [106]. Other commercial considerations may also play a role in product selection [82]. Certainly, cosmetic augmentation is not without its risks. One way to minimise risk, in common with all medical treatments, is for the procedure to be performed by an experienced, fully qualified medical practitioner in an appropriate facility with full infection control, rather than by someone of unknown capability or qualifications found through medical tourism or the self-styled beauty therapist down the back lane, third door on the left! Only then will registered materials of known provenance be used with proper infection control and with close attention to the correct anatomical location point. Even then, images on the internet suggest that not all outcomes were as expected!

While dermal augmentation was the initial and most common use for collagen-based fillers, other clinical areas have also been of interest. These include augmentation associated with laryngoplasty [107] and particularly sphincter augmentation to restore adequate function. In this aspect, treatment of stress urinary incontinence

Figure 14.1. A stabilised collagen gel as a convex lens formed by methylene blue photo-crosslinking of a collagen solution.

has received most attention, with collagen being one of the various bulking agents that have been examined [63, 108, 109]. Data indicated that collagen injections were safe and well tolerated [108], and that in selected patients, the procedure may be a suitable alternative to open surgical procedures [110].

14.1.2 Dermal repair

A particularly successful product, starting with a thermally induced collagen gel, has been the production of a dressing for severe wound and burn treatment. This product, is based on the observation that fibroblasts can cause the consolidation of a weak collagen gel and that this process could provide the basis for a 'skin equivalent' [13, 111]. A disc of native collagen gel seeded with fibroblasts in a tissue culture well will show contraction in diameter and also in depth [112]. For wound dressing it is preferable that the diameter change is minimised and contraction occurs in the depth, and anchoring technologies have been developed to enable this [113, 114]. Other studies had shown that keratinocytes could be cultured on certain synthetic materials [115] and this technology was equally applicable to the initial skin equivalent providing a dual layer construct with a lower dermal equivalent and an upper epidermal equivalent [114]. This type of material may also be considered as a tissue engineering product (see chapter 16).

14.1.3 Ophthalmic uses

The ability to form collagen solutions and clear gels into stable transparent materials (figure 14.1) has led to their proposed applications in ophthalmology. An initial proposal was to make a collagen-based vitreous replacement material [38, 40]. Enzyme solubilised and purified calf collagen was prepared at a concentration of 0.5%, resulting in a material of refractive index close to that of the human vitreous, and was crosslinked by UV irradiation in a nitrogen atmosphere. This gave a transparent, firm gel that could nevertheless still be delivered through a needle while retaining the gel format [40]. Animal studies showed that both collagen solutions and gels were well tolerated, retained viscosity, and were retained in the eye for sufficient time to allow chorioretinal adhesion [38]. It was suggested that further studies were need to optimise the materials' refractive index match to enable greater transparency [38]. Despite this progress, ideal performance is not readily achieved, although studies of synthetic polymer systems have defined the key properties that should be achieved [116, 117]. It is not clear that despite numerous materials being examined as alternatives that an ideal material has been found [118].

The use of collagen in corneal replacements has made significant progress [39], through production of suitable gel material as well as more recently through tissue engineering approaches (see chapter 16) (see [119]). A particularly interesting area of development has been through the use of recombinant collagens. Corneal replacements are a particularly suitable application for recombinant collagens, where the high uniformity and purity of the product is valuable. Initially, recombinant type I and type III collagens were examined. Transparent, firm gels could be obtained through crosslinking 13.7% collagen with 1-ethyl-3-(3-dimethylaminopropyl)carbodiimide (EDC) [120]. Recombinant type III collagen has emerged as the preferred material, as it provides better mechanical properties [120] with well characterised physical and optical properties [121]. Clinical trials have confirmed the promise of this technology showing that the cell-free devices promoted new corneal tissue and nerve formation by the host without any rejection occurring. This suggests an excellent path to supplement the supply of donor grafts in meeting the significant unmet demand for corneal tissue grafts [73].

14.1.4 Other uses of gels

A variety of other applications have been described for collagen gels, including use to form tissue adhesives. This has involved chemically modifying the collagen to activate it, so that when dissolved if self-reacts and solidifies [122]. More recently, another variation has described the formulation of an adhesive and filler, as an alternative to cyanoacrylate glue in ophthalmic surgery (e.g., [28]).

Most recently, modified collagen gels and dispersions have emerged as potential 'bioinks' and technology adapted to enable delivery [123–128]. Various approaches have been considered to stabilise the bioink during and/or after delivery, including plant-based reagents and methacrylation (see chapter 12). Methacrylated collagen is now commercially available and so could form the basis of future product developments, as this, and other formulations can be adapted to allow simultaneous inclusion of cells during delivery.

14.1.5 Hydrogels

There has been a growing interest in potential applications of collagen hydrogels, where the structure has been stabilised by crosslinking the collagen. These have been of most interest for developing tissue engineering and regenerative medicine products (see chapter 16). For chitosan, the performance of cells in hydrogels differs from those in sponges and other formats [129] and the same may be expected for collagens. One group of proposed, injectable hydrogels are those of acellular matrix derived from a wide range of tissue types [130–134]. In these materials, other non-collagenous macromolecules are retained during the processing, and these vary between tissues sources and so can also affect the performance of different cell types.

There are a range of potential applications, but the regeneration of cardiac tissue post-myocardial infarction has been a significant interest [130, 135–137].

Collagen-based hydrogels have also been proposed for cartilage repair, for example, using recombinant human type II collagen gels with chondrocytes [138]. Recombinant bacterial collagen has also been examined using peptide modified systems that enhance cell and glycosaminoglycan binding and which have been tuned to resorb through crosslink cleavage by endogenous proteases [139, 140].

14.2 Foams and sponges

The option of using collagen-based sponges, initially made from comminuted tissue, in clinical applications, emerged in the 1940s (chapter 11). These early materials addressed wound repair and haemostasis, and these still remain among the principle application, while other uses have also subsequently been described.

14.2.1 Dermal wound repair

Collagen in different formats has been used with significant success for dermal treatments, including a gel-based system [111] and also collagen-based sponges [12]. Early studies with sponges showed that pore size was a key issue, along with matrix stiffness, in providing for a good response. The porosity enables efficient transport of oxygen and nutrients and removal of cell by-products, while both parameters are important for cell behaviour and migration [141] (see chapter 11). It was shown that the pore size of the sponge should be between 20 and 125 μm for effective healing [142]. This sponge-based product design has led to commercial materials that have been used extensively in clinical practice, especially in burns treatments. Direct application to the injury site can give a slow response due to the compromised underlying tissue. An alternative strategy is to use a collagen sponge to repair the skin donor site when the patient's skin is taken to repair the principle injury site [143].

14.2.2 Haemostats

Sponges for topical haemostatic applications are produced by a range of manufacturers [144], and are particularly suitable for large surface areas, such as in liver surgery. Although intrinsically simple, each manufacturer will incorporate their own technology to improve function, typically by including some other components to form a composite (e.g., [145]). Others also look to using different species as the collagen source [146]. The alternative collagen-based format comprises 'microcrystalline' fleece material, derived from dispersed collagen fibrils from bovine hide [147, 148] (see chapter 8). Overall, the immunological properties of these materials have varied, probably due to residual non-collagen tissue immunogens [149]. In some circumstances, gelatin-based haemostats, which are widely used, may provide a better-quality clot than a collagen-based device [144, 150], perhaps due to the emergence of otherwise cryptic cell binding sites on denaturation of the collagen. A range of materials, including non-collagenous materials, are available as haemostats providing choice depending on the particular clinical need [26].

14.2.3 Orthopaedic applications

Meniscal cartilage

A widely used collagen sponge-like device is the one used for meniscal repair [52, 53], where there is high demand. It is estimated that in the US alone, meniscal injuries result in almost one million surgical procedures each year [55]. The meniscus is present in the knee to provide a degree of protection to the articular cartilage. Historically, meniscal damage was frequently treated through a total meniscectomy, which in turn could lead to the premature onset of osteoarthritis. Hence, the current trend has been to more conservative options and a meniscal implant can assist with this treatment approach. A partial meniscectomy that retains the peripheral meniscal rim allows use of the collagen implant [151].

A meniscal augmentation device is usually made from comminuted bovine tendon collagen that is formed into a meniscus-like shape, ready for use. The collagen is only partially or slowly resorbed, while the porosity allows ingrowth of the patient's own fibro-chondrocytes that subsequently augment the device [152]. Initial reports on clinical effectiveness were for 5- to 8-year implants [153, 154]. Although the number of patients was small, these data indicated that the implants gave satisfactory results, that the implant retained its structure and that the deterioration of the knee was less in the patients treated with a device. Subsequently, good clinical outcomes were reported from follow-ups at times over 10 years, with integration of the implant and incorporation of new host tissue reported [155, 156]. Recent reports and reviews have indicated that implantation of the resorbable collagen devices leads to a satisfactory clinical outcome for most patients, and provides a better outcome than a partial meniscectomy alone [157, 158].

Articular cartilage

Collagen has been used in different strategies for repair of articular cartilage, normally associated with osteoarthritis. A collagen sheet, produced from dermal collagen and containing some type III collagen as well as the predominant type I collagen, has been used without modification in combination with the microfracture technique to assist in healing [159]. The addition of the collagen matrix may protect and stabilise the blood clot that is formed prior to chondrocytes forming new tissue. It has been suggested, however, based on *in vitro* experiments, that the less readily available type II collagen, the collagen found in cartilage, may be more desirable for cell growth, showing better cell morphology and good production by cells of type I collagen and proteoglycans [160].

Other strategies have included a cellular component, where the collagen sponge acts as a delivery system [161], while in some cases the collagen is used as a supporting structure in a tissue engineering approach (see chapter 16), where signals such as mechanical forces [162] and growth factors [163] are added and a bioreactor is used to form a tissue-like implant [164, 165].

A cell-based technique, autologous chondrocyte implantation (also called transplantation) (ACI) was introduced for cartilage repair [166–168]. This approach was to expand the patient's own chondrocytes, obtained from a biopsy, over one month and then implant the cells into the knee defect in open-knee surgery, using a piece of periosteum augmented by fibrin glue for cellular retention at the defect site. One disadvantage of ACI is that under normal culture conditions in flasks, the chondrocytes dedifferentiate to give fibroblast-like cells (see figure 16.3). The method, however, can be effective in reducing pain for the patient, even though the new tissue formed is fibrocartilage rather than hyaline cartilage. A range of variations have been described, including systems that mostly retain the chondrocytic nature of the cells [169] and other systems that use a collagen support [170, 171]. This regenerative medicine (see chapter 16) approach uses the collagen support to deliver the cells to the defect. It has also been examined for lesions in other cartilages, including patellar cartilage [172].

Bone
Collagen sponge can nominally be used as a basic material for bone repair, but in practice it is used in composite materials (see chapter 18) or as a coating on devices to support formation of a strong bone-material interface (see chapter 17). Some composites for bone repair include hydroxyapatite (HAp) which can be incorporated by many different strategies, and can be osteoinductive (e.g., [173, 174]). Alternatively, or in addition, the collagen sponge can be augmented by one or more growth factor to enhance cellular activity [175–177].

14.2.4 Other applications

Collagen sponge/foam plugs have been suggested for various indications, such as for patients undergoing transthoracic needle biopsy, allowing pleural sealing to minimise the risk of pneumothorax [178]. Collagen plugs have also been proposed for sealing the femoral artery after it has been used for cardiac access [179, 180].

Collagen sponge has also been suggested for use in urinary system needs, including ureter and urethra repair, where tubes have been fabricated, with and without additional supporting polymer mesh [59, 60] and have illustrated the difficulties in making ideal materials.

It is also possible to make collagen sponges using recombinant human collagens. The costs of these materials at present may argue against use for large volume devices, but for smaller devices the benefits may be worth considering. For example, recombinant collagens have been used in gingival repair [181] and may be useful in other dental or ophthalmic situations.

14.3 Films and membranes

14.3.1 Periodontal treatment

Periodontal disease is characterised by a major reduction in connective tissue attachment and of the supporting bone, frequently leading to loss of teeth if not controlled. The bone tissue lost does not readily regenerate, in part due to migrating epithelium, so a therapy is needed to assist with regeneration [4, 182]. Hence, a therapy that interfered with the epithelial migration emerged. Initial studies examined a hydrophobic membrane made, for example, from cellulose acetate or polytetrafluorethylene [183, 184]. This approach became known as 'guided tissue regeneration'. Although the approach worked, the disadvantage was that the membrane required removal post-regeneration, giving an opportunity for infection to enter. A resorbable collagen membrane that did not require removal was proposed [4]. This approach used a membrane with very small pore sizes, much less than 1 μm, so that it was not readily infiltrated and penetrated by cells (figure 11.5). It demonstrated that in proximal and bifurcation lesions, substantial amounts of new collagen and bone were formed associated with new connective tissue attachment in established defects in a dog model [4]. Subsequently, reviews of the literature supported the use of collagen membrane for periodontal bone defects [5, 185] and a range of commercial materials are now available for this procedure.

14.3.2 Adhesion control

The principle, as above, of having a membrane that prevents cell migration and integrations of adjacent tissues provide the basis for proposals to employ resorbable collagen films or membranes to prevent tissue adhesions forming after surgery [23]. This opportunity is applicable to various procedures, including pelvic and abdominal surgery applications [24]. An area where post-operative adhesion is a significant concern is in cardiac surgery where they can significantly interfere with re-sternotomy procedures [186, 187]. Together, data from these and other studies indicated the clinical value of the opportunity. It may be of particular value for paediatric cardiac surgery, where more frequent re-interventions may be necessary [188].

14.3.3 Nerve repair

Peripheral nerve repair is a significant clinical issue. If the ends of a severed nerve cannot be re-joined, typically when the gap is >20 mm, or a nerve graft procedure cannot be done, then strategies based on medical materials become important. Collagen devices, normally as tubes, although tapes are possible, have been made as nerve guidance channels. Generally, the tube is used with other materials and/or cells included within the tube to promote regeneration, and the choice of these additional components impacts on the effectiveness of the approach [189]. In this approach, the tube seeks to isolate the repairing nerve environment from interference by external cells, such as fibroblasts, while still allowing diffusion of nutrients and other factors to assist the regeneration process. Controlled biodegradation of the collagen during the repair prevents strangulation of the new tissue [190]. Various collagen-based devices have subsequently been approved for use [191], and a range of clinical studies, generally with gaps <20 mm, showed mostly successful clinical outcomes [191]. Other studies suggested that gaps up to 25 mm could be repaired, but at a lower rate than small gaps [31]. While various materials have helped in this clinical issue, even the preferred surgical methods are limited and new and more effective device systems are still required [32].

14.3.4 Other examples

Various other applications for collagen films and membranes have been proposed that have provided effective surgical devices. These include materials as dural substitutes [192, 193], which became of greater interest after prion infection issues were associated with cadaver derived materials (see chapter 11). This led to a range of commercial materials. Clinical studies on a range of materials showed them to be compatible with the tissue requirements, to be easy to handle, and to provide an acceptable clinical outcome [35, 193–195]. Another example, is the development of collagen films for tympanic membrane repair [196], with new materials still being developed [58]. Some other suggestions, nevertheless, have been made where collagen films and membranes may not prove suitable in the long-term. These include as vascular substitutes [197] and abdominal repair (see [198]), where reconstituted materials may lack the strength and durability of tissue-based formats.

14.4 Beads and particles

The main collagen-based bead or particle application has been the use of bone particles, including processed demineralised bone, for bone filler applications [199]. One use has been in dentistry, where material is compacted into the socket of a missing or extracted tooth in order to enable new bone growth [200]. This production of new bone is undoubtedly favoured by the residual growth factors that remain with the product [201, 202], which are dependent on the donor status [203]. Overall, various commercial products are available and are widely used clinically, being regarded as safe and efficient [199]. Formation of structurally sound bone is often essential if a replacement tooth implant (see chapter 19) is able to be fixed in place.

The difficulties in making beads from mammalian collagen (see chapter 11) has limited their potential use. They could be used, however, in situations where gelatin beads have been used. Gelatin and collagen beads are excellent substrates for cell growth, for example, with chondrocytes where the cell phenotype is retained. The beads can be made resorbable [204] so can be included in an implant to deliver cells directly, without an intermediate enzyme release process, to enable tissue repair [169].

Recombinant bacterial collagen can be used to make hydrated beads for wound healing [205]. This material incorporates the benefits of being non-cytotoxic and non-immunogenic [76] as well as being non-thrombogenic [206] and inhibiting biofilm formation [207].

14.5 Reconstituted fibres

Reconstituted, wet-spun collagen fibres (see chapter 11) have not yet found great acceptance for medical products. Several groups at different times have examined their use in tendon and ligament applications, both as temporary or longer-term devices (for example, [49, 208, 209]). The issue with this type of reconstituted collagen device is that it is very hard to obtain the strength and durability of the native tissues, and so have a device that functions well in the long-term. On the other hand, the availability of material may have limited the examination of these fibres for giving strength and dimensional stability to composite wound dressings that include a collagen sponge or gel.

The other system that can produce collagen fibres, as a membrane or sponge, is electrospinning [210]. While initial laboratory studies looked very promising, it was later suggested that all or part of the collagen may have been denatured to gelatin due to solvent incompatibility [211]. Despite this report, subsequently studies have reported using the same solvent system. So, the properties of the fibres and films made from collagen may be dominated by any gelatin that is present. A few methods for electrospinning collagen using benign solvents have emerged (see chapter 11) but these are not always very convenient, nor have been described for large scale production. Large scale, commercial electrospinning of polymer fibres was achieved in the 1930s [212], so it is still possible that this can be achieved with collagens. This could be useful, for example, for wound repair materials made by cellular contraction of collagen gels [111] as a fibre mat could provide stability in two-dimensions during the contraction process.

References

[1] Ungerleider J L and Christman K L 2014 *Stem Cells Transl. Med.* **3** 1090–99

[2] Neuta P A, Rojas D M, Agredo W and Gutierrez J O 2015 *Conf. Proc. IEEE Eng. Med. Biol. Soc.* **2015** 3529–32

[3] Yaffe A, Ehrlich J and Shoshan S 1984 *J. Periodontol.* **55** 623–28

[4] Ellender G, Hammond R, Papli R, Mitrangas K, Bateman J F, Glattauer V, Thyer J M, Werkmeister J A and Ramshaw J A M 1992 *Clin. Mater.* **9** 201–9

[5] Sheikh Z, Qureshi J, Alshahrani A M, Nassar H, Ikeda Y, Glogauer M and Ganss B 2017 *Odontology* **105** 1–12

[6] Caballé-Serrano J, Munar-Frau A, Ortiz-Puigpelat O, Soto-Penaloza D, Peñarrocha M and Hernández-Alfaro F 2018 *J. Clin. Exp. Dent.* **10** e477–83

[7] Mehlisch D R, Taylor T D, Leibold D G, Hiatt R, Waite D E, Waite P D, Laskin D M, Smith S T and Koretz M M 1987 *J. Oral Maxillofac. Surg.* **45** 408–13

[8] Schnutenhaus S, Doering I, Dreyhaupt J, Rudolph H and Luthardt R G 2018 *J. Periodontal Implant Sci.* **48** 236–50

[9] Knapp T R, Kaplan E N and Daniels J R 1977 *Plast. Reconstr. Surg.* **60** 398–405

[10] Cockerham K and Hsu V J 2009 *Facial Plast. Surg.* **25** 106–13

[11] Wollina U and Goldman A 2013 *Clin. Dermatol.* **31** 731–36

[12] Yannas I V and Burke J F 1980 *J. Biomed. Mater. Res.* **14** 65–81

[13] Bell E, Ehrlich H P, Buttle D J and Nakatsuji T 1981 *Science* **211** 1052–54

[14] Chaudhari A A, Vig K, Baganizi D R, Sahu R, Dixit S, Dennis V, Singh S R and Pillai S R 2016 *Int. J. Mol. Sci.* **17** E1974

[15] Doillon C J and Silver F H 1986 *Biomaterials* **7** 3–8

[16] Doillon C J, Whyne C F, Brandwein S and Silver F H 1986 *J. Biomed. Mater. Res.* **20** 1219–28

[17] Chattopadhyay S and Raines R T 2014 *Biopolymers* **101** 821–33

[18] Kasiewicz L N and Whitehead K A 2017 *Biomater. Sci.* **5** 1962–75

[19] Remacle M, Marbaix E, Hamoir M and Bertrand B 1987 *Rev. Laryngol. Otol. Rhinol. (Bord).* **108** 39–42

[20] Hoffmann T K, Sommer P and Bier H 2006 *Laryngorhinootologie* **85** 556–58

[21] Adler R H, Pelecanos N T, Geil R G, Rosenzweig S E and Thorsell H G 1962 *Surg. Forum* **13** 29–31

[22] Limura E and Giordano P 2017 *World J. Surg.* **41** 2492–501

[23] Edwards G A, Glattauer V, Nash T J, White J F, Brock K A, Werkmeister J A and Ramshaw J A M 1997 *J. Biomed. Mater. Res.* **34** 291–97

[24] Dabrowski A, Lepère M, Zaranis C, Coelio C and Hauters P 2016 *Surg. Endosc.* **30** 2358–66

[25] Hait M R, Battista O A, Stark R B and McCord C W 1969 *Surg. Forum* **20** 51–3

[26] Forcillo J and Perrault L P 2014 *Transfus. Apher. Sci.* **50** 26–31

[27] Tardy M, Gravaagna P, Revet L, Uhlrich S, Tollier J and Tayot J L 1995 *Proc. 21st Annual Meeting Society for Biomaterials (San Francisco)* 254

[28] Samarawickrama C, Samanta A, Liszka A, Fagerholm P, Buznyk O, Griffith M and Allan B 1988 *Cornea* **37** 609–16

[29] Yannas I V 1988 *Collagen* vol 3 ed M E Nimni ((Boca Raton, FL: CRC Press) 87–115

[30] Li S T, Archibald S J, Krarup C and Madison R D 1992 *Clin. Mater.* **9** 195–200

[31] Lohmeyer J A, Kern Y, Schmauss D, Paprottka F, Stang F, Siemers F, Mailaender P and Machens H G 2014 *J. Reconstr. Microsurg.* **30** 227–34

[32] Dalamagkas K, Tsintou M and Seifalian A 2016 *Mater. Sci. Eng. C Mater. Biol. Appl.* **65** 425–32

[33] Kim M, Park S R and Choi B H 2014 *Histol. Histopathol.* **29** 1395–408

[34] Ucar B and Humpel C 2018 *Neural Regen. Res.* **13** 595–98

[35] Parlato C, di Nuzzo G, Luongo M, Parlato R S, Accardo M, Cuccurullo L and Moraci A 2011 *Acta Neurochir. (Wien)* **153** 142–47

[36] Esposito F, Grimod G, Cavallo L M, Lanterna L, Biroli F and Cappabianca P 2013 *Clin. Neurol. Neurosurg.* **115** 1735–37

[37] Tanner J C, Apanay M B, Smith J P and Rife C C 1971 *AORN J.* **13** 159–62

[38] Dunn M W, Nishihara T, Stenzel K H, Branwood A W and Rubin A L 1967 *Science* **157** 1329–30

[39] Matthyssen S, Van den Bogerd B, Dhubhghaill S N, Koppen C and Zakaria N 2018 *Acta Biomater.* **69** 31–41

[40] Stenzel K H, Dunn M W, Rubin A L and Miyata T 1969 *Science* **164** 1282–83

[41] Kleinberg T T, Tzekov R T, Stein L, Ravi N and Kaushal S 2011 *Surv. Ophthalmol.* **56** 300–23

[42] Benfer J and Struck H 1973 *Arch. Surg.* **106** 838–42

[43] Deporter D A, Komori N, Howley T P, Shiga A, Ghent A, Hansel P and Parisien K 1988 *Calcif. Tissue Int.* **42** 321–25

[44] Ferreira A M, Gentile P, Chiono V and Ciardelli G 2012 *Acta Biomater.* **8** 3191–200

[45] Kuttappan S, Mathew D and Nair M B 2016 *Int. J. Biol. Macromol.* **93** 1390–401

[46] Speer D P, Chvapil M, Volz R G and Holmes M D 1979 *Clin. Orthop. Relat. Res.* **1979** 326–35

[47] Chajra H, Rousseau C F, Cortial D, Ronzière M C, Herbage D, Mallein-Gerin F and Freyria A M 2008 *Biomed. Mater. Eng.* **18** S33–45

[48] Han L, Zhang Z W, Wang B H and Wen Z K 2018 *Cell Tissue Bank.* **19** 47–59

[49] Chvapil M, Speer D P, Holubec H, Chvapil T A and King D H 1993 *J. Biomed. Mater. Res.* **27** 313–25
[50] Kew S J, Gwynne J H, Enea D, Abu-Rub M, Pandit A, Zeugolis D, Brooks R A, Rushton N, Best S M and Cameron R E 2011 *Acta Biomater.* **7** 3237–47
[51] Rieu C, Picaut L, Mosser G and Trichet L 2017 *Curr. Pharm. Des.* **23** 3483–506
[52] Stone K R, Rodkey W G, Webber R J, McKinney L and Steadman J R 1990 *Clin. Orthop. Relat. Res.* **1990** 129–35
[53] Stone K R, Steadman J R, Rodkey W G and Li S T 1997 *J. Bone Joint Surg. Am.* **79** 1770–77
[54] Zaffagnini S, Grassi A, Marcheggiani Muccioli G M, Bonanzinga T, Nitri M, Raggi F, Ravazzolo G and Marcacci M 2015 *Knee Surg. Sports Traumatol. Arthrosc.* **23** 3228–37
[55] Merriam A R, Patel J M, Culp B M, Gatt C J and Dunn M G 2015 *Am. J. Sports Med.* **43** 2528–37
[56] Salen B and Simbach I 1965 *J. Laryngol. Otol.* **79** 159–65
[57] Patterson M E 1967 *Arch. Otolaryngol.* **86** 486–89
[58] Kim S H, Lee H J, Yoo J C, Park H J, Jeong J Y, Seo Y B, Sultan M T, Kim S H, Lee O J and Park C H 2018 *J. Biomater. Sci. Polym. Ed.* **29** 997–1010
[59] Tachibana M, Nagamatsu G R and Addonizio J C 1985 *J. Urol.* **133** 866–69
[60] Kanatani I, Kanematsu A, Inatsugu Y, Imamura M, Negoro H, Ito N, Yamamoto S, Tabata Y, Ikada Y and Ogawa O 2007 *Tissue Eng.* **13** 2933–40
[61] Tanner J C, Marcucci M A, Bradley W H and Morgan J W 1968 *J. Urol.* **99** 710–12
[62] Stenzel K H, Sullivan J F, Miyata T and Rubin A L 1969 *Trans. Am. Soc. Artif. Intern. Organs* **15** 114–17
[63] Herschorn S, Radomski S B and Steele D J 1992 *J. Urol.* **148** 1797–800
[64] Egles C, Shamis Y, Mauney J R, Volloch V, Kaplan D L and Garlick J A 2008 *J. Invest. Dermatol.* **128** 1830–37
[65] Marcelino J and McDevitt C A 1995 *Biochim. Biophys. Acta* **1249** 180–88
[66] Freyria A M, Ronzière M C, Cortial D, Galois L, Hartmann D, Herbage D and Mallein-Gerin F 2009 *Tissue Eng. Part A* **15** 1233–45
[67] Rutgers M, Saris D B, Vonk L A, van Rijen M H, Akrum V, Langeveld D, van Boxtel A, Dhert W J and Creemers L B 2013 *Tissue Eng. Part A* **19** 59–65
[68] Yurchenco P D and Furthmayr H 1984 *Biochemistry* **23** 1839–50
[69] Nakazato K, Muraoka M, Adachi E and Hayashi T 1996 *J. Biochem.* **120** 889–94
[70] Truy E, Disant F, Tiollier J, Froehlich P and Morgon A 1994 *Arch. Otolaryngol. Head Neck Surg.* **120** 1329–32
[71] Silva T H, Moreira-Silva J, Marques A L, Domingues A, Bayon Y and Reis R L 2014 *Mar. Drugs* **12** 5881–901
[72] Subhan F, Ikram M, Shehzad A and Ghafoor A 2015 *J. Food Sci. Technol.* **52** 4703–07
[73] Fagerholm P, Lagali N S, Ong J A, Merrett K, Jackson W B, Polarek J W, Suuronen E J, Liu Y, Brunette I and Griffith M 2014 *Biomaterials* **35** 2420–27
[74] Farkash U, Avisar E, Volk I, Slevin O, Shohat N, El Haj M, Dolev E, Ashraf E and Luria S 2019 *J. Shoulder Elbow Surg.* **28** 503–9
[75] Cosgriff-Hernandez E, Hahn M S, Russell B, Wilems T, Munoz-Pinto D, Browning M B, Rivera J and Höök M 2010 *Acta Biomater.* **6** 3969–77
[76] Peng Y Y, Yoshizumi A, Danon S J, Glattauer V, Prokopenko O, Mirochnitchenko O, Yu Z, Inouye M and Werkmeister J A *et al* 2010 *Biomaterials* **31** 2755–61

[77] Brodsky B and Ramshaw J A M 2017 *Subcell. Biochem.* **82** 601–29
[78] Gross J and Kirk D 1958 *J. Biol. Chem.* **233** 355–60
[79] Rosenblatt J, Devereux B and Wallace D G 1992 *Biomaterials* **13** 878–86
[80] Rosenblatt J, Devereux B and Wallace D G 1994 *Biomaterials* **15** 985–95
[81] Daniels J R and Knapp T R 1976 *US Patent* 3949073
[82] Basta S L 2015 *Facial Plast. Surg. Clin. North Am.* **23** 417–21
[83] Wichterle O and Lim D 1960 *Nature* **185** 117–18
[84] Wallace D G, McPherson J M, Ellingsworth L, Cooperman L, Armstrong R and Piez K A 1988 *Collagen* vol 3 ed M E Nimni (Boca Raton, FL: CRC Press) 117–44
[85] Ramshaw J A M, Werkmeister J A and Peters D E 1990 *Current Perspectives on Implantable Devices* vol 2 ed D F Williams (London: JAI Press) 151–220
[86] Cooperman L, Mackinnon V, Bechler G and Pharris B B 1985 *Aesth. Plastic Surg.* **9** 145–52
[87] DeLustro F, Mackinnon V and Swanson N A 1988 *J. Dermatol. Surg. Oncol.* **14** 49–56
[88] Keefe J, Wauk L, Chu S and DeLustro F 1992 *Clin. Mat.* **6** 155–62
[89] McPherson J M, Ledger P W, Sawamura S, Conti A, Wade S, Reihanian H and Wallace D G 1986 *J. Biomed. Mater. Res.* **20** 79–92
[90] McPherson J M, Sawamura S and Armstrong R 1986 *J. Biomed. Mater. Res.* **20** 93–107
[91] Wallace D G, Rhee W, Reihanian H, Ksander G, Lee R, Braun W B, Weiss B A and Pharriss B B 1989 *J. Biomed. Mater. Res.* **23** 931–45
[92] Lemperle G, Hazan-Gaúthier N and Lemperle M 1995 *Plast. Reconstr. Surg.* **96** 627–34
[93] Haneke E 2004 *Semin. Cutan. Med. Surg.* **23** 227–32
[94] Lemperle G, Knapp T R, Sadick N S and Lemperle S M 2010 *Aesthetic Plast. Surg.* **34** 264–72
[95] Cohen S, Dover J, Monheit G, Narins R, Sadick N, Werschler W P, Karnik J and Smith S R 2015 *Dermatol. Surg.* **41**(Suppl 1) S302–13
[96] Devore D, Zhu J, Brooks R, McCrate R R, Grant D A and Grant S A 2016 *J. Biomed. Mater. Res. A* **104** 758–67
[97] Marinelli E, Montanari Vergallo G, Reale G, di Luca A, Catarinozzi I, Napoletano S and Zaami S 2016 *Eur. Rev. Med. Pharmacol. Sci.* **20** 4628–34
[98] Shah A R and Kennedy P M 2018 *Med. Clin. North Am.* **102** 1041–54
[99] Attenello N H and Maas C S 2015 *Facial Plast. Surg.* **31** 29–34
[100] Haneke E 2019 *Dermatol. Ther.* **32** e12676
[101] Park T H, Seo S W, Kim J K and Chang C H 2011 *J. Plast. Reconstr. Aesthet. Surg.* **64** 892–96
[102] Bitterman-Deutsch O, Kogan L and Nasser F 2015 *Dermatol. Rep.* **7** 5851
[103] Artzi O, Loizides C, Verner I and Landau M 2016 *Dermatol. Surg.* **42** 31–7
[104] Alcântara C E P, Noronha M S, Cunha J F, Flores I L and Mesquita R A 2018 *J. Cosmet. Dermatol.* **17** 578–83
[105] Curi M M, Cardoso C L, Curra C, Koga D and Benini M B 2015 *J. Craniofac. Surg.* **26** 782–84
[106] Keizers P H J, Vanhee C, van den Elzen E M W, de Jong W H, Venhuis B J, Hodemaekers H M, Schwillens P and Lensen D G W 2018 *J. Pharm. Biomed. Anal.* **159** 173–78
[107] Lisi C, Hawkshaw M J and Sataloff R T 2013 *J. Voice* **27** 119–23
[108] Herschorn S, Steele D J and Radomski S B 1996 *J. Urol.* **156** 1305–09
[109] Li H and Westney O L 2019 *Urol. Clin. North Am.* **46** 1–15

[110] Davis N F, Kheradmand F and Creagh T 2013 *Int. Urogynecol. J.* **24** 913–19
[111] Bell E, Ivarsson B and Merrill C 1979 *Proc. Natl Acad. Sci. USA* **76** 1274–78
[112] Ramshaw J A M, Mitrangas K and Bateman J F 1991 *Connect. Tissue Res.* **25** 295–300
[113] Bell E 1984 *US Patent* 4485096
[114] Kemp P, Bell E, Kagan D T, Mason W and Cavallaro J 1996 *US Patent* 5536656
[115] Vaughan F L, Gray R H and Bernstein I A 1986 *In Vitro Cell Dev. Biol.* **22** 141–49
[116] Chirila T V, Tahija S, Hong Y, Vijayasekaran S and Constable I J 1994 *J. Biomater. Appl.* **9** 121–37
[117] Chirila T V, Hong Y E, Dalton P D, Constable I J and Refojo M F 1998 *Prog. Polym. Sci.* **23** 475–508
[118] Baino F 2011 *Acta Biomater.* **7** 921–35
[119] Connon C J 2015 *Procedia Eng.* **110** 15–20
[120] Liu W, Merrett K, Griffith M, Fagerholm P, Dravida S, Heyne B, Scaiano J C, Watsky M A and Shinozaki N *et al* 2008 *Biomaterials* **29** 1147–58
[121] Hayes S, Lewis P, Islam M M, Doutch J, Sorensen T, White T, Griffith M and Meek K M 2015 *Acta Biomater.* **25** 121–30
[122] Rhee W M, Rao P R, Chu G H, DeLustro F A, Harner C F H, Sakai N and Schroeder J A 1997 *US Patent* 5614587
[123] Lode A, Meyer M, Brüggemeier S, Paul B, Baltzer H, Schröpfer M, Winkelmann C, Sonntag F and Gelinsky M 2016 *Biofabrication* **8** 015015
[124] Kamranpour N O, Miri A K, James-Bhasin M and Nazhat S N 2016 *Biofabrication* **8** 015018
[125] Drzewiecki K E, Malavade J N, Ahmed I, Lowe C J and Shreiber D I 2017 *Technology (Singap. World Sci.)* **5** 185–95
[126] Stratesteffen H, Köpf M, Kreimendahl F, Blaeser A, Jockenhoevel S and Fischer H 2017 *Biofabrication* **9** 045002
[127] Yeo M G and Kim G H 2017 *Biofabrication* **9** 025004
[128] Marques C F, Diogo G S, Pina S, Oliveira J M, Silva T H and Reis R L 2019 *J. Mater. Sci., Mater. Med.* **30** 32
[129] Zhang J, Yang Z, Li C, Dou Y, Li Y, Thote T, Wang D A and Ge Z 2013 *Tissue Eng. Part A* **19** 2166–75
[130] Singelyn J M, DeQuach J A, Seif-Naraghi S B, Littlefield R B, Schup-Magoffin P J and Christman K L 2009 *Biomaterials* **30** 5409–416
[131] Seif-Naraghi S B, Salvatore M A, Schup-Magoffin P J, Hu D P and Christman K L 2010 *Tissue Eng. Part A* **16** 2017–27
[132] DeQuach J A, Yuan S H, Goldstein L S and Christman K L 2011 *Tissue Eng. Part A* **17** 2583–92
[133] Wassenaar J W, Boss G R and Christman K L 2015 *Biomaterials* **64** 108–14
[134] Gaetani R, Aouad S, Demaddalena L L, Straessle H, Dzieciatkowska M, Wortham M, Bender H R, Nguyen-Ngoc K V and Schmid-Schoenbein G W *et al* 2018 *Tissue Eng. Part C Methods* **24** 697–708
[135] Singelyn J M, Sundaramurthy P, Johnson T D, Schup-Magoffin P J, Hu D P, Faulk D M, Wang J, Mayle K M and Bartels K *et al* 2012 *J. Am. Coll. Cardiol.* **59** 751–63
[136] Johnson T D, Dequach J A, Gaetani R, Ungerleider J, Elhag D, Nigam V, Behfar A and Christman K L 2014 *Biomater. Sci.* **2014** 60283D

[137] Gaetani R, Yin C, Srikumar N, Braden R, Doevendans P A, Sluijter J P and Christman K L 2016 *Cell Transplant.* **25** 1653–63
[138] Pulkkinen H J, Tiitu V, Valonen P, Jurvelin J S, Rieppo L, Töyräs J, Silvast T S, Lammi M J and Kiviranta I 2013 *Osteoarthritis Cartilage* **21** 481–90.
[139] Parmar P A, Chow L W, St-Pierre J P, Horejs C M, Peng Y Y, Werkmeister J A, Ramshaw J A M and Stevens M M 2015 *Biomaterials* **54** 213–25
[140] Parmar P A, St-Pierre J P, Chow L W, Spicer C D, Stoichevska V, Peng Y Y, Werkmeister J A, Ramshaw J A M and Stevens M M 2017 *Acta Biomater.* **51** 75–88
[141] Park J S, Chu J S, Tsou A D, Diop R, Tang Z, Wang A and Li S 2011 *Biomaterials* **32** 3921–30
[142] Yannas I V, Lee E, Orgill D P, Skrabut E M and Murphy G F 1989 *Proc. Natl Acad. Sci. USA* **86** 933–37
[143] Still J, Glat P, Silverstein P, Griswold J and Mozingo D 2003 *Burns* **29** 837–41
[144] Seyednejad H, Imani M, Jamieson T and Seifalian A M 2008 *Br. J. Surg.* **95** 1197–225
[145] Toro A, Mannino M, Reale G and Di Carlo I 2011 *J. Blood Med.* **2** 31–6
[146] Luo J, Meng Y, Zheng L, Xu K and Li C 2018 *J. Biomater. Sci. Polym. Ed.* **29** 1933–48
[147] Battista O A, Erdi N Z, Ferraro C F and Karasinski N Z 1967 *J. Appl. Polym. Sci.* **11** 481–98
[148] Hait M R 1970 *Am. J. Surg.* **120** 330
[149] DeLustro F, Condell R A, Nguyen M A and McPherson J M 1986 *J. Biomed. Mater. Res.* **20** 109–20
[150] Dehen M, Niederdellmann H and Lachner J 1990 *Dtsch. Zahnarztl. Z.* **45** 553–56
[151] Buma P, van Tienen T and Veth R 2007 *Expert Rev. Med. Devices* **4** 507–16
[152] Li S-T and Stone K R 2000 *US Patent* 6042610
[153] Steadman J R and Rodkey W G 2005 *Arthroscopy* **21** 515–25
[154] Zaffagnini S, Giordano G, Vascellari A, Bruni D, Neri M P, Iacono F, Kon E, Presti M L and Marcacci M 2007 *Knee Surg. Sports Traumatol. Arthrosc.* **15** 175–83
[155] Hansen R, Choi G, Bryk E and Vigorita V 2011 *J. Long Term Eff. Med. Implants* **21** 321–37
[156] Monllau J C, Gelber P E, Abat F, Pelfort X, Abad R, Hinarejos P and Tey M 2001 *Arthroscopy* **27** 933–43
[157] Warth R J and Rodkey W G 2015 *Arthroscopy* **31** 927–41
[158] Houck D A, Kraeutler M J, Belk J W, McCarty E C and Bravman J T 2018 *Knee Surg, Sports Traumatol, Arthrosc.* **26** 2259–69
[159] Benthien J P and Behrens P 2010 *Cartilage* **1** 65–8
[160] Nehrer S, Breinan H A, Ramappa A, Shortkroff S, Young G, Minas T, Sledge C B, Yannas I V and Spector M 1997 *J. Biomed. Mater. Res.* **38** 95–104
[161] Nehrer S, Breinan H A, Ramappa A, Hsu H P, Minas T, Shortkroff S, Sledge C B, Yannas I V and Spector M 1998 *Biomaterials* **19** 2313–28
[162] Lane Smith R, Trindade M C, Ikenoue T, Mohtai M, Das P, Carter D R, Goodman S B and Schurman D J 2000 *Biorheology* **37** 95–107
[163] Fujisato T, Sajiki T, Liu Q and Ikada Y 1996 *Biomaterials* **17** 155–62
[164] Grande D A, Halberstadt C, Naughton G, Schwartz R and Manji R 1997 *J. Biomed. Mater. Res.* **34** 211–20
[165] Shahin K and Doran P M 2015 *Methods Mol. Biol.* **1340** 221–33

[166] Grande D A, Pitman M I, Peterson L, Menche D and Klein M 1989 *J. Orthop. Res.* **7** 208–18
[167] Brittberg M, Lindahl A, Nilsson A, Ohlsson C, Isaksson O and Peterson L 1994 *N. Engl. J. Med.* **331** 889–95
[168] Brittberg M, Nilsson A, Lindahl A, Ohlsson C and Peterson L 1996 *Clin. Orthop.* **326** 270–83
[169] Chiang H, Kuo T F, Tsai C C, Lin M C, She B R, Huang Y Y, Lee H S, Shieh C S and Chen M H *et al* 2005 *J. Orthop. Res.* **23** 584–93
[170] Bartlett W, Skinner J A, Gooding C R, Carrington R W, Flanagan A M, Briggs T W and Bentley G 2005 *J. Bone Joint Surg. Br.* **87** 640–45
[171] Willers C, Chen J, Wood D, Xu J and Zheng M H 2005 *Tissue Eng.* **11** 1065–76
[172] Schmiesing A M, Ridley T J and Macalena J A 2018 *JBJS Essent. Surg. Tech.* **8** e29
[173] Lickorish D, Ramshaw J A M, Werkmeister J A, Glattauer V and Howlett C R 2004 *J. Biomed. Mater. Res. A* **68** 19–27
[174] Wahl D A and Czernuszka J T 2006 *Eur. Cell. Mater.* **11** 43–56
[175] Ueda H, Hong L, Yamamoto M, Shigeno K, Inoue M, Toba T, Yoshitani M, Nakamura T, Tabata Y and Shimizu Y 2002 *Biomaterials* **23** 1003–10
[176] Shigeno K, Nakamura T, Inoue M, Ueda H, Kobayashi E, Nakahara T, Lynn A K, Toba T and Yoshitani M *et al* 2002 *Int. J. Artif. Organs* **25** 1095–102
[177] Sarban S, Senkoylu A, Isikan U E, Korkusuz P and Korkusuz F 2009 *Clin. Orthop. Relat. Res.* **467** 3113–120
[178] Engeler C E, Hunter D W, Castaneda-Zuniga W, Tashjian J H, Yedlicka J W and Amplatz K 1992 *Radiology* **184** 787–89
[179] Webb J G, Carere R A and Dodek A A 1993 *Cathet. Cardiovasc. Diagn.* **30** 314–16
[180] Dehghani P, Mohammad A, Marcuzzi D, Hong T, Holmes K, Evangelista M J, Boutin A, Leong-Poi H and Cheema A N 2010 *J. Invasive Cardiol.* **22** 505–10
[181] Yamada K, Yamaura J, Katoh M, Hata K, Okuda K and Yoshie H 2006 *J. Periodontol.* **77** 672–77
[182] Quteish D and Dolby A E 1992 *J. Clin. Periodontol.* **19** 476–84
[183] Nyman S, Lindhe J, Karring T and Rylander H 1982 *J. Clin. Periodontol.* **9** 290–96
[184] Caton J G, DeFuria E L, Polson A M and Nyman S 1987 *J. Periodontol.* **58** 546–52
[185] Stoecklin-Wasmer C, Rutjes A W, da Costa B R, Salvi G E, Jüni P and Sculean A 2013 *J. Dent. Res.* **92** 773–81
[186] Bel A, Kachatryan L, Bruneval P, Peyrard S, Gagnieu C, Fabiani J N and Menasché P 2010 *Interact. Cardiovasc. Thorac. Surg.* **10** 213–16
[187] Taksaudom N, Ketwong M, Lertprasertsuke N and Kongkaew A 2017 *Open J. Cardiovasc. Surg.* **9** 1179065217720909
[188] Armoiry X, Viprey M, Constant H, Aulagner G, Roux A S, Huot L, Roubertie F, Ninet J and Henaine R 2013 *Arch. Cardiovasc. Dis.* **106** 433–39
[189] Bellamkonda R V 2006 *Biomaterials* **27** 3515–518
[190] Mackinnon S E, Hudson A R, Bojanowski V, Hunter D A and Maraghi E 1985 *Ann. Plast. Surg.* **14** 428–36
[191] Kehoe S, Zhang X F and Boyd D 2012 *Injury* **43** 553–72
[192] McCall T D, Fults D W and Schmidt R H 2008 *Surg. Neurol.* **70** 92–6
[193] Costa B S, Cavalcanti-Mendes Gde A, Abreu M S and Sousa A A 2011 *Arq. Neuropsiquiatr.* **69** 217–20

[194] Neulen A, Gutenberg A, Takács I, Wéber G, Wegmann J, Schulz-Schaeffer W and Giese A 2011 *Acta Neurochir. (Wien)* **153** 2241–50
[195] Esposito F, Grimod G, Cavallo L M, Lanterna L, Biroli F and Cappabianca P 2013 *Clin. Neurol. Neurosurg.* **115** 1735–37
[196] Abbenhaus J I 1978 *Otolaryngology* **86** ORL485-87
[197] Chignier E, Eloy R, Huc A, Gimeno R and Gleizal C 1985 *J. Biomed. Mater. Res.* **19** 115–31
[198] Huerta S, Varshney A, Patel P M, Mayo H G and Livingston E H 2016 *JAMA Surg.* **151** 374–81
[199] Dinopoulos H T and Giannoudis P V 2006 *Expert Opin. Drug Saf.* **5** 847–66
[200] Horowitz R, Holtzclaw D and Rosen P S 2012 *J. Evid. Based Dent. Pract.* **12**(3 Suppl) 149–60
[201] Urist M R, DeLange R J and Finerman G A 1983 *Science* **220** 680–86
[202] Wildemann B, Kadow-Romacker A, Haas N P and Schmidmaier G 2007 *J. Biomed. Mater. Res. A* **81** 437–42
[203] Traianedes K, Russell J L, Edwards J T, Stubbs H A, Shanahan I R and Knaack D 2004 *J. Biomed. Mater. Res. B Appl. Biomater.* **70** 21–9
[204] Glattauer V, White J F, Tsai W B, Tsai C C, Tebb T A, Danon S J, Werkmeister J A and Ramshaw J A M 2010 *J. Biomed. Mater. Res. A* **92** 1301–09
[205] Cereceres S, Touchet T, Browning M B, Smith C, Rivera J, Höök M, Whitfield-Cargile C, Russell B and Cosgriff-Hernandez E 2015 *Adv. Wound Care* **4** 444–56
[206] Browning M B, Dempsey D, Guiza V, Becerra S, Rivera J, Russell B, Höök M, Clubb F and Miller M *et al* 2012 *Acta Biomater.* **8** 1010–21
[207] Bronk J K, Russell B H, Rivera J J, Pasqualini R, Arap W, Höök M and Barbu E M 2014 *Acta Biomater.* **10** 3354–62
[208] Goldstein J D, Tria A J, Zawadsky J P, Kato Y P, Christiansen D and Silver F H 1989 *J. Bone Joint Surg. Am.* **71** 1183–91
[209] Kew S J, Gwynne J H, Enea D, Abu-Rub M, Pandit A, Zeugolis D, Brooks R A, Rushton N, Best S M and Cameron R E 2011 *Acta Biomater.* **7** 3237–47
[210] Matthews J A, Wnek G E, Simpson D G and Bowlin G L 2002 *Biomacromolecules* **3** 232–38
[211] Zeugolis D I, Khew S T, Yew E S, Ekaputra A K, Tong Y W, Yung L Y, Hutmacher D W, Sheppard C and Raghunath M 2008 *Biomaterials* **29** 2293–305
[212] Tucker N, Stanger J J, Staiger M, Razzaq H and Hofman K 2012 *J. Eng. Fibers Fabrics* **7** 63–73

IOP Publishing

Biophysical and Chemical Properties of Collagen: Biomedical Applications

John A M Ramshaw and Veronica Glattauer

Chapter 15

Applications of biosynthetic materials

Biosynthetic materials have been a novel, niche area of collagen-based biomaterials. Nevertheless, in the particular application of vascular (artery) replacement and as an arterio-venous access device, the material has proved very effective. As such, in these areas, it has a long commercial history. While biosynthetic materials have only been used in limited indications to date, the interesting properties and versatility of the material suggest that further development of small diameter (cardiac) devices may be possible and that other non-vascular applications could also emerge.

15.1 Vascular devices

15.1.1 Background technologies

The initial biosynthetic material was designed specifically to meet the need for a more effective vascular prosthesis, especially for small diameter devices [1] where there still remains an un-met medical need. In the early 1950s, several surgeons, who were to become luminaries in the field, were looking at the development of vascular devices made from synthetic polymers [2, 3]. Over time, Dacron became the preferred fibre [4], in a variety of weights and textures [3]. Early studies showed clearly the importance of good tissue integration for effective performance [5]. In 1975, ePTFE devices were introduced, and despite an initial lack of enthusiasm, by 1992 had become the predominant device, 99%, for lower flow axillofemoral and femoropopliteal locations [3], even though they were still not ideal. Subsequently both materials have continued to be developed and improved.

Also, in the 1950s, others were looking at opportunities for biological vascular prostheses. Following on from various earlier studies that had shown that polyethylene could stimulate fibrous tissue formation [6], the principle was extended to grow fibrocollagenous tubes for vascular use [7]. Further animal studies on mandrel-grown fibrocollagenous tubes confirmed this area of opportunity, but also noted that in aortic applications an unacceptable incidence of aneurysms occurred [8]. This led to the suggested inclusion of a Dacron mesh around the mandrel [9] and the further suggestion that an autologous tube could be prepared *in situ* adjacent to the intended application site [9]. A further development was the use of a metal rather than a polymer mandrel, as a component in an assembly that could be implanted into the patient. This 'tissue die' (the Sparks Mandrel) also incorporated a polymer reinforcing mesh [10, 11]. The next development was to use a silicone rod as the mandrel, which again included either a polyester (Dacron) or polypropylene reinforcing mesh [12]. This advance was adapted and included in an updated approach to the Sparks Mandrel method for *in situ* autologous vessel production [13]. This approach gained some popularity, until in subsequent years an increasing number of adverse events were noted, mainly arising from lack of durability leading to degradation and aneurysms [14; reviewed by [15]]. Based on these data, the clinical use of autogenous mandrel-grown fibrocollagenous tubes ceased. Nevertheless, the concept with variations still appears on occasion. For example, autologous mandrel-grown tubes prepared in the patient's peritoneal cavity have been proposed [16].

At around the same time as the silicone-based Sparks Mandrel was being developed, others were examining the potential use of natural collagen vessels. This work included development of mesh surrounded (supported) human umbilical vein-based devices [17, 18]. While use of an unmodified vessel was not successful, the development of a glutaraldehyde (GA) stabilised device showed initial promise, but over longer time scales the extent of adverse events, notably aneurysms and dilatations, led to the low patencies [19]. The possibility of using GA to stabilise natural vessels from xenogeneic species was also proposed [20].

15.1.2 Biosynthetic vascular device

The background technologies had established various innovative strategies for vascular prostheses, but each showed specific difficulties that limited their use. These data had established that a silicone mandrel could be used to produce a fibrocollagenous tube, and that this tube could be made more durable by inclusion of a polyester mesh during the tissue deposition phase. This meant that the mesh was fully integrated within the collagenous structure rather than a loosely attached covering. GA fixation of collagen showed that better durability could be obtained, as well as minimising any immunological response allowing xenogeneic material to be used. The combination of all 4 of these key strategies was first shown by Ketharanathan [1, 21], using sheep as the production host, and has enabled a clinically successful, off-the-shelf vascular prosthesis product to be developed. The manufacture of this tissue-polymer composite vascular prosthesis (see chapter 11), which is best known as the Omniflow™ Vascular Prosthesis (OVP) (BioNova International, North Melbourne, now acquired by LeMaitre), has proved effective in clinical replacement of peripheral arteries for more than two decades.

Reports of early clinical studies on over 200 hundred patients using the commercial device [22, 23] suggested that if a prosthetic material were to be used, as opposed to saphenous vein, then the biosynthetic OVP performed at least as well as an ePTFE prosthesis, particularly in through and below the knee use [22]. These authors observed a low infection rate and minimal danger from aneurysm at three years. They therefore suggested that the OVP could be recommended for use when no suitable autologous vessel was available [23]. Studies reported excellent clinical performance for eight years [23]. More recently, a systematic review and analysis again reported that biosynthetic vascular prostheses are a 'promising' alternative for haemodialysis access and femoropopliteal by-pass when an appropriate autologous vein is not available [24]. Another recent report noted that biosynthetic prostheses were useful for replacing infected (polymeric) devices, with no reinfection being observed [25].

The OVP is well suited for femoral and femoropopliteal replacements; however, it is also suited to other vascular applications. For example, it can be manufactured in a U-shape format for use as an arterio-venous access device for haemodialysis. Initial studies suggested that the biosynthetic device worked well in this application, where the ready healing of a puncture wound would be of benefit [26, 27]. On-going use for this indication has confirmed the utility of the biosynthetic material, especially for high risk patients, with good patency at one year coupled to low infection and low aneurysm rates [28].

Further improvements to the present technology are possible. For example, adjustments can be made that can increase the level of fibrous tissue coverage of the polymer mesh so as to minimise risks from thrombus formation [29]. The good blood compatibility of the surface can be further improved by suitable coating with type V collagen [30], while other modified surfaces may also be applicable. Also, to enhance the blood surface characteristics, the inner surface can be successfully seeded with endothelial cells by various methods (e.g., [31]) and this cell coating can be stable under flow conditions [32].

A very significant un-met medical need is the availability of a small diameter device for use in coronary by-pass surgery. Inclusion of new technologies onto the base biosynthetic device technology may still provide a solution for this significant, persistent problem.

15.1.3 Explant analyses

When clinical explants are evaluated, cellular events are generally examined and understood. However, tissue composition changes are less frequently studied and less well characterised. And for porous materials, such as collagen-based materials, tissue changes seem to be examined even less often than for non-porous materials.

Initial studies on biosynthetic vascular prostheses were conducted in animal models, especially in dogs as an aorto-iliac by-pass graft [33]. Gross inspection of explants of these materials at various time points up to four years showed minimal intimal hyperplasia, and in the dog model showed good endothelialisation and good patency [33, 34]. The dog explants were also examined by immunohistology using collagen type and species-specific antibodies as well as antibodies to other tissue components. The availability of species-specific antibodies to collagen enabled the sheep collagen of the implanted device to be clearly differentiated from the dog collagens (or human for clinical explants) that arise from the host response [33, 34].

These data showed that the GA stabilised sheep collagen was still readily identified after four years in the dog model [33] (figure 15.1). The wall of the implant, however, was also extensively augmented throughout its full thickness with new, host derived collagen [33, 34] (figure 15.1). GA fixed collagen shows autofluorescence, so while either immunoperoxidase or immunofluorescence staining methods can be used, care is needed in certain circumstances [35].

Augmentation of the prosthesis can be seen on analyses on samples from earlier explant times. Type V collagen can be seen as part of the newly forming thin neo-intima, where it may further passivate the surface and lessen the likelihood of platelet activation and thrombosis [30], while not preventing subsequent endothelial cell growth. Type VI collagen was also observed in the early post-implant phase, marking new tissue infiltration [36]. The manufactured implant has type VI collagen mostly near the luminal surface and polyester fibres, while after one month new type VI augmentation from the adventitia is already present [34, 36] (figure 15.2). Augmentation by type III collagen, also for the adventitia, is also observed following the type VI collagen, such that by six months implant duration both have augmented the full thickness of the wall [33, 34, 36] (figure 15.2).

Clinically derived explants generally show a similar pattern of collagen augmentation. Clinical samples have generally been obtained during revisions where there has been further degradation of the natural vessel adjacent to the functional implant, giving samples from the anastomosis region. Histology showed that the explants appeared well vascularised by new capillaries formed beneath and around the polyester and at the outer margin of the vessel wall. At the anastomotic junction, pannus ingrowth and endothelialisation had occurred. Immunohistology on these samples showed that implants for up to 9 years showed retention of original sheep collagen [34]. Similarly, examination showed augmentation with recently synthesised collagens throughout the full thickness of the vessel wall at all time points [34] (figure 15.3). The neo-intimal layer at the anastomosis included collagen types I, III, VI and V, laminin, fibronectin and microfibril-associated elastic proteins.

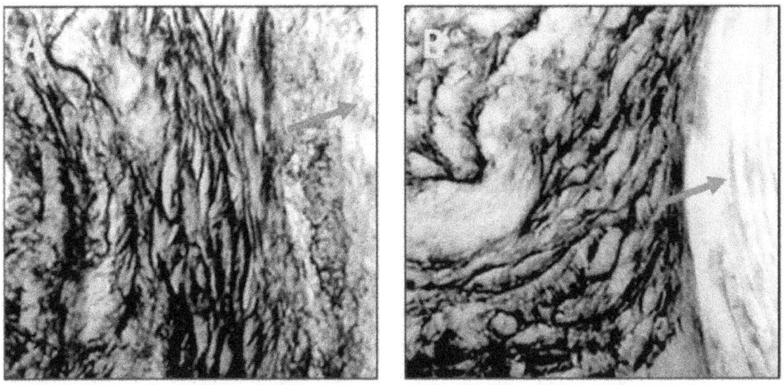

Figure 15.1. Immunohistology (peroxidase) examination of type III collagen in four year explants of OVP from a dog model (black is positive). (A) Collagen persistence, (B) Collagen augmentation. Red arrows indicate the polyester component. (Republished from [33], with permission from Begell House through the Copyright Clearance Centre Inc.)

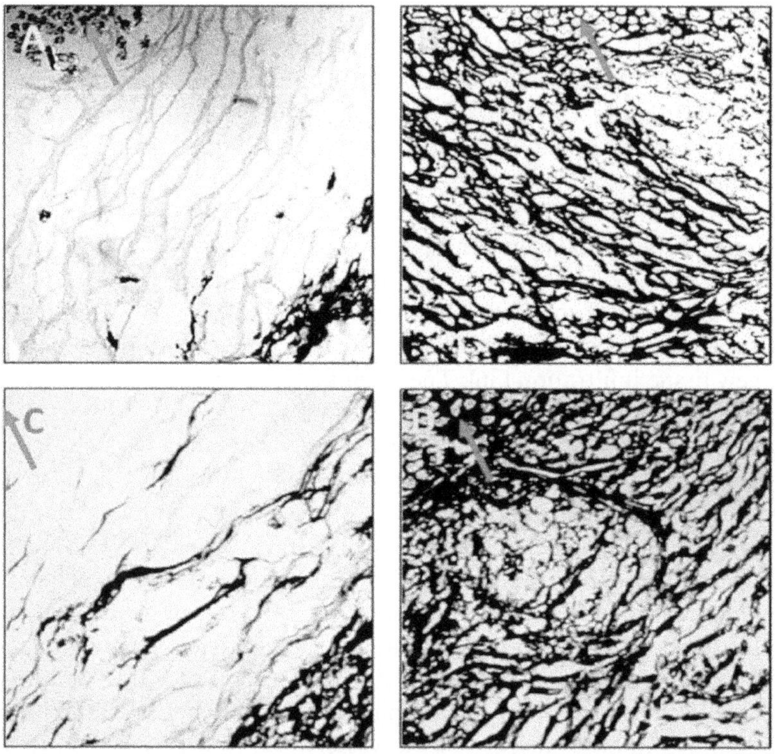

Figure 15.2. Immunohistology (peroxidase) examination of collagen augmentation of OVP explants from a dog model (black is positive). (A) Type VI collagen, after 1-month implant, (B) Type VI collagen, after six months implant, (C) Type III collagen, after three months implant, (D) Type III collagen, after 6 months implant. Red arrows indicate the polyester component. (Republished from [33], with permission from Begell House through the Copyright Clearance Centre Inc.)

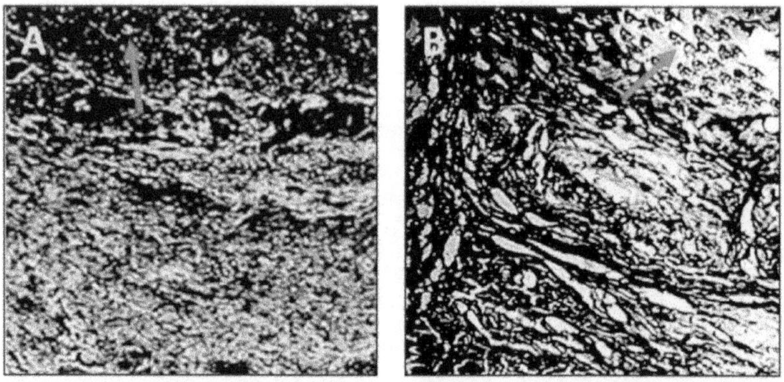

Figure 15.3. Immunohistology examination of collagen augmentation of OVP in a 2.5-year clinical explant. (A) Type III collagen (fluorescence, white is positive), (B) type VI collagen (peroxidase, black is positive). (Reprinted from [34], with permission from Springer Nature.) Arrows indicate the polyester component.

The most noticeable difference between the clinical explants and those for the dog model was the extent of endothelialisation that was seen. In the dog samples, staining for von Willebrand factor (a marker for endothelial cells) showed that a cell layer had formed along the full length of the implant, about 100 mm [34]. For the clinical samples, the extent of endothelialisation was limited. Endothelialisation was present at the anastomosis region but did not extend any further along the implanted vessel [34]. Despite this lack of endothelial cell coverage along the majority of the implant, the function did not seem to be compromised.

The various histology and immunohistology data allow a better understanding of the *in vivo* performance of the OVP. The same approach is available for other biomaterial implants. The knowledge gained for any device enables the development of strategies for further improvements to devices to enhance their functions for their specific applications.

15.2 Hernia, ligament and other options

While the biosynthetic material technology was originally developed for making vascular conduits, a range of other applications for the technology have been suggested. For example, opening up a tube leads to a flat sheet which can then be used as a patch, for example, as a vascular patch as the blood contact characteristics are already established. Changing the size, for example, a larger diameter silicone mandrel, or the shape of the silicone implant/mesh structure can provide a range of sizes and configurations.

One example for the use of a patch is for hernia repair [37]. Studies in a rabbit abdominal wall defect model have shown that the patch, with its natural porosity, led to cell infiltration, neovascularisation and new ECM deposition, giving it full tissue integration with the host. Immunohistology studies clearly showed the new tissue infiltration throughout the thickness of the patch, with type VI collagen present through the full thickness after four weeks of implant. Along with type VI collagen, other collagens including the interstitial fibril forming collagens followed and accumulated in the patch [37]. This study also compared the biosynthetic material to a polypropylene mesh implant, which has been used extensively as an implant in clinical practice. The data indicated that the biosynthetic material was superior to the mesh implant, as the polymer mesh could be associated with significant adhesions [37].

The immediate issue with using a biosynthetic material in clinical practice is higher cost. The benefit arises from the possibility of fewer complications, for example, infections, and lower adverse local tissue reactions, such that when a 'whole-of-episode' cost for extended treatment is taken into account the cost differential overall for a facility drops. While there is an increasing use of biological materials in hernia repairs [38], the cross-over point where there is a cost benefit from biologicals has probably not been reached [38], although in certain indications a biological may be preferred, and also for patients where adequate follow-up presents difficulties.

Other opportunities for biosynthetic materials, both mandrel and non-mandrel-grown could exist, for example, for ligament repair [39]. Various materials, but particularly polyester devices, have been used for ligament repair, but with mixed success. Examination of failed prostheses has typically shown that limited new ECM has been deposited [40]. Thus, while the polyester implant offers initial strength it does not seem to act as consistently as an effective scaffold. Nevertheless, it does seem that there is better performance when there is new, aligned collagen is deposited, probably in association with tensile stress [41]. Hence, one could speculate that a biosynthetic implant with a surrounding, pre-organised collagenous matrix may provide a good solution, as it may present a better scaffold for new tissue infiltration and integration, while the initial and augmented tissue integration may limit any adverse effects arising from mesh particulates that could arise from any fibre fragmentation.

References

[1] Christie B, Ketharanathan V and Perloff L J 1980 *J. Surg. Res.* **28** 519–32

[2] Voorhees A B, Jaretzki A and Blakemore A H 1952 *Ann. Surg.* **135** 332–36

[3] Sauvage L R 1993 *J. Invest. Surg.* **6** 221–25

[4] Szilagyi D E, France L C, Smith R F and Whitcomb J G 1958 *AMA Arch. Surg.* **77** 538–51

[5] Wesolowski S A 1962 *Evaluation of Tissue and Prosthetic Vascular Grafts* (Springfield, IL: Thomas)

[6] Yeager G H and Cowley R A 1948 *Ann. Surg.* **128** 509–19

[7] Pierce E C 1953 *Surgery* **33** 648–57

[8] Eiken O and Norden G 1961 *Acta Chir. Scand.* **121** 90–102

[9] Assefi I and Parsonnet V 1964 *J. Newark Beth Israel Hosp.* **15** 161–70

[10] Sparks C H 1969 *Ann. Thorac. Surg.* **8** 104–13

[11] Sparks C H 1970 *Ann. Surg.* **172** 787–94

[12] Parsonnet V, Alpert J and Brief D K 1971 *Surgery* **70** 935–39

[13] Sparks C H 1973 *Ann. Surg.* **177** 293–300

[14] Christenson J T and Eklöf B 1979 *Br. J. Surg.* **66** 514–17

[15] Ramshaw J A M, Edwards G A and Werkmeister J A 1995 *Encyclopedic Handbook of Biomaterials and Bioengineering part B* vol 2 ed D L Wise, D J Trantolo, D D Altobelli, M J Yaszemski, J D Gresser and E R Schwartz (New York: Marcel Dekker) pp 953–78

[16] Campbell J H, Efendy J L and Campbell G R 1999 *Circ. Res.* **85** 1173–78

[17] Dardik H, Ibrahim I M, Baier R, Sprayregen S, Levy M and Dardik I I 1976 *J. Am. Med. Assoc.* **236** 2859–62

[18] Dardik H D, Ibrahim I M, Sprayregen S and Dardik I I 1976 *Surgery* **79** 618–24

[19] Karkow W S, Cranley J J, Cranley R D, Hafner C D and Ruoff B A 1986 *J. Vasc. Surg.* **4** 486–92

[20] Parsonnet V, Tiro A C, Brief D K, Alpert J and Brener B J 1978 *Graft Materials in Vascular Surgery* ed H Dardik (Miami, FL: Symposia Specialists) pp 249–62

[21] Ketharanathan V 1982 *US Patent* 4319363

[22] Birchall I E, Lee V W and Ketharanathan V 2001 *Biomaterials* **22** 3139–44

[23] Koch G, Gutschi S, Pasher O, Fruhwirth J and Hauser H 1996 *Zentralbi. Chir.* **121** 761–67

[24] Koch G, Gutschi S, Pascher O, Fruhwirth H and Glanzer H 1997 *Aust. N Z J. Surg.* **67** 637–39
[25] Wilasrusmee C, Siribumrungwong B, Horsirimanont S, Poprom N, Jirasiritham J and Thakkinstian A 2017 *Ann. Med. Surg. (Lond.)* **15** 26–33
[26] Wiltberger G, Matia I, Schmelzle M, Krenzien F, Hau H M, Freitas B, Jonas S and Fellmer P T 2014 *J. Cardiovasc. Surg. (Torino)* **55** 693–98
[27] Wang S S and Chu S H 1996 *Artif. Organs* **20** 1278–81
[28] Palumbo R, Niscola P, Calabria S, Fierimonte S, Bevilacqua M, Scaramucci L, Tolu B, de Fabritiis P and Bondanini F 2009 *Nephron Clin. Pract.* **113** c76–80
[29] Al Shakarchi J, McGrogan D, Yates P J and Inston N 2016 *J. Vasc. Access* **17** 151–54
[30] Ramshaw J A M, Casagranda F, White J F, Edwards G A, Hunt J A, Williams D F and Werkmeister J A 1999 *J. Biomed. Mater. Res.* **47** 309–15
[31] Werkmeister J A, Casagranda F, Edwards G A, White J F, Glattauer V and Ramshaw J A M 2017 *JSM Cell Devel. Biol.* **5** 1022
[32] Grabenwöger M, Fitzal F, Sider J, Csekö C, Bergmeister H, Schima H, Husinsky W, Böck P and Wolner E 1998 *Ann. Thorac. Surg.* **66** S110–4
[33] Werkmeister J A, Glattauer V, Tebb T A, Ramshaw J A M, Edwards G A and Roberts G 1991 *J. Long-term Effects Med. Implants* **1** 107–19
[34] Werkmeister J A, White J F and Ramshaw J A M 1994 *Med. Prog. Technol.* **20** 231–42
[35] Werkmeister J A, Tebb T A, Peters D E and Ramshaw J A M 1990 *Clin. Mater.* **6** 13–20
[36] Werkmeister J A, Tebb T A, White J F and Ramshaw J A M 1993 *J. Histochem. Cytochem.* **41** 1701–06
[37] Werkmeister J A, Edwards G A, Casagranda F, White J F and Ramshaw J A M 1998 *J. Biomed. Mater. Res.* **39** 429–36
[38] Huerta S, Varshney A, Patel P M, Mayo H G and Livingston E H 2016 *JAMA Surg.* **151** 374–81
[39] Ketharanathan V 2001 *US Patent* 6262332 B1
[40] Marcacci M, Gubellini P, Buda R, De Pasquale V, Strocchi R, Molgora A P, Zaffagnini S, Guizzardi S and Ruggeri A 1991 *Clin. Orthop. Relat. Res.* **1991** 115–21
[41] Zaffagnini S, De Pasquale V, Montanari C, Strocchi R and Marcacci M 1997 *Knee Surg. Sports Traumatol. Arthrosc.* **5** 89–94

John A M Ramshaw and Veronica Glattauer

Chapter 16

Collagen applications in tissue engineering and regenerative medicine

The concepts of tissue engineering were initially defined around 30 years ago and suggested a different and promising new direction for materials for tissue repair and regeneration, with a product target. The research was to focus on combining cells with a supporting structure (scaffold) along with the influence of signals, biochemical and physical, to construct new tissue. The advantages of this strategy continue to emerge, providing benefits to patients. Collagen, either as a purified and fabricated material or in the form of acellular matrix, has been an important material for these developments.

In May 2000, *Time Europe* magazine predicted that the top new job to emerge by 2025 could be that of tissue engineer [1]. It was suggested that artificial cartilage would be available shortly, while by 2025 the Petri dish would be providing pancreas as well as liver, kidney and heart tissues. If, as of now, there are not the range of tissue engineered organs and tissues that were originally envisaged for medical implants, the concept has not entirely been devoid of successes. In particular, an area of significant interest has been the development of tissue-like materials for pharmaceutical development, either for developing new chemical entities or, through a path using induced pluripotent stem cells, development of patient specific strategies for medication selection.

Just as we have in our title for this chapter, tissue engineering and regenerative medicine tend to be rolled together as if a single entity. But there are distinctions, as can be seen from David Williams in his dictionary of biomaterials [2]. The definitions continue to evolve, with different versions available from difference sources [2]. Tissue engineering has been defined, explaining the key elements involved, as:

> *the persuasion of the body to heal itself, through the delivery to appropriate sites of molecular signals, cells and supporting structures* [2]

doi:10.1088/978-0-7503-2096-2ch16

Tissue engineering requires three key elements—signals, cells and support structures (figure 16.1). Frequently, these elements are brought together to form tissue through use of a bioreactor. These bioreactor systems come in various formats, normally mechanical. Presently, it is considered that in addition to molecular signals, certain other signals, such as mechanical signals, could be important, for example, when using chondrocytes, where they induce biochemical processes.

The supporting structure, which is one of the key elements, is often referred to as a scaffold. Some see scaffolds as external, whereas in tissue engineering the structure is generally internal throughout the structure, so it seen as a supporting structure. There are a growing number of reports in the literature where some form of stem cells is used as the cell source. Human mesenchymal stem cells have gained most attention to date, but, more recently, induced pluripotent stem cells are being increasingly studied. Nevertheless, in many instances an adult cell type can be relevant, even though proliferation to give sufficient numbers may be an issue. Depending on the cell source, a range of short-term and long-term safety issues are present and need to be considered. The other cell-related topic that needs to be considered is the variety of approaches that can be considered for seeding onto and particularly into the matrix.

Regeneration, the basis of regenerative medicine, is broader in sense, and has been defined as:

> *synthesis of new, natural tissue at the site of a tissue (one cell type) or organ (more than one cell type) which either has been lost due to injury or has failed due to a chronic condition* [2]

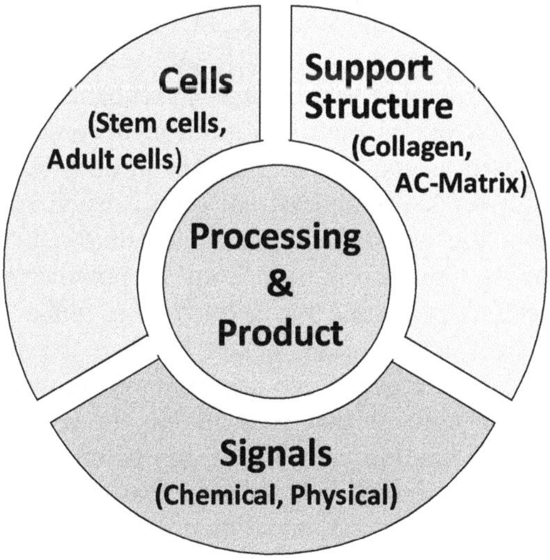

Figure 16.1. A schematic of the principle research inputs that are required for the processing and derivation of a tissue engineered material.

Hence, regenerative medicine approaches may not always include cells in a construct. Alternatively, the construct may not contain an introduced support structure, for example, if a cell pellet with new or organised matrix provided by the cells themselves is used.

In this chapter, we will briefly consider examples where collagen has been a key component of the technology. Initial studies focused on fabricated collagen supports, but more recently acellular matrix (ACM) (see chapter 8) with its abundant collagen has been in the ascendency. Collagen and ACM are ideally suited to the task. Initial and present studies have used synthetic resorbable polymers as the supporting structures. These structures define the shape and size of the resulting engineered tissue, but by themselves provide little, if any, control or input into the type of tissue that is present. If real, native tissues are examined they have complex compositions, distributions, orientations and interactions between components. They may contain numerous different collagen components each fulfilling a distinct role, as well as all the other matrix components. For example, examining a seemingly simple tissue, such as a heart valve, shows great complexity between the many components (see figure 5.5) [3]. So, it would seem much better to have a tissue engineered construct that is a close match to the native tissue, rather than poorly defined scar-like tissue, that is just the correct size and shape. The supporting structures and signals should preferably provide biological information to cells so that the cells function in the desired manner and produce a more natural tissue that is a closer match to the native tissue.

The benefit of collagen, and other proteins, as supporting structures is that they have a higher 'information content'. All molecules contain 'information' in their structures to a varying extent. This was first recognised in a very different context, that of molecular evolution, where the information in molecules was seen as a document of evolutionary history [4]. So, molecular complexity is generally associated with higher information content. In evolution, the greatest molecular complexity is found with proteins and with DNA. Initially, protein sequences, obtained with great effort, showed that changes in sequence accumulated over time, whether by natural selection or neutral mutation accumulation [5, 6] and this information could be used to gain insights into evolution, both a timeframe and a path, that was valuable for lesser understood groups, such as higher plants [7, 8]. These days, these data are much, much easier to obtain by DNA and genome sequencing. By contrast, the information seen in polymeric materials, such as carbohydrates was seen as providing less information, while small molecules that are fairly ubiquitous, such as metal ions, sugars, lipids or the individual amino acids of proteins contain little if any useful evolutionary information content. Hence, by analogy, the information content that is available for cells provided by a plain synthetic polymer is very limited, certainly if compared with collagen and other matrix proteins; for example, type I collagen alone has more than 50 individual sites of interaction [9] (see chapter 7). Also, depending on the method of fabrication, and most particularly for ACM, the topology and surface characteristics provide further information content.

Although the term tissue engineering only appeared around the mid to late 1980s [10–12], it is not necessarily consistent with today's meaning. There was also a body of work that preceded the coined phrase, including work focused on collagens [12]. For example, loose collagen gels had been transformed into more tissue-like structures through incubation with fibroblasts [13], and collagen-glycosaminoglycan materials had been used to produce artificial skin [14], along with the complementary technology to produce layers of keratinocytes that were suitable for burn patients [15].

Tissue engineering appeared to be particularly focused on patient outcomes. The field has undoubtedly progressed, but are many target products emerging? The concept of technology readiness levels (TRLs) was originally introduced in 1974 by NASA, for evaluation of space programs. This concept was more formally defined and definitions provided between 1989 and 1990, leading to the 9-level scale that is now used (see [16]). Apart from space technology, the concept has also become broadly accepted in many other industries, including the medical device industry. These other industries, and individual groups within them, have developed their own specific descriptors within the 9-levels while retaining the broad concepts. A synthesis derived from several of these is presented below (figure 16.2). If patients are to benefit, this product-oriented technology must progress to the higher levels of this scale.

A broad overview of the scientific literature shows few studies are reported that exceed TRL3 to 4. This is possibly to be expected as most of this literature comes

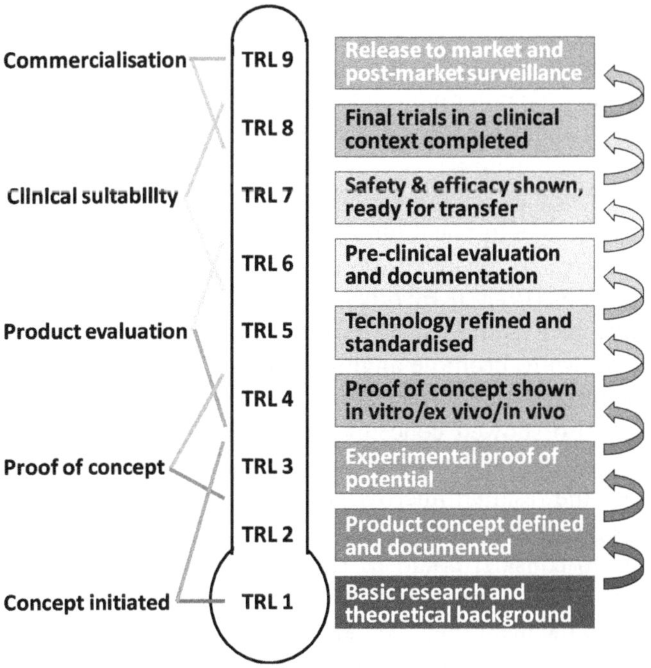

Figure 16.2. A synthesis of concepts for technology readiness levels for biomedical device production.

from academic institutions rather than from commercial endeavours, and although there is a focus now in many economies on translational research, the specialist resources that are need to move higher levels are not always available, because of cost, size or complexity. Nevertheless, more examination of concepts at an animal model stage, TRL4 to 5, would be expected if the work is to be seen as useful in the longer term. In part, a preponderance of TRL2/3 may result from the term 'tissue engineering' being used fairly loosely. Many papers describe a cell biology experiment, or more recently a stem cell experiment, rather than tissue engineering; the term 'tissue engineering' is included only as a 'may be useful in tissue engineering' context. Similarly, the biochemical description of a new collagen crosslinker is broadly useful, but is not in itself tissue engineering. At higher TRL levels, commercial issues emerge and so there is less data in the public domain as the proprietary nature of the information precludes publication. There are some reports in the patent literature, but many are lower level broad claims for collagen-based materials and processing methods, rather than the specifics of a particular tissue engineered product or process.

It is possible that the number of targets now being considered has become more limited. Whereas in the initial years when the tissue engineering concepts emerged, targets such as artificial, tissue engineered organs were being optimistically promoted, often by a tabloid press, these days the selection appears more limited. Any success in this field has, generally, been achieved through a decellularisation-recellularisation process of an existing organ [17]. A clear cause for this change is illustrated by articular cartilage repair. It has been known for centuries that once damaged or degraded articular cartilage does not repair by itself [18]. Therefore, articular cartilage repair is a frequent target for what could be termed *de novo* tissue engineering and repair, as arthritis is a significant medical issue. A key advantage of this as a target is that the potential solution is possibly less complex than many other targets, because the engineered tissue needs only to be quite thin, and, like the native tissue is avascular and aneural.

The question of how to achieve effective tissue vascularisation remains critical to success for many areas of application. In one example, vascularisation of neo-tissue has been achieved *in vivo*, through the use of an implanted chamber containing a diverted blood vessel [19–21]. In other studies, the use of systems that promote angiogenesis have been examined. For example, short chain hyaluronic acid (MW 6500 kDa) was shown to promote angiogenesis *in vivo* in collagen scaffolds [22]. In other examples, specific angiogenic factors, such as angiopoietin or VEGF, led to angiogenesis but also showed concentration dependency [23, 24]. One advantage that ACM scaffolds may have, when compared to fabricated collagen supporting structures, is that the residual micro-architecture of the tissue may provide a pro-angiogenic property to the material [25, 26].

Another issue to consider could be risk and litigation potential. If an artificial cartilage repair fails, there are proven, effective surgical options, such as total knee replacement, but in the meantime the patient will have been given a significant pain-free opportunity. Similarly, if you could make an artificial kidney, in the event of failure the patient could return, hopefully, to dialysis prior to other treatment. On the other hand, if an artificial heart or heart valve fails.... ?

16.1 Fabricated collagen as a supporting structure

Various studies have extended the use of collagen as a key component in tissue engineering studies. Many papers have included collagen by itself as a supporting material. Collagen has also been used as part of a composite material (see chapter 18), and often as a coating (see chapter 17). In these cases, it is present to provide some positive biological effect. In many reports cell studies are included, but often these have been used to assess cytotoxicity and similar effects, rather than being included as an integral component to enhance the functionality of the supporting structure. Fewer studies have followed through to assess function, but these have produced valuable results. A selection of examples is discussed below.

16.1.1 Musculoskeletal and associated tissues

A wide range of commercially available collagen-based materials have been developed for bone repair which are effective in clinical use. Most, if not all, are composite structures that involve type I collagen, the main type of collagen in bone, where the addition of hydroxyapatite (HyA), can act to stimulate bone formation. Perhaps, the success of these materials has meant that there has been less emphasis on a tissue engineering approach for bone repair than in other clinical areas.

One variable that has been considered in tissue engineering approaches has been the nature of the added cells. Initially, differentiated cells for appropriate tissues, such as periosteum were examined with positive results [27]. However, there has been a growing interest in the use of stem cells, particularly mesenchymal stem cells. These have been sourced from various tissues, including adipose tissue [28, 29], periosteum [29], tendon [30] and bone marrow [29–31], each proving suitable. Other studies indicated that there are benefits from using any cell type that is expressing CD34 [32]. Also, once safety issues have been adequately resolved, induced pluripotent stem cells may also be a useful cell source [33].

In addition to the supporting structure and the cells, the other variable to consider are physical or chemical signals. These could be physical, as shown by the increased osseous regeneration that can be obtained using very low-level, high-frequency accelerations [34]. Alternatively, chemical signals, such as appropriate growth factors [35], for example from the bone morphogenic family, also enhance osseous regeneration.

Other orthopaedic indications are possible targets for tissue engineering. However, in some cases, such as ligaments and tendons, the mechanical requirements are not readily met by a reconstituted construct and composite [36]. An acellular material may provide a better solution, in combination with appropriate cell biology studies. These have included studies of tenocytes, using fibrin to assist in phenotype retention [37] and with dermal fibroblasts to assess whether they could adapt to the application [38]. While reconstituted collagen typically has poor properties under tension, it is more suitable, though possibly not ideal, under compression. Hence, products for meniscus repair have been proposed, based of reconstituted collagen structures (see chapter 14). Although studies on suitability and the potential to culture meniscal cells from different native locations, along with the effects or oxygen tension have been discussed [39], this has not led to extensive tissue engineered construct development. Bone loss in periodontal disease is another target for bone augmentation or regrowth. In this instance, the use of a membrane to isolate the area, including by collagen membranes, has proved an effective concept [40, 41], hence not requiring use of a bioreactor and cells in preparing an implant. Similarly, insufficient bone after tooth extraction prior to dental implant placement can normally be resolved using a demineralised bone implant, with more complex strategies not being needed. Notwithstanding these existing options, the inclusion of dental pulp cells along with a collagen implant does improve clinical performance [42].

One major clinical area that has been of significant interest is repair of articular cartilage. The prevalence of osteoarthritis, which is growing with increases in life expectancy and probably with increases in obesity, will lead to ongoing demand for a satisfactory tissue regeneration. Purified and reconstituted collagen devices have been prominent in research to date.

Early research included the development of a cell-collagen gel composite for *in situ* delivery for cartilage tissue regeneration [43]. The use of external fabrication of collagen-based tissue, in tissue culture and in some cases using an independent bioreactor emerged in the mid-1990s. These included the seeding of a fibrous collagen scaffold with chondrocytes, with subsequent augmentation by soluble collagen; after 7–10 days culture the tissue substitute was implanted in animals [44]. Following the same theme, another study compared chondrocyte growth on type I collagen and on type II collagen, the principle collagen found in cartilage. After 14 days in culture, there was a significantly higher amount of DNA and glycosaminoglycan synthesis associated with the type II collagen scaffold, suggesting that it could be preferred for cartilage repair implants [45]. Further, in keeping with the concepts of tissue engineering (figure 16.1), another study grew chondrocytes, the cells, on a collagen sponge, the scaffold or supporting structure, and this was treated with epidermal growth factor, a signal, prior to implant. Comparison with a sample without the growth factor addition, showed that this signal enhanced the production of regenerated cartilage [46].

Further attention has also been paid to the quality of the cells. When autologous chondrocytes are expanded in culture from biopsy samples, the cells rapidly dedifferentiate into fibroblast-like cells (figure 16.3) that produce mainly type I collagen rather than the type II collagen required for functional cartilage [47]. The resulting fibrocartilage production is less than ideal. The introduction of microcarriers and spinner culture for chondrocyte expansion [48], showed that rapid proliferation while retaining phenotype could be achieved, giving sufficient cells for a potential therapeutic approach. The initial studies used a highly crosslinked collagen-based bead [48], while subsequently, a range of other beads and particles being shown to give similar effects [49]. Demineralised bone collagen particles were particularly effective for cell growth [47, 50]. Use of a resorbable bead

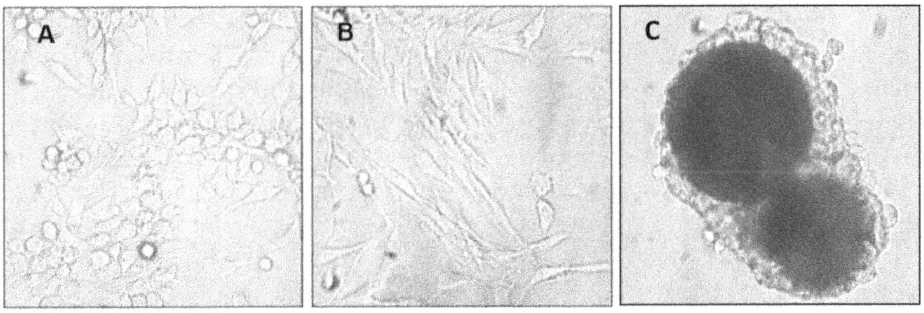

Figure 16.3. De-differentiation of chondrocytes in monolayer culture to fibroblast-like phenotype. (A) Cells after three days in culture, starting to show some elongated characteristics, (B) Cells after six days in culture, showing a predominately elongated, fibroblast-like shape. (C) Chondrocytes in spinner culture on PLGA beads coated by absorption with type II collagen [51]. After culture for 12 days, the cells have proliferated and retain their rounded morphology. (From [47], with permission of InTech Publishing.)

type, such as a polymer [51], or minimally stabilised gelatin [50], or minimally stabilised collagen [52]. The advantage of resorbable beads is that they can be directly placed in a lesion. This has been done with a collagen gel for delivery, held in location by a periosteal flap [53], which led to superior results, with good cartilage regeneration over the full aspect of the induced lesion [53]. Despite these promising data, this approach has seemingly not found favour for further clinical development, perhaps due to difficulties with spinner-culture and the lack of a cost-effective, disposable off-the-shelf reactor vessel.

A large number of studies have examined the use of a reconstituted collagen sponge for holding, stabilising and delivering cells [54, 55] as an improvement over cells alone, as used in autologous chondrocyte transplantation (ACI) [56] and cell delivery using a synthetic polymer material [57].

Various studies have continued to examine collagen scaffolds along with autologous chondrocytes, even though the dedifferentiation during cell expansion leads to fibrocartilage formation rather than hyaline cartilage (e.g., [58–62]). Although autologous, adult chondrocytes can be expanded sufficiently with retention of phenotype for use in cartilage repair, the current trend has also been to move to stem cell sources. For example, studies have used stem cells from adipose tissue [63], umbilical tissue [64] and bone marrow [65, 66]. However, whatever cell source is used, the successful integration of newly formed cartilage with adjacent tissue remains an issue that needs additional investigation [67].

Most recently bioprinting approaches have been proposed for cartilage repair, with various 'bioink' formulations proposed based on purified collagen and extracted cartilage tissue [68]. A range of different printing approaches and instruments are available for assembling an implant by printing cells and carrier materials prior to clinical implant [69]. The most recent innovation in this area is a handheld device, a 'biopen', that allows a surgeon to print directly into an arthritic lesion though either a mono, dual or coaxial print stream with *in situ* curing of the material as necessary [70].

There are other, different cartilaginous structures that have been examined as targets for repair through tissue engineering. These include nasal cartilage [71] and ear cartilage, where steady, significant progress has been made using a collagen matrix support for cells, reinforced to control shrinkage and prevent distortion [72–74]. Other tissues with a cartilaginous component that have been of interest, where there is an unmet medical need include the trachea [75] and the larynx [76], both of which need further development.

16.1.2 Cardiovascular tissues

Any tissue engineered product proposed for a cardiovascular application will need to be able to perform for an extended period under pulsatile conditions. Early endeavours to establish replacement arterial conduits clearly showed that under these pulsatile conditions aneurysms and similar durability issues presented major problems, with biosynthetic vessels providing the best biological option to date (see chapter 15). A similar situation exists for bioprosthetic heart valves where tearing, mainly due to the effects of calcification, presents a durability problem [77].

Development of a tissue engineered heart valve was an area of high initial interest in the 1990s, but 'promised more than it has delivered' [77]. This has not changed when a reconstituted collagen support is used, although the use of an acellular support (see below) may be more promising.

Treatments for myocardial infarcts have attracted attention, over an extended period. Treatments have been envisaged as providing new tissue or of delivering appropriate cell activity to adjacent tissue. A variety of different reconstituted collagen sponges and sheets have been described. It was shown that collagen was an effective substrate for myocardial cells, and seeded cells were observed to beat rapidly, up to 162 bpm, which was twice the rate observed on tissue culture plastic [78]. Initial clinical evaluation of a similar seeded material did not lead to any adverse effects, although it was recognised that further studies were needed to evaluate fully the benefits of a potential 'bioartificial myocardium' [79]. If bone marrow mesenchymal stem cells are used as a cell source, electrostimulation can be used to change cell morphology and biochemical characteristics to provide cardiac markers [80]. The stiffness of the collagen matrix is important to direct the fate and the differentiation state when stem cells are used [81]. Inclusion of VEGF bound to the collagen matrix for *in vivo* cell recruitment leads to enhanced tissue growth with an improved blood vessel density and cell survival [82]. High cell densities could be achieved through coupling Arg-Gly-Asp adhesion peptide to the collagen matrix, which also provided cell contractile properties and better cell viability [83].

16.1.3 Other tissues

Tissue engineering using a reconstituted collagen matrix has been explored for a range of other tissue repair or replacement options, although in some cases alternative, off-the shelf devices have been clinically accepted. These include tissue equivalents for ophthalmology, skin repair and bladder repair.

In ophthalmology, there is a significant shortage of suitable tissue banked corneal graft tissue in many countries, albeit that there ought to notionally be enough donors [84]. A significant demand, around 39% of procedures, arises from patients suffering from Fuchs dystrophy [84]. The structure of the cornea, with its highly ordered, orthogonal lamellar structure (see chapter 5), is critical for good optical clarity. Various approaches have been used to build this type of structure. For example, thin collagen fibre layers, where the collagen is magnetically aligned can be sequentially deposited to build a corneal tissue construct [85]. The transparency of the material was further improved by addition of glycosaminoglycans and keratocytes while the orthogonal structure was maintained during infiltration through the tissue equivalent [85]. Other approaches have used compressed collagen to achieve a translucent, hydrated structure, generally with the addition of cells [86, 87]. Both limbal cells [86] and corneal stromal cells [87] have been used to improve the quality of the devices and in both cases, successful animal trials have been reported [87, 88].

Several successful, commercial strategies for skin repair have been developed, as scaffolds alone, but notably as early examples of tissue engineered products (see above) [13–15]. Extending from these pioneering studies, other have looked at enhancing these materials. These studies have included adjustments to the scaffold architecture such that there is a gradient of interstitial space size at different depths of the collagen support [89]. Others have examined different cells types, including hair follicle dermal sheath cells that were found to provide better basal lamina establishment than fibroblasts [90] and adipose-derived cells that encouraged early vascularisation [91]. Other investigations have further developed the dermal equivalent materials for study of diseases and their treatment [92].

Bladder repair has led to a several studies, although recent results may suggest that an acellular matrix may be better as a supporting structure than reconstituted collagen (see below). Nevertheless, the studies using reconstituted collagen have proved valuable, for example, for providing a tissue platform for studying various disease and treatment options [93]. A selection of preliminary reports has also been published covering the development of other tissue engineering products that use a reconstituted collagen supporting structure. These include studies on nerve repair [94], nucleus pulposus repair [95] and fat regeneration. Despite the huge global business that encourages and helps some people to lose body fat and reduce obesity, there are some clinical instances where a fat transplant is required. This need for a fat transplant has been approached using stem cells in a collagen matrix [96], with studies tracking the fate of the implanted cells [97].

16.2 ACM as a supporting structure

ACM is gaining in popularity as a supporting structure. In part, the methods for production from various tissue sources are now well established; also the benefits of its potential pro-angiogenic properties have been recognised [25, 26]. In some cases, it is used as a generic support for regeneration of unrelated tissues, for example, bladder ACM for skeletal muscle repair [98] or dermis ACM for neural repair [99] or for tympanic membrane replacement [100]. Alternatively, in other cases, it is used for regeneration of a similar tissue, for example, acellular porcine heart valve has been used for new heart valves [101] or bladder ACM for bladder regeneration [102, 103].

16.2.1 Bladder and urologic tissues

There have been many studies reporting on the use of bladder ACM, frequently porcine, for the development of engineered tissue for the repair or replacement of diseased or damaged bladder tissue. These studies cover a broad range of activities. They include as a starting point the development of decellularization strategies that extend from the use of detergents and enzymes [104] to include a process of dynamic, cyclic distention of the tissue that increases the extraction efficiency of cellular components [105]. Various cell sources have been used for cells to seed into the wall of prototype devices. These include stem cells derived from adipose tissue [106, 107], human umbilical tissue [108], bone marrow [109] and autologous smooth muscle cells [110]. Specific attention has been given to culture and seeding of urothelial surface cells [111] whose presence on the seeded tissue facilitates the recruitment and differentiation of host cells to the implant [112, 113]. Various molecules have been added to enhance the cellular process, including hyaluronic acid [113]. Mostly, these molecules have been growth factors, including platelet-derived growth factor, vascular endothelial growth factor, fibroblast growth factor and epidermal growth factor, individually and in combination [114–117], each providing benefits. Many studies have progressed to successful animal trials, using mouse, dog and pig models [107, 115, 118, 119]. The animal trials have led to limited clinical studies that can be compared to other treatments (see [120, 121]).

The extensive development of bladder ACM has led to it being used in other potential applications. Notably, this has been in urethral reconstruction [122, 123], where the use of an acellular bladder structure, as an alternative to an acellular urethral-based implant [124], has been developed [125–127], and has proved favourable in clinical application [128].

16.2.2 Other tissues

Other clinical areas where there has been some interest in the development of ACM-based materials include heart valves, where the acellular tissue could be useful in assisting with producing the long-term durability that is essential. Initial studies showed the potential for this approach, including the results of 3-month trials in a sheep model [129, 130]. In oesophageal applications, it includes the development of appropriate scaffold structures using different sources of tissue [131] and regeneration of the muscle layer [132]. More recently, studies in mini-pig models have continued to show progress [133]. Other investigations have looked at systems that allow regeneration of muscle, especially for body wall repair, with promising results obtained in animal models after appropriate cell seeding. However, these data indicated that additional studies were needed for better structural and functional integrity of implants over longer periods [134–136].

Often, ACM has only been used to a limited extent for tissue engineering in areas where well-established materials and/or protocols have been established. When ACM materials have been examined, they have performed well for cartilage repair [137, 138] and for skin repair, particularly using acellular amniotic membrane [139, 140]. There have also been some studies based on ACM to produce small diameter vascular implants, when difficulties related to thrombogenicity [141] and to aneurysms may occur. When natural vessels were proposed, even after addition of external supporting fabric, these problems emerged [142, 143]. It has been suggested that the use of a completely unrelated ACM, such as tendon may provide a solution [144].

16.2.3 Organ replacements

The most important application of like-to-like use is likely to be the use of ACM in the development of functional, effective whole organs. While this area still remains in its infancy, it is the most probable path to a successful device that will bring benefits to patients. An understanding of critical factors for success is starting to emerge, based on results for selected organs and tissues [145].

An example is liver. Initial studies used a polymer support structure [146, 147]. These, and other studies, suggested that the observed efficiency of the initial constructs may preclude the development of a suitably sized organ-mimic to allow implant. Hence the concepts of extra-corporeal liver-assist devices gained attention, with various topologies being possible [148, 149]. A clear positive of these initial studies was that it could be shown that effective tissue mimics could be made that could be used for drug testing [150, 151]. Recent progress in the development of functional liver tissue has been achieved through the use of ACM [26]. These data have showed that appropriate cell infiltration was possible with production of a range of hepatic markers, that pro-angiogenic effects were present and that the material could be implanted in a mouse model [26]. The potential to use autologous cells in regenerated materials, for drug testing or eventually possibly implant has been demonstrated via an induced pluripotent stem cell (iPS) approach [152].

Progress towards the artificial heart has been through ACM, using using decellularisation-recellularisation on existing organs. Some functionality could be regained, but it was limited [17, 153].

Some other organ systems for which there is a clear clinical need and where ACM could be used as the supporting structure have also received attention. These include the development of an artificial pancreas [154], an artificial kidney [155, 156] and development of acellular lung tissue for studies of disease and potentially as a replacement tissue [157].

References

[1] Rawe J 2000 *Time Europe* **155** 52

[2] Williams D F 1999 *The Williams Dictionary of Biomaterials* (Liverpool: Liverpool University Press)

[3] White J F, Werkmeister J A, Hilbert S L and Ramshaw J A M 2010 *J. Heart Valve Dis.* **19** 766–71
[4] Zuckerkandl E and Pauling L 1965 *J. Theor. Biol.* **8** 357–66
[5] Kimura M 1968 *Nature* **217** 624–6
[6] Kimura M 1983 *The Neutral Theory of Molecular Evolution* (Cambridge: Cambridge University Press)
[7] Boulter D, Ramshaw J A M, Thompson E W, Richardson M and Brown R H 1972 *Proc. Roy. Soc. Lond.* B **181** 441–55
[8] Ramshaw J A M, Richardson D L, Meatyard B T, Brown R H, Richardson M, Thompson E W and Boulter D 1972 *New Phytol.* **71** 773–9
[9] Sweeney S M, Orgel J P, Fertala A, McAuliffe J D, Turner K R, Di Lullo G A, Chen S, Antipova O and Perumal S *et al* 2008 *J. Biol. Chem.* **283** 21187–97
[10] Wolter J R and Meyer R F 1984 *Trans. Am. Ophthalmol. Soc.* **82** 187–202
[11] Matsuda T, Akutsu T, Kira K and Matsumoto H 1989 *ASAIO Trans.* **35** 553–5
[12] Vacanti C A 2006 *J. Cell Mol. Med.* **10** 569–76
[13] Bell E, Ivarsson B and Merrill C 1979 *Proc. Natl Acad. Sci. USA* **76** 1274–8
[14] Yannas I V and Burke J F 1980 *J. Biomed. Mater. Res.* **14** 65–81
[15] Green H, Kehinde O and Thomas J 1979 *Proc. Natl Acad. Sci. USA* **76** 5665–8
[16] Banke J 2010 *Technology Readiness Levels Demystified* (Washington DC: NASA)
[17] Ott H C, Matthiesen T S, Goh S K, Black L D, Kren S M, Netoff T I and Taylor D A 2008 *Nat. Med.* **14** 213–21
[18] Hunter W 1742-3 *Phil. Trans.* **42** 514–21
[19] Cronin K J, Messina A, Knight K R, Cooper-White J J, Stevens G W, Penington A J and Morrison W A 2004 *Plast. Reconstr. Surg.* **113** 260–9
[20] Chan E C, Kuo S M, Kong A M, Morrison W A, Dusting G J, Mitchell G M, Lim S Y and Liu G S 2016 *PLoS One* **11** e0149799
[21] Yap K K, Yeoh G C, Morrison W A and Mitchell G M 2018 *Trends Biotechnol.* **36** 1011–24
[22] Perng C K, Wang Y J, Tsi C H and Ma H 2011 *J. Surg. Res.* **168** 9–15
[23] Yin J, Gong G, Sun C, Yin Z, Zhu C, Wang B, Hu Q, Zhu Y and Liu X 2018 *Biomed. Pharmacother.* **105** 932–9
[24] Jiang X, Xiong Q, Xu G, Lin H, Fang X, Cui D, Xu M, Chen F and Geng H 2015 *Ann. Biomed. Eng.* **43** 2577–86
[25] Maghsoudlou P, Georgiades F, Tyraskis A, Totonelli G, Loukogeorgakis S P, Orlando G, Shangaris P, Lange P and Delalande J M *et al* 2013 *Biomaterials* **34** 6638–48
[26] Agarwal T, Maiti T K and Ghosh S K 2019 *Mater. Sci. Eng. C Mater. Biol. Appl.* **98** 939–48
[27] Ryu Y M, Hah Y S, Park B W, Kim D R, Roh G S, Kim J R, Kim U K, Rho G J, Maeng G H and Byun J H 2011 *Mol. Biol. Rep.* **38** 2887–94
[28] Kakudo N, Shimotsuma A, Miyake S, Kushida S and Kusumoto K 2008 *J. Biomed. Mater. Res. A* **84** 191–7
[29] Stockmann P, Park J, von Wilmowsky C, Nkenke E, Felszeghy E, Dehner J F, Schmitt C, Tudor C and Schlegel K A 2012 *J. Craniomaxillofac. Surg.* **40** 310–20
[30] Kunkel N, Wagner A, Gehwolf R, Heimel P, Tempfer H, Korntner S, Augat P, Resch H and Redl H *et al* 2017 *J. Tissue Eng. Regen. Med.* **11** 2014–23
[31] Todo M and Arahira T 2013 *Conf. Proc. IEEE Eng. Med. Biol. Soc.* **2013** 409–12

[32] Hertweck J, Ritz U, Götz H, Schottel P C, Rommens P M and Hofmann A 2018 *J. Biomed. Mater. Res. B Appl. Biomater.* **106** 1505–16
[33] Ozasa R, Matsugaki A, Isobe Y, Saku T, Yun H S and Nakano T 2018 *J. Biomed. Mater. Res. A* **106** 360–9
[34] Hwang S J, Lublinsky S, Seo Y K, Kim I S and Judex S 2009 *Clin. Orthop. Relat. Res.* **467** 1083–91
[35] Qian Y, Yao G, Lin Z, Chen J, Fan Y, Davey T, Xu J and Zheng M 2009 *J. Biomed. Mater. Res. B Appl. Biomater.* **90** 778–88
[36] Ran J, Hu Y, Le H, Chen Y, Zheng Z, Chen X, Yin Z, Yan R and Jin Z *et al* 2017 *Acta Biomater.* **53** 307–17
[37] Pawelec K M, Wardale R J, Best S M and Cameron R E 2015 *J. Mater. Sci., Mater. Med.* **26** 5349
[38] Tremblay P, Cloutier R, Lamontagne J, Belzil A M, Larkin A M, Chouinard L, Chabaud S, Laverty S, Lussier B and Goulet F 2011 *Cell Transplant.* **20** 535–42
[39] Croutze R, Jomha N, Uludag H and Adesida A 2013 *BMC Musculoskelet. Disord.* **14** 353
[40] Ellender G, Hammond R, Papli R, Mitrangas K, Bateman J F, Glattauer V, Thyer J M, Werkmeister J A and Ramshaw J A M 1992 *Clin. Mater.* **9** 201–9
[41] Sheikh Z, Qureshi J, Alshahrani A M, Nassar H, Ikeda Y, Glogauer M and Ganss B 2017 *Odontology* **105** 1–12
[42] Ferrarotti F, Romano F, Gamba M N, Quirico A, Giraudi M, Audagna M and Aimetti M 2018 *J. Clin. Periodontol.* **45** 841–50
[43] Wakitani S, Kimura T, Hirooka A, Ochi T, Yoneda M, Yasui N, Owaki H and Ono K 1989 *J. Bone Joint Surg. Br.* **71** 74–80
[44] Sams A E and Nixon A J 1995 *Osteoarth. Cartilage* **3** 47–59
[45] Nehrer S, Breinan H A, Ramappa A, Shortkroff S, Young G, Minas T, Sledge C B, Yannas I V and Spector M 1997 *J. Biomed. Mater. Res.* **38** 95–104
[46] Fujisato T, Sajiki T, Liu Q and Ikada Y 1996 *Biomaterials* **17** 155–62
[47] Glattauer V, Tsai W-B, White J F, Nigro J, Tebb T A, Werkmeister J A and Ramshaw J A M 2011 *Regenerative Medicine and Tissue Engineering – Cells and Biomaterials* ed D Eberli (Rijeka: InTech) pp 261–86
[48] Frondoza C, Sohrabi A and Hungerford D 1996 *Biomaterials* **17** 879–88
[49] Malda J and Frondoza C G 2006 *Trends Biotechnol.* **24** 299–304
[50] Glattauer V, White J F, Tsai W B, Tsai C C, Tebb T A, Danon S J, Werkmeister J A and Ramshaw J A M 2010 *J. Biomed. Mater. Res. A* **92** 1301–09
[51] Thissen H, Chang K Y, Tebb T A, Tsai W B, Glattauer V, Ramshaw J A M and Werkmeister J A 2006 *J. Biomed. Mater. Res. A* **77** 590–8
[52] Tebb T A, Tsai S W, Glattauer V, White J F, Ramshaw J A M and Werkmeister J A 2006 *Cytotechnology* **52** 99–106
[53] Chiang H, Kuo T F, Tsai C C, Lin M C, She B R, Huang Y Y, Lee H S, Shieh C S and Chen M H *et al* 2005 *J. Orthop. Res.* **23** 584–93
[54] Frenkel S R, Toolan B, Menche D, Pitman M I and Pachence J M 1997 *J. Bone Joint Surg.* **79B** 831–6
[55] Lee C R, Grodzinsky A J, Hsu H P and Spector M 2003 *J. Orthop. Res.* **21** 272–81
[56] Brittberg M, Lindahl A, Nilsson A, Ohlsson C, Isaksson O and Peterson L 1994 *N. Engl. J. Med.* **331** 889–94

[57] Chu C R, Coutts R D, Yoshioka M, Harwood F L, Monosov A Z and Amiel D 1995 *J. Biomed. Mater. Res.* **29** 147–54
[58] Zheng M H, Willers C, Kirilak L, Yates P, Xu J, Wood D and Shimmin A 2007 *Tissue Eng.* **13** 737–46
[59] Grigolo B, Desando G, Cavallo C, Zini N, Ghisu S and Facchini A 2011 *J. Biol. Regul. Homeost. Agents* **25** S53–62
[60] Nürnberger S, Meyer C, Ponomarev I, Barnewitz D, Resinger C, Klepal W, Albrecht C and Marlovits S 2013 *Anat. Histol. Embryol.* **42** 332–43
[61] Ma N, Wang H, Xu X, Wan Y, Liu Y, Wang M, Yu W, Dai Y and Peng J *et al* 2017 *Trials* **18** 519
[62] Wang K H, Wan R, Chiu L H, Tsai Y H, Fang C L, Bowley J F, Chen K C, Shih H N and Lai W T 2018 *PLoS One* **13** e0196779
[63] Kang H, Lu S, Peng J, Yang Q, Liu S, Zhang L, Huang J, Sui X and Zhao B *et al* 2015 *Cell Tissue Bank.* **16** 123–33
[64] Sadlik B, Jaroslawski G, Gladysz D, Puszkarz M, Markowska M, Pawelec K, Boruczkowski D and Oldak T 2017 *Adv. Exp. Med. Biol.* **1020** 113–22
[65] Araki S, Imai S, Ishigaki H, Mimura T, Nishizawa K, Ueba H, Kumagai K, Kubo M and Mori K *et al* 2015 *Acta Orthop.* **86** 119–26
[66] Chou C L, Rivera A L, Williams V, Welter J F, Mansour J M, Drazba J A, Sakai T and Baskaran H 2017 *Acta Biomater.* **60** 210–9
[67] Pabbruwe M B, Esfandiari E, Kafienah W, Tarlton J F and Hollander A P 2009 *Biomaterials* **30** 4277–86
[68] Jung C S, Kim B K, Lee J, Min B H and Park S H 2017 *Tissue Eng. Regen. Med.* **15** 155–62
[69] Francis S L, Di Bella C, Wallace G G and Choong P F M 2018 *Front. Surg.* **5** 70
[70] Di Bella C, Duchi S, O'Connell C D, Blanchard R, Augustine C, Yue Z, Thompson F, Richards C and Beirne S *et al* 2018 *J. Tissue Eng. Regen. Med.* **12** 611–21
[71] Andrews S H J, Kunze M, Mulet-Sierra A, Williams L, Ansari K, Osswald M and Adesida A B 2017 *Sci. Rep.* **7** 6490
[72] Zhou L, Pomerantseva I, Bassett E K, Bowley C M, Zhao X, Bichara D A, Kulig K M, Vacanti J P, Randolph M A and Sundback C A 2011 *Tissue Eng. Part A* **2011** 1573–81
[73] Bichara D A, Pomerantseva I, Zhao X, Zhou L, Kulig K M, Tseng A, Kimura A M, Johnson M A and Vacanti J P *et al* 2014 *Tissue Eng. Part A* **20** 303–12
[74] Zhao X, Bichara D A, Zhou L, Kulig K M, Tseng A, Bowley C M, Vacanti J P, Pomerantseva I, Sundback C A and Randolph M A 2015 *J. Craniomaxillofac. Surg.* **43** 382–9
[75] Lin C H, Hsu S H, Huang C E, Cheng W T and Su J M 2009 *Biomaterials* **30** 4117–26
[76] Zhang H, Voytik-Harbin S, Brookes S, Zhang L, Wallace J, Parker N and Halum S 2018 *Laryngoscope* **128** E123–9
[77] Vesely I 2005 *Circ. Res.* **97** 743–55
[78] Guo Z, Iku S, Zheng X, Sammons R L and Kuboki Y 2012 *Artif. Organs* **36** 816–9
[79] Chachques J C, Trainini J C, Lago N, Masoli O H, Barisani J L, Cortes-Morichetti M, Schussler O and Carpentier A 2007 *Cell Transplant.* **16** 927–34
[80] Haneef K, Lila N, Benadda S, Legrand F, Carpentier A and Chachques J C 2012 *Heart Int.* **7** e14
[81] Lin Y L, Chen C P, Lo C M and Wang H S 2016 *J. Biomed. Mater. Res. A* **104** 2234–42

[82] Miyagi Y, Chiu L L, Cimini M, Weisel R D, Radisic M and Li R K 2011 *Biomaterials* **32** 1280–90

[83] Schussler O, Coirault C, Louis-Tisserand M, Al-Chare W, Oliviero P, Menard C, Michelot R, Bochet P and Salomon D R *et al* 2009 *Nat. Clin. Pract. Cardiovasc. Med.* **6** 240–9

[84] Gain P, Jullienne R, He Z, Aldossary M, Acquart S, Cognasse F and Thuret G 2016 *JAMA Ophthalmol.* **134** 167–73

[85] Torbet J, Malbouyres M, Builles N, Justin V, Roulet M, Damour O, Oldberg A, Ruggiero F and Hulmes D J S 2007 *Biomaterials* **28** 4268–76

[86] Mi S, Chen B, Wright B and Connon C J 2010 *Tissue Eng. Part A* **16** 2091–100

[87] Cui Z, Zeng Q, Liu S, Zhang Y, Zhu D, Guo Y, Xie M, Mathew S and Cai D *et al* 2018 *Acta Biomater.* **75** 183–99

[88] Xiao X, Pan S, Liu X, Zhu X, Connon C J, Wu J and Mi S 2014 *J. Biomed. Mater. Res. A* **102** 1782–7

[89] Wang Y, Xu R, Luo G, Lei Q, Shu Q, Yao Z, Li H, Zhou J and Tan J *et al* 2016 *Acta Biomater.* **30** 246–57

[90] Higgins C A, Roger M F, Hill R P, Ali-Khan A S, Garlick J A, Christiano A M and Jahoda C A B 2017 *Br. J. Dermatol.* **176** 1259–69

[91] Foubert P, Barillas S, Gonzalez A D, Alfonso Z, Zhao S, Hakim I, Meschter C, Tenenhaus M and Fraser J K 2015 *Burns* **41** 1504–16

[92] Jannasch M, Groeber F, Brattig N W, Unger C, Walles H and Hansmann J 2015 *Exp. Parasitol.* **150** 22–30

[93] Janssen D A, Geutjes P J, Odenthal J, van Kuppevelt T H, Schalken J A, Feitz W F and Heesakkers J F 2013 *J. Urol.* **190** 341–9

[94] Nakada A, Fukuda S, Ichihara S, Sato T, Itoi S, Inada Y, Endo K and Nakamura T 2009 *Cells Tissues Organs* **190** 326–35

[95] Zhou X, Tao Y, Chen E, Wang J, Fang W, Zhao T, Liang C, Li F and Chen Q 2018 *J. Biomed. Mater. Res. A* **106** 1258–68

[96] Lequeux C, Oni G, Wong C, Damour O, Rohrich R, Mojallal A and Brown S A 2012 *Plast. Reconstr. Surg.* **130** 1208–17

[97] Venugopal B, Fernandez F B, Harikrishnan V S and John A 2017 *J. Mater. Sci., Mater. Med.* **28** 28

[98] Corona B T, Ward C L, Baker H B, Walters T J and Christ G J 2014 *Tissue Eng. Part A* **20** 705–15

[99] Syu W Z, Chen S G, Chan J Y, Wang C H, Dai N T and Huang S M 2019 *Ann. Plast. Surg.* **82** S108–18

[100] Deng Z, Wu J, Qiu J, Wang J, Tian Y, Li Y and Jin Y 2009 *Tissue Eng. Part A* **15** 3729–39

[101] Bader A, Schilling T, Teebken O E, Brandes G, Herden T, Steinhoff G and Haverich A 1998 *Eur. J. Cardiothorac. Surg.* **14** 279–84

[102] Song L, Murphy S V, Yang B, Xu Y, Zhang Y and Atala A 2014 *Tissue Eng. Part B Rev.* **20** 163–72

[103] Pokrywczynska M, Gubanska I, Drewa G and Drewa T 2015 *Biomed. Res. Int.* **2015** 613439

[104] Bolland F, Korossis S, Wilshaw S P, Ingham E, Fisher J, Kearney J N and Southgate J 2007 *Biomaterials* **28** 1061–70

[105] Consolo F, Brizzola S, Tremolada G, Grieco V, Riva F, Acocella F, Fiore G B and Soncini M 2016 *J. Tissue Eng. Regen. Med.* **10** E101–12

[106] Zhu W D, Xu Y M, Feng C, Fu Q, Song L J and Cui L 2010 *World J. Urol.* **28** 493–8
[107] Pokrywczynska M, Jundzill A, Rasmus M, Adamowicz J, Balcerczyk D, Buhl M, Warda K, Buchholz L and Gagat M *et al* 2018 *Stem Cell Res. Ther.* **9** 328
[108] Yuan H, Zhuang Y, Xiong J, Zhi W, Liu L, Wei Q and Han P 2013 *PLoS One* **8** e80959
[109] Coutu D L, Mahfouz W, Loutochin O, Galipeau J and Corcos J 2014 *PLoS One* **9** e111966
[110] Zhou L, Xia J, Wang P, Jia R, Zheng J, Yao X, Chen Y, Dai Y and Yang B 2018 *Tissue Eng. Part A* **24** 1066–81
[111] Kurzrock E A, Lieu D K, deGraffenried L A and Isseroff R R 2005 *J. Urol.* **173** 281–5
[112] Master V A, Wei G, Liu W and Baskin L S 2003 *J. Urol.* **170** 1628–32
[113] Brown A L, Ringuette M J, Prestwich G D, Bagli D J and Woodhouse K A 2006 *Biomaterials* **27** 3825–35
[114] Kanematsu A, Yamamoto S, Noguchi T, Ozeki M, Tabata Y and Ogawa O 2003 *J. Urol.* **170** 1633–8
[115] Loai Y, Yeger H, Coz C, Antoon R, Islam S S, Moore K and Farhat W A 2010 *J. Biomed. Mater. Res. A* **94** 1205–15
[116] Zhou L, Yang B, Sun C, Qiu X, Sun Z, Chen Y, Zhang Y and Dai Y 2013 *Tissue Eng. Part A* **19** 264–76
[117] Roelofs L A, Oosterwijk E, Kortmann B B, Daamen W F, Tiemessen D M, Brouwer K M, Eggink A J, Crevels A J and Wijnen R M *et al* 2016 *Tissue Eng. Part A* **22** 83–92
[118] Farhat W A, Chen J, Sherman C, Cartwright L, Bahoric A and Yeger H 2006 *Can. J. Urol.* **13** 3000–8
[119] Hou X, Shi C, Chen W, Chen B, Jia W, Guo Y, Ma C, Ye G, Kang J and Dai J 2016 *Biomed. Mater.* **11** 031001
[120] Pokrywczynska M, Adamowicz J, Sharma A K and Drewa T 2014 *Exp. Biol. Med. (Maywood)* **239** 264–71
[121] Gasanz C, Raventós C and Morote J 2018 *Actas Urol. Esp.* **42** 435–41
[122] Liu Y, Bharadwaj S, Lee S J, Atala A and Zhang Y 2009 *Biomaterials* **30** 3865–73
[123] Ribeiro-Filho L A and Sievert K D 2015 *Adv. Drug Deliv. Rev.* **82–83** 38–46
[124] Osman Y, Shokeir A, Gabr M, El-Tabey N, Mohsen T and El-Baz M 2004 *J. Urol.* **172** 1151–4
[125] Li C, Xu Y M, Song L J, Fu Q, Cui L and Yin S 2008 *J. Urol.* **180** 1538–42
[126] Liao W, Yang S, Song C, Li X, Li Y and Xiong Y 2013 *Transplant. Proc.* **45** 730–4
[127] Huang J W, Xie M K, Zhang Y, Wei G J, Li X, Li H B, Wang J H, Zhu W D and Li C *et al* 2014 *Urology* **84** 1499–505
[128] Ansari S and Karram M 2017 *Int. Urogynecol. J.* **28** 1257–60
[129] Steinhoff G, Stock U, Karim N, Mertsching H, Timke A, Meliss R R, Pethig K, Haverich A and Bader A 2000 *Circulation* **102** III50–55
[130] Wilcox H E, Korossis S A, Booth C, Watterson K G, Kearney J N, Fisher J and Ingham E 2005 *J. Heart Valve Dis.* **14** 228–36
[131] Bhrany A D, Lien C J, Beckstead B L, Futran N D, Muni N H, Giachelli C M and Ratner B D 2008 *J. Tissue Eng. Regen. Med.* **2** 365–72
[132] Wang F, Maeda Y, Zachar V, Ansari T and Emmersen J 2018 *Biochem. Biophys. Res. Commun.* **503** 271–7
[133] Catry J, Luong-Nguyen M, Arakelian L, Poghosyan T, Bruneval P, Domet T, Michaud L, Sfeir R and Gottrand F *et al* 2017 *Cell Transplant.* **26** 1831–9

[134] Marzaro M, Conconi M T, Perin L, Giuliani S, Gamba P, De Coppi P, Perrino G P, Parnigotto P P and Nussdorfer G G 2002 *Int. J. Mol. Med.* **10** 177–82
[135] Conconi M T, De Coppi P, Bellini S, Zara G, Sabatti M, Marzaro M, Zanon G F, Gamba P G, Parnigotto P P and Nussdorfer G G 2005 *Biomaterials* **26** 2567–74
[136] Machingal M A, Corona B T, Walters T J, Kesireddy V, Koval C N, Dannahower A, Zhao W, Yoo J J and Christ G J 2011 *Tissue Eng. Part A* **17** 2291–303
[137] Yang Q, Peng J, Guo Q, Huang J, Zhang L, Yao J, Yang F, Wang S and Xu W *et al* 2008 *Biomaterials* **29** 2378–87
[138] Meng Q *et al* 2017 *Acta Biomater.* **53** 279–92
[139] Reyzelman A, Crews R T, Moore J C, Moore L, Mukker J S, Offutt S, Tallis A, Turner W B and Vayser D *et al* 2009 *Int. Wound J.* **6** 196–208
[140] Sanluis-Verdes A, Yebra-Pimentel Vilar M T, García-Barreiro J J, García-Camba M, Ibáñez J S, Doménech N and Rendal-Vázquez M E 2015 *Cell Tissue Bank.* **16** 411–23
[141] Yuan J M, Xiong S H, Liu Z, Wen Y, Dang R S, Shen M R, Zhang Y Z, Zhang X, Yang X Q and Zhang C S 2016 *J. Tissue Eng. Regen. Med.* **10** 554–63
[142] Yang D, Guo T, Nie C and Morris S F 2009 *Ann. Plast. Surg.* **62** 297–303
[143] Li Q, Huang C, Xu Z, Liu G, Liu Y, Xiao Z, Nie C, Zheng B and Yang D 2013 *Aesthetic Plast. Surg.* **37** 822–32
[144] Ghazanfari S, Alberti K A, Xu Q and Khademhosseini A 2019 *J. Biomed. Mater. Res. A* **107** 1225–34
[145] Yannas I V 2013 *Biomaterials* **34** 321–30
[146] Kim S S, Utsunomiya H, Koski J A, Wu B M, Cima M J, Sohn J, Mukai K, Griffith L G and Vacanti J P 1998 *Ann. Surg.* **228** 8–13
[147] Kulig K M and Vacanti J P 2004 *Transpl. Immunol.* **12** 303–10
[148] Tilles A W, Berthiaume F, Yarmush M L, Tompkins R G and Toner M 2002 *J. Hepatobiliary Pancreat. Surg.* **9** 686–96
[149] Powers M J, Janigian D M, Wack K E, Baker C S, Beer Stolz D and Griffith L G 2002 *Tissue Eng.* **8** 499–513
[150] Sivaraman A, Leach J K, Townsend S, Iida T, Hogan B J, Stolz D B, Fry R, Samson L D, Tannenbaum S R and Griffith L G 2005 *Curr. Drug Metab.* **6** 569–91
[151] Dash A, Inman W, Hoffmaster K, Sevidal S, Kelly J, Obach R S, Griffith L G and Tannenbaum S R 2009 *Expert Opin. Drug Metab. Toxicol.* **5** 1159–74
[152] Wang B, Jakus A E, Baptista P M, Soker S, Soto-Gutierrez A, Abecassis M M, Shah R N and Wertheim J A 2016 *Stem Cells Transl. Med.* **5** 1257–67
[153] Weymann A, Loganathan S, Takahashi H, Schies C, Claus B, Hirschberg K, Soós P, Korkmaz S and Schmack B *et al* 2011 *Circ. J.* **75** 852–60
[154] De Carlo E, Baiguera S, Conconi M T, Vigolo S, Grandi C, Lora S, Martini C, Maffei P and Tamagno G *et al* 2010 *Int. J. Mol. Med.* **25** 195–202
[155] Figliuzzi M, Remuzzi G and Remuzzi A 2014 *Nephron Exp. Nephrol.* **126** 113–8
[156] Liu R F, Gao J S, Yang Y F and Zeng W X 2015 *Urol. J.* **12** 2457–61
[157] Wagner D E, Bonenfant N R, Sokocevic D, DeSarno M J, Borg Z D, Parsons C S, Brooks E M, Platz J J and Khalpey Z I *et al* 2014 *Biomaterials* **35** 2664–79

IOP Publishing

Biophysical and Chemical Properties of Collagen: Biomedical Applications

John A M Ramshaw and Veronica Glattauer

Chapter 17

Coating of biomedical materials with collagen

As collagen is the most abundant component of the extracellular matrix it was a natural expectation that coating other materials with collagen would improve the interface that was formed between host tissue and implants. This proved true with better cell adhesion, higher cell proliferation and new, appropriate tissue formation observed. Attachment methods have ranged from simple absorption from solutions, coating with fibrils to covalent attachment, with higher levels of collagen often leading to the best results.

The development of surface modifications that incorporate biomolecules, including collagen, is targeted at giving non-biological medical devices a more biologically compatible surface in order to influence the biocompatibility of materials for specific applications. This involves enhancing localised signalling and recognition processes and active involvement, including stimulation, in the tissue response at a host-implant interface [1].

Initial studies predominantly targeted introduction of heparin onto device surfaces so as to improve their blood compatibility [2], showing clearly that immobilised biological molecules could influence the local response. Commercially available devices were treated, but with mixed results [3]. Other significant directions included the modification of surfaces using synthetic peptides [4] with a particular interest in the fibronectin derived sequence Arg-Gly-Asp [5]. Peptides based on this motif show a broad range of binding activities, so the emerging focus is in modifications using peptides and peptide assemblies that could enable production of functional, cell-specific surfaces [6].

In the application of collagen as a coating, the collagen is essentially a minor component of the overall device but is there for the significant biological and performance advantages that the coating can provide. In its specific location it can provide significant benefits even though the amount present can be very small.

doi:10.1088/978-0-7503-2096-2ch17

Collagen coatings are usually used on non-biological implants, such as polymers and metals, as these materials frequently lack good cell and tissue interaction properties. Studies on collagen coating of synthetic polymeric materials were well under way in the late 1980s [7, 8]. Rarely, collagen coatings have been described for collagen tissue-based devices. An example is the use of a collagen with poor platelet binding properties, such as type V collagen [9], to enhance the blood compatibility of a material. Collagen tissue-based devices are normally rich in type I and type III collagens and these collagens interact strongly with platelets [10]. Thus, coating a tissue-based, biosynthetic device (Omniflow™ Vascular Prosthesis, see chapter 15) with type V collagen led to a major reduction in the extent of platelet attachment (figure 17.1) [11]. Thus, in the search for better small diameter vascular prosthesis to meet the un-met need for a suitable off-the-shelf coronary by-pass device, the use of an appropriate collagen coating could provide an improved lumen surface with poor thrombogenic properties that could be one of the tools that could be combined to assist in development of an effective coronary by-pass device.

17.1 Coating of synthetic polymers

Collagen coatings have now been described for a wide range of synthetic polymers, including, for example, soft silicone rubber [7, 8], polypropylene [12, 13], poly (L-lactide-*co*-glycolide) (PLGA) [14, 15], polyurethane [16, 17], and poly(hydroxyethylmethacrylate-*co*-methylmethacrylate) [18]. In some instances, such as polypropylene, these polymers are used in a variety of medical devices. However, much of the data obtained has been on disks in cell culture plates and coating efficiencies for more complex topologies are less well discussed.

Collagen and gelatin coatings have been used on polyester (Dacron) vascular prostheses [19], but the aim here has frequently been as a sealant pre-surgery rather than a functional coating for the polymer thread.

In most examples, the coating has been with purified type I collagen (e.g., [16, 20, 21]. However, in some cases coating mixtures containing collagen were used. These include an extracellular matrix preparation that was predominantly type I collagen,

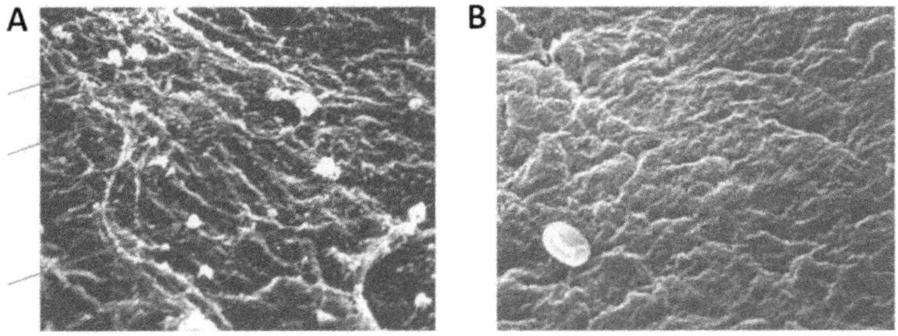

Figure 17.1. Collagen type V coating to limit platelet adhesion to a collagen-based device, the Omniflow™ Vascular Prosthesis. (A) Without further modification, with adherent platelets shown by arrows, and (B) after coating with type V collagen, with no platelets seen binding. (Reprinted from [11], with permission from SciMedCentral.)

where washed and processed porcine dermal tissue had been solubilised by proteolysis, but not further fractionated [22]. This type of preparation would, therefore, contain potentially variable amounts of other matrix components and collagen types. In other cases, defined mixtures of purified components were used, including collagen with chondroitin sulfate [15].

A range of formats for the collagen used in the coating procedure have been used. These have included collagen solutions [23], collagen fibril suspensions [16], collagen sponges [20], and collagen in the form of an acellular dermal matrix [24]. Similarly, a range of different methods have been used for the coating process. These include absorption from solution [23]. In addition, three further approaches, gas-plasma modification, chemical modification, and layer-by-layer methods are described in the following sections.

17.1.1 Plasma modification prior to collagen coating

The surfaces of many, possibly most, materials need modifying prior to collagen coating to allow an efficient adhesion or bonding of the collagen to the substrate material. A frequently described approach to this activation of the material has been the use of a plasma technologies, plasma being a confined ionised gas which is generally produced by applying an electric field to a neutral state gas or vapour. There are various possible equipment designs and formats described for plasma methods, which are included in various reviews (e.g., [25, 26]). The advantages of these techniques are that they change the surface properties without affecting the bulk properties of the material [27] and that the coating is robust and can be tailored to meet specific needs [28].

The possibly most popular approach is to create a new functional surface through attachment of new chemical entities [25]. Often, this approach is used to change the hydrophilicity or hydrophobicity of a sample, giving changes in the water contact angle that provides a measure of the balance of these parameters. In general, the production of a hydrophilic surface will provide options for immobilisation of collagen to the surface, as the hydrophilicity can arise, for example, from the presence of new carboxyl or amino functions. Thus, use of an O_2 or N_2, or NH_3 plasma can be used to modify the surface. However, it is important to minimise the extent of degradation of the surface, so for this reason O_2 is best avoided and either N_2 or NH_3 are preferable [25]. These will lead to amino functions on the surface which can be used as binding sites. The parameters of the process will usually need optimising. For this, various methods for determining the nature and extent of the modification are available [29] of which the use of x-ray photoelectron spectroscopy (XPS), also known as electron spectroscopy for chemical analysis (ESCA), is the preferred method as it can measure the elemental and bonding composition at parts per thousand in the top few nanometres of the surface. The other approach is to produce plasma polymerised surfaces. In this case the plasma is generated containing a suitable small organic molecule, such as allylamine [30], which leads to a polymerised coating on the surface with reactive functional groups. The subsequent attachment of collagen to a modified surface can be by absorption, but is preferably covalent using reaction strategies such as those developed for crosslinking collagen materials (see chapter 12).

17.1.2 Chemical modification of surfaces

Although gas-plasma methods have been widely used to activate a substrate surface for collagen coating, they have the disadvantage of requiring specialist equipment. Consequently, a selection of other, fairly simple physical and chemical solution methods has been described that can be used to activate a surface, or provide attachment, mostly for synthetic polymeric substrates.

A simple physical technique is irradiation of a polymer surface with strong UV light [31]. This energetic irradiation causes bond cleavage at the polymer surface, often at –C–H bonds. This results in short-lived radicals that react with proximal molecules, which can be added for the purpose. This can be used to generate a range of new reactive groups. The nature of the groups will depend on the type of polymer and any added chemical. The new groups can allow direct reaction with a coating material such as collagen, or attachment of a linker group to enable subsequent coating.

Another approach, which has been used for silicone rubber substrates is a direct adhesive bonding of a preformed collagen sponge layer [20]. In this approach, polymer sheets were treated with fresh cure mix. This could be a mixture of vinyl-derivatised polydimethylsiloxane with addition of 10% crosslinker, hydride-derivatised polydimethylsiloxane plus platinum-vinyl methylsiloxane catalyst, which cures over 24+ h at room temperature [20]. The stabilised collagen sheet, e.g., about 2 mm thick, was pressed onto the cure mix layer and was adherent after the 24 h cure [20]. In *in vivo* implant studies using these coated materials, a significantly lower extent of fibrous capsule formation was observed around the implant [20].

A further method is to chemically modify the substrate surface to leave a reactive group that can then be used for collagen attachment from collagen solutions or from fibril slurries. The chemistry that is used here resembles the numerous reactions that are available for crosslinking collagen (see chapter 12). An example was the modification of a carboxyl group enriched polyurethane with epoxide groups using a range of linear bis-epoxy compounds [16]. These data indicated that a greater extent of modification was achieved by using aliphatic bis-epoxides rather than those with aromatic groups in the linker chain [16]. The epoxide-polyurethane substrate was reacted with a collagen solution in phosphate buffered saline at 37 °C, leading to fibril formation during the process. The more extensive collagen coating was found on the surfaces with the higher aliphatic epoxide concentration [16].

17.1.3 Layer-by-layer coating

A variety of methods for producing films for coatings have been proposed, including use of Langmuir–Blodgett films [32–34] and self-assembling systems [35, 36]. However, these approaches are often limited for biomedical applications, for example, due to costs, long processing times and poor physiological stability of the film [37]. However, the development of layer-by-layer (LbL) coatings, initially as inorganic assemblies [38], has led to the stacked, alternating polyelectrolyte technology [39] that is now quite widely employed, particularly for cardiovascular and bone applications [37]. The LbL approach is easily achieved, and uses alternating exposure of the sample to solutions of negatively and positively charged polyelectrolytes [37] (figure 17.2). The approach allows coating of charged substrates of varied shape and size, as well as allowing development of designed functionality for the surface [37]. Various parameters need to be optimised for efficient and effective coating, including pH and ionic strength so that the polyelectrolytes are appropriately charged [40]. While electrostatic interactions have been most widely used, other approaches including hydrophobic interactions [41] and host-guest interactions [42] have also been used. The development of the LbL approach, the experimental considerations and potential applications have been reviewed (e.g., [37, 40]).

For collagen coatings, the interstitial collagens provide excellent polycations. Although type I collagen has been the preferred collagen to date, it is possible that other members of this collagen group may find use in specific applications. For example, type II collagen may be useful in cartilage applications, while type V collagen could be useful in providing more thromboresistant surfaces [11]. As coatings typically consume only very small amounts of materials, other collagen types that cannot be readily be purified in larger quantity could be useful. For example, type IV collagen, possibly in combination with a laminin, could be used to finish a coating to produce a mimic of a basement membrane.

Type I collagen has been used for LbL coatings, with several different polyanion components. In many cases, the sample, whether polymer or metal, is first activated to give it an appropriate surface charge to initiate the LbL process, using either chemical or plasma approaches (see above). These non-collagenous polyanions include heparin [43], where a heparin rather than collagen terminal layer gave better endothelial cell coverage on stents [43]. A collagen/heparin combination coating has been shown to improve the blood compatibility of titanium [44, 45]. Another polyanion that has been used in combination with collagen is chondroitin sulfate, which has been used with collagen to provide a surface on a polymer surface so as to enhance chondrogenesis [46]. Further polyanions that could be used include various

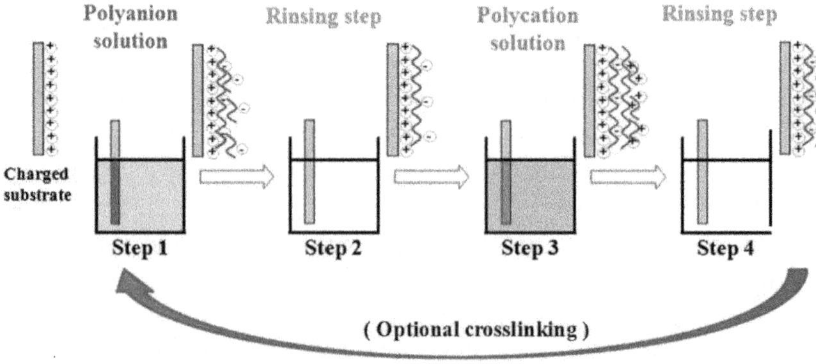

Figure 17.2. A schematic representation of the LbL process, based on exposure of the charged substrate alternating solutions of oppositely charged polyelectrolytes with intervening washing steps, with the coating cycle repeated to for the desired thickness of coating. (Based on [37], reprinted with permission from IOP Publishing.)

synthetic polyelectrolytes [37], plus other biological polymers such as alginate [47] and hyaluronic acid [37].

The option of producing a homogeneous, collagen only layer [48], rather than the heterogeneous layers [43, 45], provides a further option. In this case, however, in addition to using a native collagen as the polycation, a modified collagen with an acidic isoelectric point can provide the polyanion [48]. There are various approaches to provide the modified collagen. Treatment of collagen with alkali leads to loss of amides from the Asn and Gln residues to give the acidic Asp and Glu, dropping the pI [49]. Alternatively, Lys side chains can be chemically modified to provide carboxylic acid side chains, for example, by succinylation [49]. This reaction is not reversible, but other reagents can allow reversible modification. These include maleylation, which can then be reversed over days in acidic conditions [50] and citraconylation which can be reversed over hours, also in acidic conditions [51]. Particularly if reversible reagents are used, very mild crosslinking such as with minimal glutaraldehyde (GA) (see chapter 12) could be considered after washing steps to provide a stable LbL coating; this step may also be valuable for other coating strategies (figure 17.2).

17.2 Collagen on metals and inorganic materials

17.2.1 Metallic surfaces

Collagen has been used to coat a variety of metallic and ceramic materials, mainly titanium [52–55], but also including tantalum [56], steel [57], various alloys, including Ti6Al4V [57–59], X2CrNiMo18 [57] and Mg-rare earth alloys [60] as well as various calcium phosphate ceramics [61, 62] and bioactive glass [63].

In these cases, type I collagen has usually been used as the coating, with the aim of enhancing the cell-material interactions and cell function and proliferation. This leads to better adhesion between the substrate and new collagen fibrils for more effective osteointegration in orthopaedic and dental implant applications. Another interstitial collagen, type II in a fibrillar form, has also been examined, together with proteoglycan additives, for coating titanium, where the results were found to vary with different collagen preparations [64]. In addition, type IV collagen coating has been used for coating titanium used on stents to promote smooth muscle cell contractile phenotype [65]. An interesting example that has emerged is the use of bacterial collagen as a coating for titanium [66]. The *S. pyogenes* Scl2 protein, which is normally a 'blank slate' (see chapter 2) was engineered to include an integrin binding domain. This modification led to selective activation of the $\alpha 2\beta 1$ integrin, leading to enhanced osteoblast differentiation, osteoid formation and mineralisation [66]. Furthermore, this coating inhibited bacterial colonisation and biofilm formation [66], suggesting that further development of bacterial collagens may be a valuable tool for improving medical device performance.

As well as this interesting result with the bacterial collagen, overall, in all papers, beneficial results from the collagen coatings, albeit with variations depending on the coating method, have been consistently reported.

Various methods have been used to enable collagen coatings on metallic surfaces. The simplest has been passive absorption on un-modified materials, such as titanium, from collagen solutions (e.g., [54]). The effectiveness of this approach, however, was variable with some studies showing positive results, including higher proliferation rates and up-regulation of cell adhesion markers [54]. On the other hand, others reported that the collagen coating did not extensively affect cell proliferation [59]. The key variable among these and other studies is that often the amount of coating is not optimal nor well quantified. Responses have been reported to be greater when the coating consists of more collagen [67].

More studies have used coatings of titanium and tantalum with collagen fibrils [56, 68]. In these cases, the collagen solution is formed into a fibrillar mass, for example, by heating to 37 °C (see chapter 4) [68] and the fibrils used as the coating, potentially giving a greater amount of collagen on the surface of the implant. The good binding to tantalum and its oxide, suggested that it could be used as an intermediate coating to significantly improve collagen coating of steel [56, 69]. Another approach to coating with collagen fibrils that has been suggested is through use of electrospinning [70].

Various other surface modification reactions to the metal substrates have been used to enhance collagen binding and performance. These have included studies that used gas-plasma modification of the surface giving a more robust collagen coating [57, 67, 71], which, in some cases, covalent attachment was used with further improvement of function [67]. The gas-plasma treatment significantly improved the otherwise poor coating of collagen on steel devices [57]. In *in vivo* studies, this combination approach led to better vascularisation and a reduced inflammatory response for material implanted in mice [71]. Other surface modifications prior to the collagen coating have included HF treatment of magnesium-rare earth alloys [60].

In some cases, the collagen coating material has been augmented with additional biological components, which typically improve the performance of the solely collagen coated material. For example, including decorin or biglycan in the coating enhanced the effect of collagen [64]. Similarly, inclusion of RGD peptide [52] or phospholipid [72] also had positive effects.

17.2.2 Ceramic surfaces

A selection of bioceramics, generally containing hydroxyapatite and tricalcium phosphate have been proposed for orthopaedic repairs. In some cases, particles of these materials are formed into composites (see chapter 18), but in other cases the large implant has been coated with collagen (e.g., [61, 62, 73]. These larger scale materials can be either porous (or microporous) or non-porous to the collagen.

Coating of calcium phosphate ceramics with collagen, through gelation of a collagen solution on the surface [74], indicated that collagen coating could increase the extent of osseointegration. The collagen coating led to enhanced cell proliferation by 14 days for most sample compositions, excepting those with a 50:50 mixture of hydroxyapatite and tricalcium phosphate [73]. However, the strength of the materials of different compositions was often reduced by the collagen coating, so adaptations to stabilise and maintain mechanical strength may be need for an ideal coated material [61].

In another example, the effects of collagen coating on a microporous surface, with pores <10 μm, was examined [62]. The microporous surface of the ceramic was formed by dipping in a slurry of particles that aggregated on drying prior to sintering. The collagen was then coated from a suspension and samples washed then dried under vacuum [62]. The modified and coated surface showed enhanced cell-material interactions including better spreading, proliferation and differentiation compared to un-coated and un-modified samples [62].

References

[1] Morra M 2007 *Expert Rev. Med. Devices* **4** 361–72

[2] Lagergren H, Olsson P and Swedenborg J 1974 *Surgery* **75** 643–50

[3] Bianchi J J, Swartz M T, Raithel S C, Braun P R, Illes M Z, Barnett M G and Pennington D G 1992 *ASAIO J.* **38** M143–6

[4] Massia S P and Hubbell J A 1990 *Ann. N Y Acad. Sci.* **589** 261–70

[5] Pierschbacher M D and Ruoslahti E 1984 *Nature* **309** 30–3

[6] Fairman R and Akerfeldt K S 2005 *Curr. Opin. Struct. Biol.* **15** 453–63

[7] Ksander G A and Gray L 1985 *Ann. Plast. Surg.* **14** 351–60

[8] Ksander G A 1988 *Ann. Plast. Surg.* **20** 215–24

[9] Parsons T J, Haycraft D L, Hoak J C and Sage H 1983 *Arteriosclerosis* **3** 589–98

[10] Barnes M J, Knight C G and Farndale R W 1998 *Curr. Opin. Hematol.* **5** 314–20

[11] Werkmeister J A, Casagranda F, Edwards G A, White J F, Glattauer V and Ramshaw J A M 2017 *JSM Cell Devel. Biol.* **5** 1022

[12] van't Riet M, Burger J W, Bonthuis F, Jeekel J and Bonjer H J 2004 *Surg. Endosc.* **18** 681–5

[13] Ackermann M, Wang X, Wang S, Neufurth M, Schröder H C, Isemer F E and Müller W E G 2017 *J. Biomed. Mater. Res. B Appl. Biomater.* **106** 2109–21

[14] Vandrovcová M, Douglas T, Hauk D, Grössner-Schreiber B, Wiltfang J, Bačáková L and Warnke P H 2011 *Physiol. Res.* **60** 797 813

[15] Wojak-Cwik I M, Hintze V, Schnabelrauch M, Moeller S, Dobrzynski P, Pamula E and Scharnweber D 2013 *J. Biomed. Mater. Res. A* **101** 3109–22

[16] Huang L L H, Lee P C, Chen L W and Hsieh K H 1998 *J. Biomed. Mater. Res.* **39** 630–6

[17] Douglas T and Haugen H J 2008 *J. Mater. Sci., Mater. Med.* **19** 2713–9

[18] Yan T, Sun R, Li C, Tan B, Mao X and Ao N 2010 *J. Mater. Sci., Mater. Med.* **21** 2425–33

[19] Hirt S W, Aoki M, Demertzis S, Siclari F, Haverich A and Borst H G 1993 *J. Vasc. Surg.* **17** 538–45

[20] Wallace D G, Rosenblatt J and Ksander G A 1992 *J. Biomed. Mater. Res.* **26** 1517–34

[21] Siniscalchi R T, Melo M, Palma P C, Dal Fabbro I M, Vidal Bde C and Riccetto C L 2013 *Int. Urogynecol. J.* **24** 1747–54

[22] Faulk D M, Londono R, Wolf M T, Ranallo C A, Carruthers C A, Wildemann J D, Dearth C L and Badylak S F 2014 *Biomaterials* **35** 8585–95

[23] Adamczak M, Scisłowska-Czarnecka A, Genet M J, Dupont-Gillain C C and Pamuła E 2011 *Acta Bioeng. Biomech.* **13** 63–75

[24] Komorowska-Timek E, Oberg K C, Timek T A, Gridley D S and Miles D A 2009 *Plast. Reconstr. Surg.* **123** 807–16

[25] Poncin-Epaillard F and Legeay G 2003 *J. Biomater. Sci. Polym. Ed.* **14** 1005–28
[26] Minati L, Migliaresi C, Lunelli L, Viero G, Dalla Serra M and Speranza G 2017 *Biophys. Chem.* **229** 151–64
[27] Zhao Y, Yeung K W K and Chu P K 2014 *Appl. Surf. Sci.* **310** 11–8
[28] Chu P K, Chem J Y, Wang L P and Huang N H 2002 *Mat. Sci. Eng.: R: Reports* **36** 143–206
[29] Werner C and Jacobasch H J 1999 *Int. J. Artif. Organs* **22** 160–76
[30] Thissen H, Johnson G, Hartley P G, Kingshott P and Griesser H J 2006 *Biomaterials* **27** 35–43
[31] Ahad I U, Bartnik A, Fiedorowicz H, Kostecki J, Korczyc B, Ciach T and Brabazon D 2014 *J. Biomed. Mater. Res. A* **102** 3298–310
[32] Blodgett K B 1935 *J. Am. Chem. Soc.* **57** 1007–22
[33] Langmuir I and Schaefer V J 1937 *J. Am. Chem. Soc.* **59** 2075–6
[34] Punkka E and Rubner M F 1992 *J. Electron. Mater.* **21** 1057–63
[35] Bain C D, Evall J and Whitesides G M 1989 *J. Am. Chem. Soc.* **111** 7155–64
[36] Ulman A 1996 *Chem. Rev.* **96** 1533–54
[37] Gentile P, Carmagnola I, Nardo T and Chiono V 2015 *Nanotechnology* **26** 422001
[38] Kirkland J J 1965 *Anal. Chem.* **37** 1458–61
[39] Decher G and Hong J-D 1991 *Makromol. Chem. Macromol. Symp.* **46** 321–7
[40] Xiao F X, Pagliaro M, Xu Y J and Liu B 2016 *Chem. Soc. Rev.* **45** 3088–121
[41] Kotov N A 1999 *Nanostruct. Mater.* **12** 789–96
[42] van der Heyden A, Wilczewski M, Labbé P and Auzély R 2006 *Chem. Commun. (Camb).* **2006** 3220–2
[43] Lin Q, Yan J, Qiu F, Song X, Fu G and Ji J 2011 *J. Biomed. Mater. Res. A* **96** 132–41
[44] Chen J L, Li Q L, Chen J Y, Chen C and Huang N 2009 *Appl. Surf. Sci.* **255** 6894–900
[45] Chou C-C, Zeng H-J and Yeh C-H 2013 *Thin Solid Films* **549** 117–22
[46] Gong Y, Zhu Y, Liu Y, Ma Z, Gao C and Shen J 2007 *Acta Biomater.* **3** 677–85
[47] Chaubaroux C, Vrana E, Debry C, Schaaf P, Senger B, Voegel J C, Haikel Y, Ringwald C and Hemmerlé J *et al* 2012 *Biomacromol.* **13** 2128–35
[48] Truong Y B, Glattauer V, Briggs K L, Zappe S and Ramshaw J A M 2012 *Biomaterials* **33** 9198–204
[49] Rauterberg J and Kühn K 1968 *Hoppe-Seylers Z. Physiol. Chem.* **349** 611–22
[50] Butler P J, Harris J I, Hartley B S and Leberman R 1967 *Biochem. J.* **103** 78P–9P
[51] Dixon H B F and Perham R N 1968 *Biochem. J.* **109** 312–4
[52] Rammelt S, Illert T, Bierbaum S, Scharnweber D, Zwipp H and Schneiders W 2006 *Biomaterials* **27** 5561–71
[53] Sartori M, Giavaresi G, Parrilli A, Ferrari A, Aldini N N, Morra M, Cassinelli C, Bollati D and Fini M 2015 *Int. Orthop.* **39** 2041–52
[54] Ritz U, Nusselt T, Sewing A, Ziebart T, Kaufmann K, Baranowski A, Rommens P M and Hofmann A 2017 *Clin. Oral Investig.* **21** 255–65
[55] Sharan J, Koul V, Dinda A K, Kharbanda O P, Lale S V, Duggal R, Mishra M, Gupta G and Singh M P 2018 *Colloids Surf. B. Biointerfaces* **161** 1–9
[56] Li Y, Zhang S, Guo L, Dong M, Liu B and Mamdouh W 2012 *Colloids Surf. B Biointerfaces* **95** 10–5
[57] Hauser J, Koeller M, Bensch S, Halfmann H, Awakowicz P, Steinau H U and Esenwein S 2010 *J. Biomed. Mater. Res. A* **94** 19–26

[58] Geissler U, Hempel U, Wolf C, Scharnweber D, Worch H and Wenzel K 2000 *J. Biomed. Mater. Res.* **51** 752–60
[59] Becker D, Geissler U, Hempel U, Bierbaum S, Scharnweber D, Worch H and Wenzel K W 2002 *J. Biomed. Mater. Res.* **59** 516–27
[60] Zhao N, Workman B and Zhu D 2014 *Int. J. Mol. Sci.* **15** 5263–76
[61] Brodie J C, Merry J and Grant M H 2006 *J. Mater. Sci., Mater. Med.* **17** 43–8
[62] Lee M H, You C and Kim K H 2015 *Materials* **8** 1150–61
[63] Magyari K, Vanea E, Baia L and Simon V 2016 *Biomed. Mater. Eng.* **27** 63–74
[64] Douglas T, Heinemann S, Hempel U, Mietrach C, Knieb C, Bierbaum S, Scharnweber D and Worch H 2008 *J. Mater. Sci., Mater. Med.* **19** 1653–60
[65] Li J, Zhang K, Chen H, Liu T, Yang P, Zhao Y and Huang N 2014 *Mater. Sci. Eng. C Mater. Biol. Appl.* **38** 235–43
[66] Bronk J K, Russell B H, Rivera J J, Pasqualini R, Arap W, Höök M and Barbu E M 2014 *Acta Biomater.* **10** 3354–62
[67] Ao H, Xie Y, Tan H, Wu X, Liu G, Qin A, Zheng X and Tang T 2014 *J. Biomed. Mater. Res. A* **102** 204–14
[68] Rammelt S, Schulze E, Bernhardt R, Hanisch U, Scharnweber D, Worch H, Zwipp H and Biewener A 2004 *J. Orthop. Res.* **22** 1025–34
[69] Müller R, Abke J, Schnell E, Macionczyk F, Gbureck U, Mehrl R, Ruszczak Z, Kujat R and Englert C *et al* 2005 *Biomaterials* **26** 6962–72
[70] Iafiscol M, Quirici N, Foltran I and Rimondini L 2013 *J. Nanosci. Nanotechnol.* **13** 4720–6
[71] Hauser J, Ring A, Schaffran A, Henrich L, Esenwein S A, Steinau H U, Stricker I and Langer S 2009 *Eur. Surg. Res.* **43** 262–8
[72] Ruiz G C M, Cruz M A E, Faria A N, Zancanela D C, Ciancaglini P and Ramos A P 2017 *Mater. Sci. Eng. C. Mater. Biol. Appl.* **77** 102–10
[73] Brodie J C, Goldie E, Connel G, Merry J and Grant M H 2005 *J. Biomed. Mater. Res. A* **73** 409–21
[74] Elsdale T and Bard J 1972 *J. Cell Biol.* **54** 626–37

John A M Ramshaw and Veronica Glattauer

Chapter 18

Composites of collagen with other materials

The production of composites of collagen with other components provides many opportunities for the development of materials with novel properties. These arise from the synergy between the components. The composites could, for example, provide additional biological activities and/or have enhanced mechanical properties. Composites are possible with other biological polymers, synthetic polymers, mineral phases and with bioactive entities as a delivery method.

Composites containing collagen comprise a significantly increasing number of reports in the scientific literature. Most reports are of composites of two components, giving a large number of potential combinations, but when three or more components are involved, the number of options becomes particularly large. This can be even further increased if a crosslinking step is included. As such, the following section is not comprehensive, but rather chooses examples to illustrate the concepts and opportunities, where the synergy in the bulk composition provides new or enhanced properties. In some cases, the synergies are less obvious, and some may consider them as mixtures or as coatings rather than true composites.

Various aims are discussed for the development of these materials. Normally, it is to exploit and/or augment the biological properties of collagen to provide materials with functions that would not occur naturally. In other cases, it can be to obtain better value for the collagen. Medical collagen is typically an expensive material, so composites with other, cheaper materials that allow retention of the biological properties of collagen can reduce the overall cost of production. Although the use of a composite may seem a good idea, in some cases the benefits of the mixture over the individual components are not always demonstrated, and often experiments to optimise the proportions are limited.

In some instances, collagen preparations contain more than one collagen type, such as the mixture of types I and III collagens that can be obtained from skin prior

doi:10.1088/978-0-7503-2096-2ch18

to further fractionation. These composites are coincidental in origin, and generally show compositional variations, and are not dealt with further in this chapter. This chapter looks at the options for deliberate composites with defined compositions.

18.1 Composites with other biopolymers

18.1.1 Proteins

The choices of proteins that are used in composites is limited by the possibility that the additional protein may introduce an immunogenic response. The protein that is frequently associated with collagen in tissues is elastin. Elastin, like collagen is available from native tissues either as an insoluble fibrous material or a chemically solubilised material [1, 2]. It is also available as a full-length protein as a recombinant human sequence product [3]. In addition, recombinant elastin-like polypeptides are available and have been frequently used in composites, often for similar indications [4].

Collagen-elastin composites have generally been focused on a limited number of clinical applications. These include cardiovascular applications where the inclusion of elastin in a composite aims to assist with adapting to the pulsatile requirements of these tissues. This includes heart valves, where a 5-fold excess of soluble collagen over soluble elastin has been used to form thermal gels that were then transformed to a more tissue-like matrix through incorporation of appropriate cells [5]. Layered collagen–elastin composites have also been examined for blood vessel fabrication, as elastin is a major component in such tissues. In one example, a three-layer collagen–elastin–collagen was constructed on a mandrel using fibrous protein component, crosslinked using *N*-ethyl-*N'*-(3-dimethylaminopropyl)carbodiimide (EDC) and incorporating heparin [6].

Another target area for collagen–elastin composites has been for dermal and similar applications where a sheet of material is required. For example, insoluble bovine fibrils have been prepared as composites with soluble elastin for dermal repair, with the elastin intended to provide a more compliant material that could conform to surface undulations and movement [7]. As with collagen alone, the pore size and orientation can be regulated by the freeze-drying process and can be optimised for clinical efficiency [8]. Electrospinning can also be used with soluble collagen–elastin mixtures to provide a membrane structure [9], although use of a non-denaturing solvent system instead of hexafluoroisopropanol (HFIP) may be preferable for retaining the collagen structure [10]. In addition to dermal use, these composite membranes may also be useful as adhesion prevention barriers [11].

The other protein system that has been regularly examined in composite materials with collagen has been silk fibroin, principally from *Bombyx mori* but also from other, related species. Another source is from spiders, where recombinant variations are also available [12]. *Bombyx mori* silk is a commodity material that is readily obtained and relatively cheap, with around 70 000 tonnes of finished silk fibre produced annually [13]. Silk fibroin solution, free of sericin, can be readily prepared and has been used to fabricate various structures for biomedical applications [14]. Soluble spider silk, on the other hand, is best obtained as a recombinant protein [15].

The availability of collagen and silk protein solutions allows a range of mixtures to be made and used in different fabrication approaches. As silks generally lack active biological motifs, they provide the strength and stiffness to the composite material; the collagen provides the biological interactions. The mixtures can be used to form hydrogels, which can be used for cell encapsulation with tailored strength through variation in composition [16]. Further tuning of the hydrogel strength can be achieved by crosslinking, for example, using EDC [17, 18]. The protein mixtures can also be used to fabricate films through electrospinning [19]. Initial studies used HFIP as the solvent and included subsequent crosslinking to give strength [19], prior to the denaturing effect on collagen being identified [10]. Despite this issue, subsequent studies [20, 21] have continued to use this solvent, which could lead to a material with different properties. On the other hand, non-denaturing aqueous solvents have been used for 3D-printing of a collagen-silk composite onto a low temperature stage with associated freeze-drying [22].

18.1.2 Carbohydrates

The extracellular matrix contains carbohydrates that interact or are associated with collagens, especially the glycosaminoglycan (GAG) chains, which are components of the proteoglycan structures, and hyaluronic acid (HyA).

There is a large amount of literature concerning the fabrication, properties and applications of collagen-GAG composites. The majority concerns materials that combine acetic acid treated fibrous collagen formed into a composite by addition of GAG (normally chondroitin sulfate); these studies have also been the subject of various reviews (e.g., [23–26]. Several materials have received US Federal Drug Administration product approvals. In particular, a material for dermal repair has achieved a significant success and has been a widely used product for more than two decades [24].

Collagen–GAG composites have undergone extensive development to optimise the properties for dermal applications. For example, control of pore size solely by physical rather than chemical approaches was examined [27]. The pore size and its uniformity were shown to be important for the effectiveness of cell seeding, with matrices being effective provided that the average pore diameter ranged between 20 and 125 μm [28]. Factors that can affect pore size include freezing temperature and freezing rate [29, 30], while pore uniformity can be influenced by the uniformity of heat transfer during freezing and drying [29]. Further, the method of drying, for example, freeze-drying or critical point drying, presents another variable that alters the pore composition of the resulting material [27]. In addition, cell adhesion can depend on the surface area associated with the pore size and conformation [30]. This wealth of information has allowed the collagen-GAG dermal matrix to be changed and adapted for other different clinical applications [24, 26], such as for the development of a device for peripheral nerve repair [26, 31].

Another carbohydrate, hyaluronic acid (HyA), has generated recent interest because it is a critical component of tissues. Although it is difficult to obtain from tissue sources [32], it is now readily available as a medical grade product from bacterial sources, such as *Streptococcus zooepidermicus* [33, 34]. In addition to its biological interactions, HyA can also contribute to the control of hydration in clinical applications [35]. HyA derivatives, such as sulfated hyaluronan, enhanced cell interactions [36], and addition of a hydrazide function can be used to provide crosslinked stability [37].

Although freeze-dried sponges have been a frequently used format, other fabrication methods have been explored. These include electrospun membranes, using HFIP and formic acid as the solvent [38], although, as noted previously, HFIP may not be an ideal solvent. Also, constructs have been described based on layers of collagen and HyA, or chondroitin sulfate (CS), built by layer-by-layer coating (see chapter 17) on a petri dish, with collagen type I as the polycation and HyA (or CS) as the polyanion [39]. An injectable composition, which can be used as a dermal filler, has been described and suggested as having superior properties compared to the individual components [40]. Type II collagen has been used in composites with HyA for potential repair of damaged articular cartilage [41]. Inclusion of HyA is useful as it increased the compression modulus [42].

In many cases, crosslinking has been used to provide better stability and mechanical properties. While EDC has been used more frequently [35, 43] than other crosslinkers, glyoxal, dialdehyde starch [33], 1,4- butanediol-diglycidyl ether [44] and genipin [37] have also be used. In some examples, a third component has been added to the collagen-HyA composite mixture. These have included chitosan [45] and the natural tissue components, fibrin [46] and chondroitin sulfate [47].

Several studies have described composites of chitosan and collagen. Chitosan is cheap and readily available as a by-product of the seafood industry based on shrimp and other crustacean production, although other sources are possible [48]. Chitin forms the exoskeleton and is a linear polysaccharide, made up of β-(1,4)-linked N-acetyl-D-glucosamine units. It can be converted to chitosan by enzymatic or chemical methods [49, 50]. The chemical method is cheaper and allows large scale production [49]. Chemical treatment, for example, uses a high concentration of NaOH at elevated temperatures. This method leads to rapid deacetylation of the N-acetyl-D-glucosamine; when around 50% deacetylation has occurred, the chitin becomes soluble [51] although products with a higher degree of deacetylation are generally used.

Chitosan, when used alone, is an effective dressing material [52]. The benefits of chitosan in a composite with collagen include its possible anti-microbial activity [53], and that it can add mechanical strength through non-covalent interactions with collagen, while contributing to control of porosity [54]. As chitosan is very cheap, especially when compared to collagen, it can be used as an appropriate diluent in composites with collagen, where it can replace collagen as the bulk material, while sufficient collagen is present to ensure the appropriate biological properties are retained. However, chitosan is possibly less readily degraded *in vivo* than collagen, so may still be retained, in part, when the collagen has been lost. This will be different again if crosslinking agents or other modifications to the chitosan have been introduced. Chitosan-collagen composites have been suggested as scaffolds for various applications, as well as being further augmented in various multicomponent composites.

Agarose has also been used in composites with collagen, but in this case not so much for synergistic biological activities, but rather as an aid in fabrication [55, 56]. In cases such as printing or bead fabrication, where collagen can be difficult to handle, the agarose mixture can be more readily fabricated. For example, introduction into a Ca^{2+} solution leads to an insoluble form of agarose. The collagen can then be stabilised, for example, with GA, and the agarose substantially removed afterwards, by solubilising in an Na^{+} solution, leaving a stable, porous collagen structure [55].

18.2 Composites with synthetic polymers

Composites of collagen with synthetic polymers come in two forms. One form, as a hydrogel, appears macroscopically to be a single phase. The other contains distinct elements where the two components are clearly evident with the collagen augmenting a polymer network.

18.2.1 Hydrogel and related composites

Combination of collagen with water soluble or miscible polymers is a basis for hydrogel formation. These polymers could include, for example, PEGs which do not form any strong associations with collagens. Care is needed, however, with these and other polymers, that the quantity added for a given molecular weight does not lead to collagen precipitation [57]. Collagen–PEG hydrogels will typically need to be stabilised. However, PEGs of different molecular weights can be readily obtained in modified, reactive forms, including as dendrimers, so that the composite hydrogel is formed through crosslinking of a collagen solutions [58].

Possibly the best hydrogels are those that can be made by creating a double network, an interpenetrating network, as this leads to a composite where the mechanical properties can considerably exceed those of the individual primary component, a collagen gel [59]. Thus, for example, an interpenetrating network between collagen with a small amount of added poly(*N,N′*-dimethylacrylamide) leads to a greater than 10-fold increase in the gel fracture stress compared with a gel from collagen alone [59].

Another synthetic polymer that has been used in hydrogels with collagen is polyvinyl alcohol (PVA). Uniform, composite scaffolds of collagen with PVA can be prepared, for example, by mixing selected amounts of collagen and PVA solutions, followed by GA stabilisation and drying sufficiently to give a gel [60]. Alternatively, 2 phase composites can be prepared by mixing collagen particles into a PVA solution followed by air drying [61]. For gels, properties such as compression strength depend on the composition of the mixture (figure 18.1), as well as the extent of stabilisation and the pH [60]. The properties of PVA composites depend on the degree of crystallinity of the polymer. This can be increased by taking the material through a series of freeze–thaw steps, giving extra strength without the inclusion of crosslinks [62].

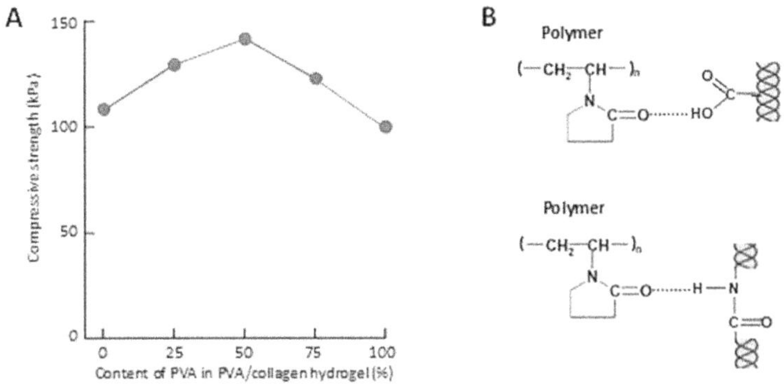

Figure 18.1. Interactions between collagen and soluble polymers in hydrogels. (A) The effects of different ratios of collagen and PVA on compressive strength of GA stabilised hydrogels, (redrawn from the data given in [60] and (B) proposed [63] hydrogen bonding schemes between collagen and PVP.

A further interesting collagen-soluble polymer composite system is when polyvinylpyrrolidone (PVP) is used. Studies of collagen-PVP mixtures using viscometry, DSC and FTIR showed strong interactions between these components, mainly through hydrogen bonds (figure 18.1) [63]. Examination of collagen-PVP composites *in vitro* and *in vivo* indicated that the material did not lead to any cellular damage nor any adverse cellular or humoral immunological effects [64]. The mixture can be further stabilised by crosslinking. A collagen-PVP mixture (1:16), with low molecular weight PVP, can be converted into a crosslinked co-polymer system by gamma irradiation, with almost 60% of the PVP involved in crosslinks [65]. Microscopy indicated that the PVP had formed a coating around the collagen fibres, possibly masking the collagen as the collagens cell interactions were lost [65].

18.2.2 Two phase systems

Composites have been made of collagen and different resorbable polymers. These polymers are generally insoluble in water. These include polymers such as polyglycolic acid (PGA), polylactic acid (PLA), poly-4-hydroxybutyrate (P4HB) and polycaprolactone (PCL), as well as copolymers such as PGA-PLA copolymers. The selected polyesters are slowly hydrolysed in contact with water, such as in tissue, so as to be resorbed in weeks up to years, depending on compositions, while losing strength in shorter time frames [66, 67]. These polymers can readily be fabricated into porous structures that define or control the size and shape of the material, for example, using fibres for weaving or knitting, by 3D-printing, by electrospinning methods and so forth [68]. These porous structures can then be further processed to incorporate a collagen network as a distinct, separate component in the structure [69–71]. The concept used for these materials can also be applied to constructs that combine collagen with non-resorbable polymers, for implants that need long-term durability [72].

18.3 Composites with inorganic materials

18.3.1 Hydroxyapatite and calcium phosphate phases

Hydroxyapatite (HAp) is the principle mineral components of bone, which is a natural composite of HAp and collagen (see chapter 5), and so has been the main, obvious choice for a long time for use in composites for bone repair. Various commercial collagen-HAp composites have been developed, and are a significant part of the market [73, 74]. The main issues that emerge in making suitable composites include the potential immunogenicity of xenogeneic collagen and the difficulty in establishing a biomineralization approach that mimics the natural processes [75]. Processes using co-precipitation [76] through mixed suspensions [77] or through immersion [77] and other methods [78] have been developed. The collagen component is normally crosslinked to provide stability, for example, using glutaraldehyde (GA) [79, 80], EDC [81] and dehydrothermal [76] treatments. The effect of the composite structure increased the compressive strength of the materials more than 10-fold when compared to collagen alone [77]. A special case of an immersion approach was introduced in which a GA stabilised sponge was immersed in 'simulated body fluid' [79], using a 2-stage protocol [82].

A variety of successful commercial products have emerged [73, 74] and remain readily available as the clinical performance is seen as particularly good. The results obtained were comparable with those obtained by bone grafts [83–85], but without any of the issues that can emerge with the use of autografts, such as the procurement and processing of allograft and banked cadaveric bone tissue.

In addition to HAp, various other calcium phosphate phases have been used, sometimes in combination with HAp [74, 83, 86], including in commercial materials. Tricalcium phosphate (TCP) is the most frequently used (e.g., [87]), although others, such as octacalcium phosphate have also been examined [88]. In these and other cases of calcium phosphate materials, it is the surface contact properties and porosity that are key variables [89]. In addition, next generation materials are emerging with additional components, such as growth factors and cells being part of the composition coupled with different fabrication approaches [78].

18.3.2 Other inorganic materials

Collagen has been added to bioactive glasses (bioglass), to provide a material with improved properties. The bioglass materials are variants of a CaO-P_2O_5-SiO_2 system [90, 91] and show osteogenic potential [89, 92], with nanosized bioglass being much better than micrometre glass, due to its greater surface area [93], particularly when used as a collagen composite [94]. Various composite systems have been reported [91]. In those prepared by freeze-drying the collagen–bioglass material, which gives high porosity [95], crosslinking of the collagen with EDC has generally been used [91]. More recently, vacuum impregnation of bioglass has been used, providing about 20% w/w impregnation by the collagen, and still providing the composite advantages seen with other methods [96].

18.4 Composites with bioactive entities

Collagen has been identified as a useful material for drug delivery for a long time (e.g., [97]). Initially examined for use in drug delivery (compounds with a molecular weight typically <900) [98, 99], subsequent studies also saw collagen examined for delivery of higher molecular weight bioactive materials, especially growth factors [100]. It is not surprising that prepared and fabricated collagen materials are good vehicles for growth factor delivery as these proteins are often associated with collagens in natural tissues [101, 102]. There is an extensive literature of biomaterials for drug delivery, principally for synthetic polymers, that covers many aspects of kinetics of release, efficacy, requisite physical properties, degradation and so forth. The following section is brief, and does not discuss this literature, but considers only a limited number of examples. The reader should look elsewhere for these details.

18.4.1 Growth factors

The most common growth factors for successful delivery by collagen are bone morphogenic proteins (BMP), although various others, such as nerve growth factor (NGF), and basic fibroblast growth factor (bFGF) have also been examined [100]. There are at least 15 distinct BMPs [103], which are members of the transforming growth factor beta (TGF-β) super family [104], with BMP-2 and recombinant human rhBMP-2 having been the more extensively examined [104], and leading to regulatory approval of certain products for clinical application.

In this case composites were initially formed by addition of BMP to preformed collagen sponges [103]. Data has shown that use of collagen increases the time of availability of BMP [105], with binding to collagen occurring at near neutral pH, with the extent dependent on ionic strength [103, 106]. Stabilisation of the collagen prior to composite formation by mild crosslinking, such as with formaldehyde, reduces BMP binding [106]. Stabilisation, however, can improve the handling properties of the wet material [103]. A range of fabrication and stabilisation strategies as well as the effects of different sterilisation methods have been explored [103]. Clinical efficacy in small animal models seemed good, but was less convincing in larger animal models. The approach has been used to examine fracture repair [107], particularly aiming at non-union fracture repair. Other surgical areas that have been of interest include spinal fusion [108], dental and craniofacial repair [109]. For bone repair, the addition of a mineral component (e.g., HAp) to the composite proteins provides an extra benefit [110, 111]. Overall, this type of composite may be particularly valuable in diabetic patients, where fracture healing can be impaired [112]. While BMP has been the object of most studies of collagen and growth factor composites, others have been examined (see the literature) including basic FGF for vocal fold repair [113].

18.4.2 Delivery of drugs, including antibiotics

A wide variety of fabricated collagen options have been examined for drug delivery, including sponges, films, particles, gels, etc, and a much wider selection of drugs [98].

While antibiotics are the main compounds that have been examined, various others including steroids, anti-rejection drugs, anti-cancer drugs and pain control agents have been studied [98]. The range of clinical indications that have been considered is equally large [98]. It is quite likely that the optimum formulation will vary in each case.

A useful product has been the development of corneal shields for use after various ophthalmic procedures and various commercial products have emerged. The conforming collagen film can be loaded with drug, with the collagen acting as a reservoir and to retard drug release. The lower dose required limits any possible systemic toxicity [98, 114]. The collagen film is thin enough to allow acceptable oxygen permeability and can dissolve and lubricate the eye surface [115]. Examples of manufacturing approaches and the wide range of antibiotics included in composites have been described in various reviews [98, 116].

Many of the examples use collagen sponges or membranes for delivery, with modifications such as crosslinking used to control the release rate. For example, several trials using gentamicin-collagen sponges have been described. A compilation of reports of randomised controlled trials of surgical site infections, covering nearly 7000 patients, concluded that these data show that gentamicin-collagen composites are good at decreasing the rate of these infections [117]. Similarly, chloramphenicol-collagen sponges were useful in dentistry, where the persistent bacterial environment may be more challenging [118]. Formats other than sponges are necessary for certain indications. Thus, collagen-based nanoparticles have been developed as drug carriers for pulmonary delivery [119].

References

[1] Partridge S M, Davis H F and Adair G S 1955 *Biochem. J.* **61** 11–21

[2] Robert L and Hornbeck W 1976 *Methodology of Connective Tissue Research* ed D A Hall (Oxford: Joynson Bruvvers) 81–104

[3] Martin S L, Vrhovski B and Weiss A S 1995 *Gene* **154** 159–66

[4] Gurumurthy B, Griggs J A and Janorkar A V 2018 *J. Mech. Behav. Biomed. Mater.* **84** 116–25

[5] Wang X, Ali M S and Lacerda C M R 2018 *Bioengineering* **5** 69

[6] Koens M J W, Faraj K A, Wismans R G, van der Vliet J A, Krasznai A G, Cuijpers V M, Jansen J A, Daamen W F and van Kuppevelt T H 2010 *Acta Biomater.* **6** 4666–74

[7] Haslik W, Kamolz L P, Manna F, Hladik M, Rath T and Frey M 2010 *J. Plast. Reconstr. Aesthet. Surg.* **63** 360–64

[8] Boekema B K H L, Vlig M, Olde Damink L, Middelkoop E, Eummelen L, Bühren A V and Ulrich M M W 2014 *J. Mater. Sci., Mater. Med.* **25** 423–33

[9] Rnjak-Kovacina J, Wise S G, Li Z, Maitz P K, Young C J, Wang Y and Weiss A S 2012 *Acta Biomater.* **8** 3714–22

[10] Zeugolis D I, Khew S T, Yew E S, Ekaputra A K, Tong Y W, Yung L Y, Hutmacher D W, Sheppard C and Raghunath M 2008 *Biomaterials* **29** 2293–305

[11] Wichelhaus D A, Beyersdoerfer S T, Gierer P, Vollmar B and Mittlmeier T 2016 *Arch. Orthop. Trauma Surg.* **136** 1021–29

[12] Werkmeister J A and Ramshaw J A M 2012 *Biomed. Mater.* **7** 012002

[13] Tomita M 2011 *Biotechnol. Lett.* **33** 645–54

[14] Jin H J, Fridrikh S V, Rutledge G C and Kaplan D L 2002 *Biomacromolecules* **3** 1233–39

[15] Schacht K and Scheibel T 2011 *Biomacromolecules* **12** 2488–95

[16] Buitrago J O, Patel K D, El-Fiqi A, Lee J H, Kundu B, Lee H H and Kim H W 2018 *Acta Biomater.* **69** 218–33

[17] Lv Q, Hu K, Feng Q and Cui F 2008 *J. Biomed. Mater. Res. A* **84** 198–207

[18] Long K, Liu Y, Li W, Wang L, Liu S, Wang Y, Wang Z and Ren L 2015 *J. Biomed. Mater. Res. A* **103** 1159–68

[19] Yeo I S, Oh J E, Jeong L, Lee T S, Lee S J, Park W H and Min B M 2008 *Biomacromolecules* **9** 1106–16

[20] Zhu B, Li W, Lewis R V, Segre C U and Wang R 2015 *Biomacromolecules* **16** 202–13

[21] Zhu B, Li W, Chi N, Lewis R V, Osamor J and Wang R 2017 *ACS Omega* **2** 2439–50

[22] Sun K, Li R, Jiang W, Sun Y and Li H 2016 *Biochem. Biophys. Res. Commun.* **477** 1085–91

[23] Orgill D P, Straus F H and Lee R C 1999 *Ann. N Y Acad. Sci.* **888** 233–48

[24] Yannas I V, Tzeranis D S, Harley B A and So P T 2010 *Philos. Trans. Roy. Soc. A* **368** 2123–39

[25] Yannas I V, Orgill D P and Burke J F 2011 *Plast. Reconstr. Surg.* **127**(Suppl 1) 60S–70S

[26] Yannas I V 2018 *Regen. Biomater.* **5** 189–95

[27] Dagalakis N, Flink J, Stasikelis P, Burke J F and Yannas I V 1980 *J. Biomed. Mater. Res.* **14** 511–28

[28] Yannas I V, Lee E, Orgill D P, Skrabut E M and Murphy G F 1989 *Proc. Natl. Acad. Sci. USA* **86** 933–37

[29] O'Brien F J, Harley B A, Yannas I V and Gibson L 2004 *Biomaterials* **25** 1077–86

[30] O'Brien F J, Harley B A, Yannas I V and Gibson L J 2005 *Biomaterials* **26** 433–41

[31] Yannas I V, Zhang M and Spilker M H 2007 *J. Biomater. Sci. Polym. Ed.* **18** 943–66

[32] Boeriu C G, Springer J, Kooy F K, van den Broek L A and Eggink G 2013 *Int. J. Carbohyd. Chem.* **2013** 624967

[33] Reháková M, Bakos D, Vizárová K, Soldán M and Jurícková M 1996 *J. Biomed. Mater. Res.* **30** 369–72

[34] Sze J H, Brownlie J C and Love C A 2016 *3 Biotech* **6** 67

[35] Tang S, Vickers S M, Hsu H P and Spector M 2007 *J. Biomed. Mater. Res. A* **82** 323–35

[36] Kliemt S, Lange C, Otto W, Hintze V, Möller S, von Bergen M, Hempel U and Kalkhof S 2013 *J. Proteome Res.* **12** 378–89

[37] Zhang L, Xiao Y, Jiang B, Fan H and Zhang X 2010 *J. Biomed. Mater. Res. B Appl. Biomater.* **92** 307–16

[38] Hsu F Y, Hung Y S, Liou H M and Shen C H 2010 *Acta Biomater.* **6** 2140–47

[39] Dai B, Pan Q, Li Z, Zhao M, Liao X, Wu K and Ma X 2016 *Stem Cells Int.* **2016** 4796578

[40] Kim Z H, Lee Y, Kim S M, Kim H, Yun C K and Choi Y S 2015 *J. Microbiol. Biotechnol.* **25** 399–406

[41] Taguchi T, Ikoma T and Tanaka J 2002 *J. Biomed. Mater. Res.* **61** 330–36

[42] Yeh H Y, Lin T Y, Lin C H, Yen B L, Tsai C L and Hsu S H 2013 *Differentiation* **86** 171–83

[43] Park S N, Park J C, Kim H O, Song M J and Suh H 2002 *Biomaterials* **23** 1205–12

[44] Maier V, Lefter C M, Maier S S, Butnaru M, Danu M, Ibanescu C, Popa M and Desbrieres J 2014 *Mater. Sci. Eng. C Mater. Biol. Appl.* **42** 243–53

[45] Lin Y C, Tan F J, Marra K G, Jan S S and Liu D C 2009 *Acta Biomater.* **5** 2591–600

[46] Filová E, Jelínek F, Handl M, Lytvynets A, Rampichová M, Varga F, Cinátl J, Soukup T, Trc T and Amler E 2008 *J. Biomed. Mater. Res. B Appl. Biomater.* **87** 415–24
[47] Ko C S, Huang J P, Huang C W and Chu I M 2009 *J. Biosci. Bioeng.* **107** 177–82
[48] Philibert T, Lee B H and Fabien N 2017 *Appl. Biochem. Biotechnol.* **181** 1314–37
[49] Younes I and Rinaudo M 2015 *Mar. Drugs* **13** 1133–74
[50] Muxika A, Etxabide A, Uranga J, Guerrero P and de la Caba K 2017 *Int. J. Biol. Macromol.* **105** 1358–68
[51] Chang K L B, Tsai G, Lee J and Fu W R 1997 *Carbohydr. Res.* **303** 327–32
[52] Park J U, Song E H, Jeong S H, Song J, Kim H E and Kim S 2018 *Adv. Exp. Med. Biol.* **1077** 527–37
[53] Cazón P, Velázquez G, Ramirez J A and Vázquez M 2017 *Food Hydrocoll.* **68** 136–48
[54] Martínez A, Blanco M D, Davidenko N and Cameron R E 2015 *Carbohydr. Polym.* **132** 606–19
[55] Tebb T A, Tsai S W, Glattauer V, White J F, Ramshaw J A M and Werkmeister J A 2006 *Cytotechnology* **52** 99–106
[56] Köpf M, Campos D F, Blaeser A, Sen K S and Fischer H 2016 *Biofabrication* **8** 025011
[57] Ramshaw J A M, Bateman J F and Cole W G 1984 *Anal. Biochem.* **141** 361–65
[58] Sargeant T D, Desai A P, Banerjee S, Agawu A and Stopek J B 2012 *Acta Biomater.* **8** 124–32
[59] Gong J P, Katsuyama Y, Kurokawa T and Osada Y 2003 *Adv. Mater.* **15** 1155–58
[60] Peng Z, Li Z, Zhang F and Peng X 2012 *J. Macromol. Sci. B* **51** 1934–41
[61] Hameed N, Glattauer V and Ramshaw J A M 2015 *J. Mech. Behav. Biomed. Mater.* **48** 38–45
[62] Peppas N A and Stauffer S R 1991 *J. Controlled Rel.* **16** 305–10
[63] Sionkowska A 2003 *Eur. Polym. J.* **39** 2135–40
[64] Furuzawa-Carballeda J, Rojas E, Valverde M, Castill I, Diaz de León L and Krötzsch E 2003 *Can. J. Physiol. Pharmacol.* **81** 1029–35
[65] Leyva-Gómez G, Lima E, Krötzsch G, Pacheco-Marin R, Rodríguez-Fuentes N, Quintanar-Guerrero D and Krötzsch E 2014 *J. Phys. Chem. B* **118** 9272–89273
[66] Miller R A, Brady J M and Cutright D E 1977 *J. Biomed. Mater. Res.* **11** 711–19
[67] Amecke B, Bendix D and Entenmann G 1992 *Clin. Mater.* **10** 47–50
[68] Chen G, Sato T, Ohgushi H, Ushida T, Tateishi T and Tanaka J 2005 *Biomaterials* **26** 2559–66
[69] Lu H, Oh H H, Kawazoe N, Yamagishi K and Chen G 2012 *Sci. Technol. Adv. Mater.* **13** 064210
[70] Chou Y C, Yeh W L, Chao C L, Hsu Y H, Yu Y H, Chen J K and Liu S J 2016 *Int. J. Nanomedicine* **11** 4173–86
[71] Mozdzen L C, Vucetic A and Harley B A C 2017 *J. Mech. Behav. Biomed. Mater.* **66** 28–36
[72] Sawyer M A J 2018 *JSLS* **22** e2018.00057
[73] Yunus Basha R, Sampath Kumar T S and Doble M 2015 *Mater. Sci. Eng. C Mater. Biol. Appl* **57** 452–63
[74] Kuttappan S, Mathew D and Nair M B 2016 *Int. J. Biol. Macromol.* **93** 1390–401
[75] Uskoković V 2015 *Mater. Sci. Eng. C Mater. Biol. Appl* **57** 434–51
[76] Yunoki S, Ikoma T, Monkawa A, Marukawa E, Sotome S, Shinomiya K and Tanaka J 2007 *J. Biomater. Sci. Polym. Ed.* **18** 393–409

[77] Cunniffe G M, Dickson G R, Partap S, Stanton K T and O'Brien F J 2010 *J. Mater. Sci., Mater. Med.* **21** 2293–98
[78] Turnbull G, Clarke J, Picard F, Riches P, Jia L, Han F, Li B and Shu W 2017 *Bioact. Mater* **3** 278–314
[79] Lickorish D, Ramshaw J A M, Werkmeister J A, Glattauer V and Howlett C R 2004 *J. Biomed. Mater. Res. A* **68** 19–27
[80] Panda N N, Jonnalagadda S and Pramanik K 2013 *J. Biomater. Sci. Polym. Ed.* **24** 2031–44
[81] Villa M M, Wang L, Huang J, Rowe D W and Wei M 2014 *J. Biomed. Mater. Res. B Appl. Biomater.* **103** 243–53
[82] Kokubo T, Hata K, Nakamura T and Yamamuro T 1991 *Bioceramics: Proc. of the 4th Int. Symp. on Ceramics in Medicine (London)* pp 113–20
[83] Cornell C N, Lane J M, Chapman M, Merkow R, Seligson D, Henry S, Gustilo R and Vincent K 1991 *J. Orthop. Trauma* **5** 1–8
[84] Habibovic P, Kruyt M C, Juhl M V, Clyens S, Martinetti R, Dolcini L, Theilgaard N and van Blitterswijk C A 2008 *J. Orthop. Res.* **26** 1363–70
[85] Giorgi P, Capitani D, Sprio S, Sandri M, Tampieri A, Canella V, Nataloni A and Schirò G R 2017 *J. Appl. Biomater. Funct. Mater.* **15** e262–70
[86] Grimes J S, Bocklage T J and Pitcher J D 2006 *Orthopedics* **29** 145–48
[87] Takahashi K 2009 *Bull. Tokyo Dent. Coll.* **50** 1–11
[88] Kawai T, Anada T, Honda Y, Kamakura S, Matsui K, Matsui A, Sasaki K, Morimoto S, Echigo S and Suzuki O 2009 *Tissue Eng. Part A* **15** 23–32
[89] Davies J E 2007 *Biomaterials* **28** 5058–67
[90] Will J, Gerhardt L C and Boccaccini A R 2012 *Adv. Biochem. Eng. Biotechnol.* **126** 195–226
[91] Sarker B, Hum J, Nazhat S N and Boccaccini A R 2015 *Adv. Healthc. Mater* **4** 176–94
[92] Hench L L, Splinter R J, Allen W C and Greenlee T K 1971 *J. Biomed. Mater. Res.* **5** 117–41
[93] Misra S K, Mohn D, Brunner T J, Stark W J, Philip S E, Roy I, Salih V, Knowles J C and Boccaccini A R 2008 *Biomaterials* **29** 1750–61
[94] Marelli B, Ghezzi C E, Mohn D, Stark W J, Barralet J E, Boccaccini A R and Nazhat S N 2011 *Biomaterials* **32** 8915–26
[95] Xu C, Su P, Chen X, Meng Y, Yu W, Xiang A P and Wang Y 2011 *Biomaterials* **32** 1051–58
[96] Kido H W, Gabbai-Armelin P R, Avanzi I R, da Silva A C, Fernandes K R, Fortulan C A and Rennó A C M 2019 *J. Biomed. Mater. Res. B Appl. Biomater.* **107** 211–22
[97] Bloomfield S E, Miyata T, Dunn M W, Bueser N, Stenzel K H and Rubin A L 1978 *Arch. Ophthalmol.* **96** 885–87
[98] Friess W 1998 *Eur. J. Pharm. Biopharm.* **45** 113–36
[99] Lee C H, Singla A and Lee Y 2001 *Int. J. Pharm.* **221** 1–22
[100] Malafaya P B, Silva G A and Reis R L 2007 *Adv. Drug Deliv. Rev.* **59** 207–33
[101] Urist M R 1965 *Science* **150** 893–99
[102] Nienhuijs M E L, Poulsen K, van der Zande M, Briest A, Merkx M A W, Stoelinga P J W, Jensen O N, Jansen J A and Walboomers X F 2009 *J. Biomed. Mater. Res. B Appl. Biomater.* **89** 300–5
[103] Geiger M, Li R H and Friess W 2003 *Adv. Drug Deliv. Rev.* **55** 1613–29

[104] Begam H, Nandi S K, Kundu B and Chanda A 2017 *Mater. Sci. Eng. C Mater. Biol. Appl* **70** 856–69
[105] Li R H and Wozney J M 2001 *Trends Biotechnol.* **19** 255–65
[106] Friess W, Uludag H, Foskett S, Biron R and Sargeant C 1999 *Int. J. Pharm.* **187** 91–9
[107] Welch R D, Jones A L, Bucholz R W, Reinert C M, Tjia J S, Pierce W A, Wozney J M and Li X J 1998 *J. Bone Miner. Res.* **13** 1483–90
[108] Schimandle J H, Boden S D and Hutton W C 1995 *Spine* **20** 1326–37
[109] Nevins M, Kirker-Head C, Nevins M, Wozney J A, Palmer R and Graham D 1996 *Int. J. Periodon. Restor. Dent.* **16** 8–19
[110] Koo K H, Ahn J M, Lee J M, Kim B S, Kim C S and Im G I 2014 *Artif. Organs* **38** 893–99
[111] Quinlan E, Thompson E M, Matsiko A, O'Brien F J and López-Noriega A 2015 *J. Control. Release* **207** 112–19
[112] de Santana R B and Trackman P C 2015 *Int. J. Oral Maxillofac. Implants* **30** 707–14
[113] Hiwatashi N, Hirano S, Mizuta M, Tateya I, Kanemaru S, Nakamura T, Ito J, Kawai K and Suzuki S 2015 *Ann. Otol. Rhinol. Laryngol.* **124** 116–25
[114] Kaufman H E 1988 *J. Cataract Refract. Surg.* **14** 487–88
[115] Poland D E and Kaufman H E 1988 *J. Cataract Refract. Surg.* **14** 489–91
[116] Willoughby C E, Batterbury M and Kaye S B 2002 *Surv. Ophthalmol.* **47** 174–82
[117] Chang W K, Srinivasa S, MacCormick A D and Hill A G 2013 *Ann. Surg.* **258** 59–65
[118] Tihan G T, Ungureanu C, Barbaresso R C, Zgârian R G, Rău I, Meghea A, Albu M G and Ghica M V 2015 *C. R. Chimie* **18** 986–92
[119] Carter K C and Puig-Sellart M 2016 *Curr. Pharm. Des.* **22** 3324–31

Part V

Conclusions

Chapter 19

Concluding remarks

Collagen-based biomaterials are successful and are readily available for a wide range of clinical applications. Despite the advances in biomedical materials and devices there still remain a significant number of un-met medical needs. In many cases, collagen could have an impact in the solution to these problems, directly or as acellular matrix. New advances, the leaps not the steps, will likely turn on individuals who show imagination and lateral thinking.

Collagen-based biomaterials have proved successful in a wide range of products that have been demonstrated to be effective in clinical application and which have achieved strong consumer acceptance. Indeed, collagen derived from animal sources has been used with great success, and has led the way in the use of extracellular matrix proteins in biomedical applications. New research and knowledge on collagen structure and function is driving a wide range of new initiatives that will see new products and opportunities in the future. Of course, one must be realistic and recognise that collagen will not be the single answer for all medical problems, for example, where compressive strength may be needed. However, with the emerging interest in composite materials and tissue engineering, features of collagens including the information content and designed recombinant products, will see new applications and uses emerge.

In these conclusions we will raise just a few points that may merit consideration. Our first, is on identifying the un-met medical needs that drive the demand for new materials and new devices. In 1990, the publication *Biomedical Business International* [1] published a list of the leading un-met medical needs that lacked a satisfactory solution (table 19.1). Looking at this list, now, nearly 30 years later, it is clear that several (most) of these items still remain as un-met needs. So, there are still many prime targets for research. To this can be added the demands for further improvement on performance and reliability of existing materials and devices.

doi:10.1088/978-0-7503-2096-2ch19

Table 19.1. Some of the un-met needs for biomaterials and medical devices in 1990, suggested by *Biomedical Business International* [1].

Application	Un-met medical need
Burn and wound cover	Material(s) for epithelial regeneration
Cardiovascular system	Antithrombogenic surfaces
Cardiovascular system	Small diameter arterial prosthesis
Dental treatment	Stable peri-gingival junction for implant
Neural repair	Materials that induce nerve regeneration
Ophthalmology	Artificial cornea
Ophthalmology	Vitreous augmentation or replacement

A second point is what type of education and environment will lead to the major advances that are needed in certain areas? Steady advances may be sufficient to resolve the incremental improvements that may be needed for existing products. But, will this lead to the intellectual leaps forward that may be needed to resolve the main un-met needs? Over 150 years ago, Louis Pasteur (1822–1895), in a Lecture at the University of Lille (7 December 1854), stated that *Dans les champs de l'observation le hasard ne favorise que les esprits préparés* (In the field of observation chance favours only the prepared mind). The extent to which so called serendipity has progressed science is fascinating [2]. Synthetic polymers provide several examples. One example was the discovery of polytetrafluoroethylene by Roy Plunkett. On 6 April, 1938, when he was trying to use a pressurised cylinder of tetrafluoroethylene, no gas came out, even though the cylinder weight indicated that gas was present. So, he cut the cylinder open and discovered that the gas had polymerised, giving polytetrafluoroethylene. It was later found that rust in the old cylinder had probably catalysed the reaction.

Serendipity arises from the prepared mind and has certainly happened in biomaterials science. For example, intraocular lenses emerged from observations during the Second World War by Harold Ridley, an ophthalmologist, who was performing surgery on Royal Air Force pilots with eye injuries. In combat, if the pilots' Perspex/Plexiglass canopy was hit and shattered, many fragments of the poly (methyl methacrylate) material could end up in the pilot's eyes. In August 1940, Ridley observed that the body was not reacting adversely against foreign material, and realised that the poly(methyl methacrylate) could be used for implantable artificial lenses for treatment of cataracts. This concept was realised in practice later that decade (see [3]). Shortly after, in a different clinical area of interest, an orthopaedic surgeon, Per-Ingvar Brånemark, was studying titanium chamber implants, which had been developed in Cambridge for soft tissue implantation. In his work he implanted them in rabbit femurs. After the initial experiments, because of their value, he tried to retrieve the implants but found this impossible as the bone had strongly adhered to the metal implant. Whence, from this observation, he, and also Leonard Linkow, moved to the development of dental implants and other devices, such as for finger joints [4].

Following from this, the next point is the assembling of knowledge. A broad knowledge, as well as a specialised knowledge on collagen, will be essential for new biomaterial development and innovation. The practitioner will need not just an understanding of the material and its fabrication, but will also need to understand a wealth of biological and clinical information. With the explosion of science and technology in society, finding people who possess the depth of knowledge across all the necessary fields will become increasingly difficult. Hence, the important trend of building interactions and collaborations will become essential. But it is never expected that a biomedical materials or device team will need a team of thousands of active members, as in particle physics! Changes in grant fund strategies, including for big item equipment, will help team building. There is also a growing pressure for translational research and especially in obtaining even more funding, but care is needed to ensure that a balance between fundamental, strategic and tactical research is maintained. Certainly, further fundamental studies on collagens will undoubtedly lead to new insights, especially relating to the *in vivo* performance of biomedical implants. Also, we can expect a range of new, unexpected scientific treasures to emerge that can lead to new approaches to development of successful materials and therapies.

Research in many laboratories has clearly demonstrated that collagen can be used in a wide range of different medical devices, and this will continue. These medical devices are suitable for many distinct clinical applications, and in many cases have proven very successful in clinical use. Indeed, animal derived collagen is expected to retain its dominant position in the development of new materials as a result of its long-term proven safety profile and its adaptability to fulfil many different applications. This adaptability includes its ready use in the development of various coated and composite materials with new functions. There are many opportunities in the future for further commercial products, particularly in the emerging field of tissue engineering where collagen is a logical choice as the scaffold structures that are necessary for tissue regeneration. Undoubtedly, one area of growing importance will be the inclusion of collagen in composite materials from which many new materials and solutions to un-met needs will emerge. The other emerging direction is that of personalised medicine. Most countries see the benefit of providing good health care to all their citizens, to go with good education and other areas of support. Personalised medicine will be seen to emerge most rapidly in well-resourced environments, but will probably still be limited by costs and cost recovery as well as the amortisation and effective use of specialist manufacturing capabilities. Elsewhere, covering the vast majority of people worldwide, the availability of low cost, effective devices enabling full access for all to medical device advances will undoubtedly be, and should be, the first priority.

References

[1] Biomedical Business International 13(10) October 1990

[2] Roberts R M 1989 *Serendipity: Accidental Discoveries in Science* (New York: Wiley)

[3] Auffarth G U and Apple D J 2001 *Ophthalmologe* **98** 1017–28

[4] Brånemark P I 1986 *Tissue-Integrated Prostheses - Osseointegration in Clinical Dentistry* ed P I Brånemark, G Zarb, T Albrektsson and T (Chicago, IL: Quintessence) pp 11–76

Lightning Source UK Ltd.
Milton Keynes UK
UKHW032034191219
355695UK00006B/289/P